HEAT TRANSFER AND HYDRAULIC RESISTANCE AT SUPERCRITICAL PRESSURES IN POWER–ENGINEERING APPLICATIONS

I.L. PIORO AND R.B. DUFFEY

ASME PRESS

Library of Congress Cataloging-in-Publication Data

Pioro, I. L. (Igor* Leonardovich)
Heat transfer and hydraulic resistance at supercritical pressures in power engineering applications / I.L. Pioro and R.B. Duffey.
p. cm.
ISBN 0-7918-0252-3
1. Heat--Transmission. 2. Frictional resistance (Hydrodynamics) 3. Supercritical fluids. I. Duffey, R. B. (Romney B.) II. Title.

TJ260.P54 2006
621.402'2—dc22

Dedicated to our families,
Natalia Kozioura, Violetta Pioro, Anastasia Pioro, and Victoria Duffey.
Thank you for your patience, love, and understanding.

TABLE OF CONTENTS

PREFACE

This monograph summarizes the findings from 650 references devoted to heat transfer and hydraulic resistance of fluids flowing inside channels of various geometries at critical and supercritical pressures. The objectives are to assess the work that was done for the last fifty years in these areas, to understand the specifics of heat transfer and hydraulic resistance, and to propose the most reliable correlations to calculate the heat transfer coefficient and total pressure drop at these conditions.

Analysis of the open literature sources showed that the majority of studies were devoted to the heat transfer of fluids at near-critical and supercritical pressures flowing inside circular tubes. Three major working fluids are involved: water, carbon dioxide, and helium. The main objective of these studies was the development and design of supercritical "steam" generators for fossil-fired power plants (utilizing supercritical water as a working fluid) in the 1950s, 1960s, and 1970s. Supercritical carbon dioxide was usually used as the modeling fluid because it had lower values of the critical parameters. Supercritical helium and carbon dioxide were considered as possible working fluids in some special designs of nuclear reactors and heat exchangers.

Some applications of supercritical water such as supercritical "steam" generators and supercritical water-cooled nuclear reactor concepts are also discussed.

This monograph is intended for scientists, researchers, engineers, specialists and students working in power, nuclear, mechanical, aerospace and chemical engineering areas and dealing with the heat transfer, thermalhydraulics and design of advanced thermal systems and various equipment working at critical and supercritical pressures.

ACKNOWLEDGMENTS

Support from Chalk River Laboratories AECL, Canada, for this work is gratefully acknowledged.

Also, the authors express their great appreciation to Dr. S. Bushby (Chalk River Laboratories, AECL, Chalk River, Canada), Professor J. Buongiorno (Massachusetts Institute of Technology, Cambridge, MA, USA), Professor Q. Bi (Xi'an University, Xi'an, China), Professor I. Catton (University of California, Los Angeles, CA, USA), Professor V. Chatoorgoon (University of Manitoba, Winnipeg, Canada), Professor T. Chen (Xi'an University, Xi'an, China), Professor G.A. Dreitser (Technical University "Moscow State Aviation Institute," Moscow, Russia), Dr. H.F. Khartabil (Chalk River Laboratories, AECL, Chalk River, Canada), Professor J.D. Jackson (University of Manchester, Manchester, UK), Professor P.L. Kirillov (State Scientific Centre "Institute of Physics and Power Engineering," Obninsk, Russia), Dr. V.A. Kurganov (Institute of High Temperatures Russian Academy of Sciences, Moscow, Russia), Professor Yu.N. Kuznetsov (N.A. Dollezhal' Scientific-Research and Design Institute of Power Engineering, Moscow, Russia), Professor Yo. Oka (University of Tokyo, Tokyo, Japan), and Dr. V.A. Silin (Russian Research Centre "Kurchatov Institute," Institute of Nuclear Reactors, Moscow, Russia) for their valuable comments, information and advice during the preparation of this monograph.

SYMBOLS AND ABBREVIATIONS

A	area, m^2
A_{fl}	flow area, m^2
c_p	specific heat at constant pressure, J/kg·K
$\overline{c}_p$	averaged specific heat within the range of $(T_w - T_b)$; $\left(\dfrac{H_w - H_b}{T_w - T_b}\right)$, J/kgK
D	inside diameter, m
D_{ext}	external diameter, m
D_{hy}	hydraulic diameter, m; $\left(\dfrac{4\,A_{fl}}{P_{wetted}}\right)$
f	friction factor; $\left(\dfrac{\sigma_w}{\dfrac{G^2}{8\,\rho}}\right)$
f_d	drag coefficient
G	mass flux, kg/m^{2}s; $\left(\dfrac{m}{A_{fl}}\right)$
g	gravitational acceleration, m/s^2
H	specific enthalpy, J/kg
h	heat transfer coefficient, W/m^{2}K
HL	heat loss, W
I	current, A
k	thermal conductivity, W/mK
L	heated length, m
L_{tot}	total length, m
m	mass-flow rate, kg/s; (ρV)
p	pressure, MPa
POW	power, W
Q	heat-transfer rate, W
q	heat flux, W/m^2; $\left(\dfrac{Q}{A_h}\right)$
q_v	volumetric heat flux, W/m^3; $\left(\dfrac{Q}{V_h}\right)$
R	molar gas constant, 8.31451 J/mol·K
R_a	arithmetic average surface roughness, μm
R_{bend}	radius of bending (for tube), mm

R_{el}	electrical resistance, Ohm
r	radial coordinate or radius, m; regression coefficient
T	temperature, K
t	temperature, °C
U	voltage, V
u	axial velocity, m/s
V	volume, m^3 or volumetric flow rate, m^3/s
V_m	molar volume, m^3/mol
v	radial velocity, m/s
x	axial coordinate, m
y	radial distance; $(r_0 - r)$, m
z	axial coordinate, m

Greek Letters

α	thermal diffusivity, m^2/s; $\left(\frac{k}{c_p\,\rho}\right)$
β	volumetric thermal expansion coefficient, 1/K
Δ	difference
Δ_{HB}	error in heat balance, %
δ	thickness, mm
ε	dissipation of turbulent energy
μ	dynamic viscosity, Pa·s
π	reduced pressure; $\left(\frac{p}{p_{cr}}\right)$
P	perimeter, m
ρ	density, kg/m^3
ρ_{el}	electrical resistivity, Ohm·m
σ	dispersion
σ_w	viscous stress, N/m^2
υ	kinematic viscosity, m^2/s
ξ	friction coefficient

Non-dimensional Numbers

Ga	Galileo number; $\left(\frac{g\,D^3}{v^2}\right)$
Gr	Grashof number; $\left(\frac{g\,\beta\,\Delta T\,D^3}{v^2}\right)$
$\mathbf{Gr_q}$	modified Grashof number; $\left(\frac{g\,\beta\,q_w\,D^4}{k\,v^2}\right)$
Nu	Nusselt number; $\left(\frac{h\,D}{k}\right)$
Pr	Prandtl number; $\left(\frac{\mu\,c_p}{k}\right)=\left(\frac{\upsilon}{\alpha}\right)$

$\overline{\mathbf{Pr}}$	averaged Prandtl number within the range of $(t_w - t_b)$; $\left(\frac{\mu\ \bar{c}_p}{k}\right)$
Re	Reynolds number; $\left(\frac{G\,D}{\mu}\right)$
Ra	Raleigh number; (**Gr Pr**)
St	Stanton number; $\left(\frac{\mathbf{Nu}}{\mathbf{Re\ Pr}}\right)$

Symbols with an overline at the top denote average or mean values (e.g., $\overline{\mathbf{Nu}}$ denotes average (mean) Nusselt number).

Subscripts or superscripts

ac	acceleration
amb	ambient
ave	average
b	bulk
cal	calculated
cr	critical
cr sect	cross-section
dht	deteriorated heat transfer
el	electrical
ext	external
f	fluid
fl	flow
fm	flowmeter
fr	friction
g	gravitational
h	heated
HB	Heat Balance
hor	horizontal
hy	hydraulic
in	inlet
int	internal
iso	isothermal
ℓ	liquid or local
m	molar
max	maximum
meas	measured
min	minimum
nom	nominal or normal
0	constant properties, scale, reference, characteristic, initial, or axial value
out	outlet or outside
OD	outside diameter

pc	pseudocritical
T	value of turbulent flow
TS	test section
th	threshold value
tot	total
v	volumetric
vert	vertical
w	wall

Abbreviations and acronyms widely used in the text and list of references

AC	Alternating Current
A/D	Analog-to-Digital (conversion)
A/I	Analog Input
AECL	Atomic Energy of Canada Limited (Canada)
AERE	Atomic Energy Research Establishment (UK)
AGR	Advanced Gas-cooled Reactor
AIAA	American Institute of Aeronautics and Astronautics
AIChE	American Institute of Chemical Engineers
ANS	American Nuclear Society
ASME	American Society of Mechanical Engineers
ASHRAE	American Society of Heating, Refrigerating and Air-conditioning Engineers
AWG	American Wire Gauge
BWR	Boiling Water Reactor
CANDU®	CANada Deuterium Uranium nuclear reactor
CFD	Computational Fluid Dynamics
CHF	Critical Heat Flux
CRL	Chalk River Laboratories, AECL (Canada)
DAS	Data Acquisition System
DC	Direct Current
DOE	Department Of Energy (USA)
DP	Differential Pressure
emf	electromagnetic force
ENS	European Nuclear Society
EU	European Union
EXT	EXTernal
FA	Fuel Assembly
FBR	Fast Breeder Reactor
FM	FlowMeter
F/M	Ferritic-Martensitic (steel)
FR	Fuel Rod
f.s.	full scale
FT	Flow Transducer
GIF	Generation IV International Forum
HMT	Heat Mass Transfer
HT	Heat Transfer

HTC	Heat Transfer Coefficient
HTD	Heat Transfer Division
HTR	High Temperature Reactor
HVAC & R	Heating Ventilating Air-Conditioning and Refrigerating
IAEA	International Atomic Energy Agency (Vienna, Austria)
ID	Inside Diameter
INEEL	Idaho National Engineering and Environmental Laboratory (USA)
IP	Intermediate-Pressure (turbine)
IPPE	Institute of Physics and Power Engineering (Obninsk, Russia)
JAERI	Japan Atomic Energy Research Institute
JSME	Japan Society of Mechanical Engineers
KAERI	Korea Atomic Energy Research Institute (South Korea)
KPI	Kiev Polytechnic Institute (nowadays National Technical University of Ukraine "KPI") (Kiev, Ukraine)
KP-SKD	Channel Reactor of Supercritical Pressure (in Russian abbreviations)
LP	Low-Pressure (turbine)
LOCA	Loss Of Coolant Accident
LOECC	Loss Of Emergency Core Cooling
Ltd.	Limited
LWR	Light Water Reactor
MEI	Moscow Power Institute (Moscow, Russia) (In Russian abbreviations)
MIT	Massachusetts Institute of Technology (Cambridge, MA, USA)
MOX	Mixed OXide (nuclear fuel)
NASA	National Aeronautics and Space Administration (USA)
NIST	National Institute of Standards and Technology (USA)
NPP	Nuclear Power Plant
OD	Outside Diameter
PC	Personal Computer
PDT	Pressure Differential Transducer
Ph.D.	Philosophy Degree
PLC	Programmable Logic Controller
ppb	parts per billion
ppm	parts per million
PT	Pressure Tube or Pressure Transducer
PWAC	Pratt & Whitney AirCraft
PWR	Pressurized Water Reactor
R	Refrigerant
RAS	Russian Academy of Sciences
RBMK	Reactor of Large Capacity Channel type (in Russian abbreviations)
RDIPE	Research and Development Institute of Power Engineering (Moscow, Russia) (NIKIET in Russian abbreviations)
R&D	Research and Development
RMS	Root-Mean-Square (error or surface roughness)
RPV	Reactor Pressure Vessel
RSC	Russian Scientific Centre

RT	propulsion fuel (in Russian abbreviations)
RTD	Resistance Temperature Detector
SCP	SuperCritical Pressure
SCR	SuperCritical Reactor
SCW	SuperCritical Water
SCWO	SuperCritical Water Oxidation (technology)
SCWR	SuperCritical Water-cooled Reactor
SFL	Supercritical Fluid Leaching
SFR	Sodium Fast Reactor
SKD	SuperCritical Pressure (in Russian abbreviations)
SMR	Steam-Methane-Reforming (process)
SS	Stainless Steel
T	fuel (in Russian abbreviation)
TC	ThermoCouple
TE	TEmperature
TECO	TEmperature of CO_2
TS	Test Section
TsKTI	Central Boiler-Turbine Institute (St.-Petersburg, Russia) (in Russian abbreviations)
UCG	Uranium-Carbide Grit pored over with calcium (nuclear fuel)
UK	United Kingdom
U.K.A.E.A.	United Kingdom Atomic Energy Authority (UK)
UNESCO	United Nations Educational, Scientific and Cultural Organization (Paris, France)
US or USA	United States of America
USSR	Union of Soviet Socialist Republics
VHTR	Very High-Temperature Reactor
VNIIAM	All-Union Scientific-Research Institute of Atomic Machine Building (Russia) (in Russian abbreviations)
VTI	All-Union Heat Engineering Institute (Moscow, Russia) (in Russian abbreviations)
wt	weight
WWPR	Water-Water Power Reactor (“VVER” in Russian abbreviations)

GLOSSARY AND DEFINITIONS OF SELECTED TERMS AND EXPRESSIONS RELATED TO CRITICAL AND SUPERCRITICAL REGIONS SPECIFICS

Compressed fluid is a fluid at a pressure above the critical pressure but at a temperature below the critical temperature.

Critical point (also called a *critical state*) is the point where the distinction between the liquid and gas (or vapor) phases disappears, i.e., both phases have the same temperature, pressure and volume. The *critical point* is characterized by the phase state parameters T_{cr}, p_{cr} and V_{cr}, which have unique values for each pure substance.

Deteriorated heat transfer is characterized with lower values of the wall heat transfer coefficient compared to those at the normal heat transfer; and hence has higher values of wall temperature within some part of a test section or within the entire test section.

Improved heat transfer is characterized with higher values of the wall heat transfer coefficient compared to those at the normal heat transfer; and hence lower values of wall temperature within some part of a test section or within the entire test section. In our opinion, the improved heat-transfer regime or mode includes peaks or "humps" in the heat transfer coefficient near the critical or pseudocritical regions.

Near-critical point is actually a narrow region around the critical point where all the thermophysical properties of a pure fluid exhibit rapid variations.

Normal heat transfer can be characterized in general with wall heat transfer coefficients similar to those of subcritical convective heat transfer far from the critical or pseudocritical regions, when are calculated according to the conventional single-phase Dittus-Boelter type correlations.

Pseudo-boiling is a physical phenomenon similar to subcritical pressure nucleate boiling, which may appear at supercritical pressures. Due to heating of the supercritical fluid with a bulk-fluid temperature below the pseudocritical temperature (high-density fluid, i.e., "liquid"), some layers near a heating surface may attain temperatures above the pseudocritical temperature (low-density fluid, i.e., "gas"). This low-density "gas" leaves the heating surface in the form of variable density (bubble) volumes. During the pseudo-boiling, the wall heat transfer coefficient usually increases (improved heat-transfer regime).

Pseudocritical point (characterized with p_{pc} and t_{pc}) is a point at a pressure above the critical pressure and at a temperature ($t_{pc} > t_{cr}$) corresponding to the maximum value of the specific heat for this particular pressure.

Pseudo-film boiling is a physical phenomenon similar to subcritical pressure film boiling, which may appear at supercritical pressures. At pseudo-film boiling, a low-density fluid (a fluid at temperatures above the pseudocritical temperature, i.e., "gas") prevents a high-density fluid (a fluid at temperatures below the pseudocritical temperature, i.e., "liquid") from contacting ("rewetting") a heated surface. Pseudo-film boiling leads to the deteriorated heat-transfer regime.

Supercritical fluid is a fluid at pressures and temperatures that are higher than the critical pressure and critical temperature. However, in the present monograph, the term *supercritical fluid* includes both terms—a *supercritical fluid* and *compressed fluid*.

Supercritical steam is actually supercritical water because at supercritical pressures there is no difference between phases. However, this term is widely (and incorrectly) used in the literature in relation to supercritical steam generators and turbines.

Superheated steam is a steam at pressures below the critical pressure but at temperatures above the critical temperature.

ABOUT THE AUTHORS

Igor L. Pioro – Ph.D. (1983), Doctor of Technical Sciences (1992), is an internationally recognized scientist in areas of nuclear engineering (thermalhydraulics of nuclear reactors, Generation IV nuclear reactor concepts and high-level radioactive-wastes management); thermal sciences (boiling, forced convection including supercritical pressures, etc.); and heat engineering (two-phase thermosyphons, heat exchangers, heat-recovery systems, submerged-combustion melters, etc.).

He is author/co-author of 6 state-of-the-art monographs, more than 100 papers, 24 patents, and more than 30 major technical reports. His latest monographs are L.S. Pioro, I.L. Pioro, T.O. Kostyuk, and B.S. Soroka, *Industrial Application of Submerged Combustion Melters*, Fact Publ. House, Kiev, Ukraine, 2006, 240 pages; M.K. Bezrodny, I.L. Pioro and T.O. Kostyuk, *Transfer Processes in Two-Phase Thermosyphon Systems. Theory and Practice*, (In Russian), 2nd Edition, Augmented and Revised, Fact Publ. House, Kiev, Ukraine, 2005,

704 pages; and L.S. Pioro and I.L. Pioro, *Industrial Two-Phase Thermosyphons*, Begell House, Inc., New York, NY, USA, 1997, 288 pages.

Dr. Pioro graduated from the National Technical University of Ukraine "Kiev Polytechnic Institute", Heat Engineering Department with Master of Science in Thermal Physics (Diploma of Honor) in 1979. After that he worked on various positions including an engineer, scientist, senior scientist and deputy director at the Institute of Engineering Thermal Physics of the National Academy of Sciences of Ukraine. Last 6.5 years Dr. Pioro was a senior scientist in the Thermalhydraulics Branch of Chalk River Laboratories, Atomic Energy of Canada Limited. Currently, he is associated with the School of Power Systems and Nuclear Engineering University of Ontario Institute of Technology.

Dr. Pioro is an active member of the American Society of Mechanical Engineers and the American Nuclear Society.

In 1990, Dr. Pioro was awarded with the Medal of the National Academy of Sciences of Ukraine for the best scientific work of a young scientist and with the Badge "Inventor of the USSR".

Romney B. Duffey, Principal Scientist with AECL (Canada), is a leading expert in commercial nuclear reactor studies; is active in global environmental and energy studies and in advanced system design; and is currently leading work on advanced energy concepts. In addition to consulting positions, previously he was Department Chair at Brookhaven National Laboratory, New York, and Deputy Department Manager at Idaho National Laboratory, Idaho, having held senior technical positions at the Electric Power Research Institute, California and with the Central Electricity Generating Board, UK. He has an extensive technology background, including energy, environment waste, safety, risk, two-phase flow and heat transfer, simulation, physical modeling, and uncertainty analysis. Dr. Duffey was awarded his B/Sc. (Honors) in Physics and Ph.D. in Physics/Geophysics from the University of Exeter, UK.

Dr. Duffey is the author of the original text about learning from errors in technology (*"Know the Risk"*, Butterworth-Heinemann, 2002), and has authored and contributed to more than 200 published technical papers and reports. He is a past Chair of the American Society of Mechanical Engineers' Nuclear Engineering Division, an active Member of the American and Canadian Nuclear Societies, and a past Chair of the American Nuclear Society's Thermal Hydraulics Division. He has been a co-chair of the Generation IV International Forum Steering Committee on SCWR, and of many International Conferences. In 2004, he was elected a Fellow of ASME for exceptional engineering achievements and contributions to the engineering profession.

Chapter 1

INTRODUCTION

The objective of this monograph is to show what work has been done worldwide in the area of heat transfer and hydraulic resistance of fluids at critical and supercritical pressures for the last five decades, including achievements in development concepts of SuperCritical Water-cooled nuclear Reactors (SCWRs) and in operation of supercritical power-plant "steam" generators.

This monograph is the most exhaustive literature survey on this topic published so far. However, some reviews found in the open literature may complement materials presented in this monograph providing in-depth analysis on some specific topics. Therefore, short descriptions of these reviews are listed in Appendix A.

This monograph will help the reader to understand the main directions of investigations conducted so far, the problems encountered during the operation of supercritical units and development concepts of SCWRs, and will provide with practical recommendations for the calculation of heat transfer and hydraulic resistance at critical and supercritical pressures.

With the objective mentioned above, the monograph is divided into chapters according to the main problems of heat transfer and hydraulic resistance at critical and supercritical pressures, with subdivisions, i.e., sections, created for better understanding of the specifics of the underlying problems. Special attention was paid to the problems related to operating supercritical "steam" generators and developing concepts of future nuclear reactors working at supercritical pressures.

In general, the following topics associated with the heat transfer and hydraulic resistance at critical and supercritical pressures can be noted:

- thermophysical properties near critical and pseudocritical points;
- forced convection (mainly turbulent) in upward, downward and horizontal flows at low, intermediate and high heat fluxes;
- special heat-transfer regimes such as deteriorated heat transfer, improved heat transfer, pseudo-boiling, pseudo-film boiling, etc.;
- mixed convection in upward and downward flows;
- free convection (mainly near vertical plate);
- hydraulic resistance; and
- special problems at critical and supercritical pressures: oscillations of temperature and pressure, dissolved gases, deposits on the inner surface of tubes, corrosion of materials, etc.

The use of supercritical fluids in different processes is not new and, actually, is not a human invention. Nature has been processing minerals in aqueous solutions at near or above the critical point of water for billions of years (Levelt Sengers 2000). In the late 1800s, scientists started to use this natural process in their labs for creating various crystals. During the last 50 – 60 years, this process, called hydrothermal processing (operating parameters: water pressure from 20 to 200 MPa and temperatures from 300 to 500°C), has been widely used in the industrial production of high-quality single crystals (mainly gem stones) such as sapphire, tourmaline, quartz, titanium oxide, zircon and others (Levelt Sengers 2000).

The first works devoted to the problem of heat transfer at supercritical pressures started as early as the 1930s (Pioro and Pioro 1997; Hendricks et al. 1970). Schmidt and his associates investigated free convection heat transfer of fluids at the near-critical point with the application to a new effective cooling system for turbine blades in jet engines. They found (Schmidt 1960; Schmidt et al. 1946) that the free convection heat transfer coefficient (HTC) at the near-critical state was quite high and decided to use this advantage in single-phase thermosyphons with an intermediate working fluid at the near-critical point (Pioro and Pioro 1997). (In general, thermosyphons are used to transfer heat flux from a heat source to a heat sink located at some distance.)

In the 1950s, the idea of using supercritical steam-water appeared to be rather attractive for "steam" generators. At supercritical pressures, there is no liquid-vapour phase transition; therefore, there is no such phenomenon as critical heat flux or dryout. Only within a certain range of parameters a deterioration of heat transfer may occur. The objective of operating "steam" generators at supercritical pressures was to increase the total efficiency of a power plant. Work in this area was mainly done in the USA and former USSR in the 1950s – 1980s (International Encyclopedia of Heat & Mass Transfer 1998).

At the end of the 1950s and the beginning of the 1960s, some studies were conducted to investigate the possibility of using supercritical fluids in nuclear reactors (Oka 2000; Wright and Patterson 1966; Bishop et al. 1962; Skvortsov and Feinberg 1961; Marchaterre and Petrick 1960; Supercritical pressure power reactor 1959). Several designs of nuclear reactors using water as the coolant at supercritical pressures were developed in the USA and USSR. However, this idea was abandoned for almost 30 years and regained support in the 1990s.

Use of supercritical water in power-plant "steam" generators is the largest application of a fluid at supercritical pressures in industry. However, other areas exist where supercritical fluids are used or will be implemented in the near future.

The latest developments in these areas focus on

- increasing the efficiency of the existing ultra-supercritical and supercritical "steam" generators (Smith 1999);
- developing supercritical water-cooled nuclear reactors (Kirillov 2001a,b; Oka 2000);
- using supercritical water in the Rankine cycle for lead-cooled nuclear reactors (Boehm et al. 2005) and in the Brayton cycle (Sohn et al. 2005) including the Brayton cycle for future Sodium Fast Reactors (SFRs);
- using supercritical carbon dioxide in an indirect cycle of the gas cooled fast reactors (Hejzlar et al. 2005; Kato et al. 2005);

- using supercritical carbon dioxide for cooling of a printed circuit (Ishizuka et al. 2005);
- the use of near-critical helium to cool the coils of superconducting electromagnets, superconducting electronics and power-transmission equipment (Hendricks et al. 1970a);
- the use of supercritical hydrogen as a fuel for chemical and nuclear rockets (Hendricks et al. 1970a);
- the use of methane as a coolant and fuel for supersonic transport (Hendricks et al. 1970a);
- the use of liquid hydrocarbon coolants and fuels at supercritical pressures in the cooling jackets of liquid rocket engines and in fuel channels of air-breathing engines (Altunin et al. 1998; Kalinin et al. 1998, Dreitser 1993, Dreitser et al. 1993);
- the use of supercritical carbon dioxide as a refrigerant in air-conditioning and refrigerating systems (Lorentzen 1994; Lorentzen and Pettersen 1993);
- the use of a supercritical cycle in the secondary loop for transformation of geothermal energy into electricity (Abdulagatov and Alkhasov 1998);
- the use of supercritical water oxidation technology (SCWO) for treatment of industrial and military wastes (Levelt Sengers 2000; Lee 1997);
- the use of carbon dioxide in the supercritical fluid leaching (SFL) method for removal uranium from radioactive solid wastes (Tomioka et al. 2005) and in decontamination of surfaces (Shadrin et al. 2005); and
- the use of supercritical fluids in chemical and pharmaceutical industries in such processes as supercritical fluid extraction, supercritical fluid chromatography, polymer processing and others (Supercritical Fluids 2002; Levelt Sengers 2000).

Experiments at supercritical pressures are very expensive and require sophisticated equipment and measuring techniques. Therefore, some of these studies (for example, heat transfer in bundles) are proprietary and hence were not published in the open literature.

The majority of the studies deal with heat transfer and hydraulic resistance of working fluids, mainly *water* (Pioro and Duffey 2005, 2003a; Pioro et al. 2004), *carbon dioxide* (Duffey and Pioro 2005b), *and helium, in circular tubes.* In addition to these fluids, forced- and free-convection heat transfer experiments were conducted at supercritical pressures, using[1]

- *liquefied gases* such as air and argon (Budnevich and Uskenbaev 1972), hydrogen (International Encyclopedia of Heat & Mass Transfer 1998; Hess and Kunz 1965; Thompson and Geery 1962); nitrogen (Popov et al. 1977; Akhmedov et al. 1974; Uskenbaev and Budnevich 1972), nitrogen tetra-oxide (Nesterenko et al. 1974; McCarthy et al. 1967), oxygen (Powell 1957) and sulphur hexafluoride (Tanger et al. 1968);

[1] For additional information, see Chapter 8.

- *alcohols* such as ethanol and methanol (Kafengauz 1983; Alad'yev et al. 1967, 1963);
- *hydrocarbons* such as n-heptane (Isayev 1983; Alad'ev et al. 1976; Kaplan and Tolchinskaya 1974a, 1971), n-hexane (Isaev et al. 1995), di-iso-propyl-cyclo-hexane (Kafengauz 1983, 1969, 1967; Kafengauz and Fedorov 1970, 1968, 1966), n-octane (Yanovskii 1995), iso-butane, iso-pentane and n-pentane (Abdulagatov and Alkhasov 1998; Bonilla and Sigel 1961);
- *aromatic hydrocarbons* such as benzene and toluene (Rzaev et al. 2003; Abdullaeva et al. 1991; Kalbaliev et al. 1983, 1978; Isaev and Kalbaliev 1979; Mamedov et al. 1977; 1976), and poly-methyl-phenyl-siloxane (Kaplan et al. 1974b);
- *hydrocarbon coolants* such as kerosene (Kafengauz 1983), TS-1 and RG-1 (Altunin et al. 1998), jet propulsion fuels RT and T-6 (Kalinin et al. 1998; Yanovskii 1995; Valueva et al. 1995; Dreitser et al. 1993); and
- *refrigerants* (Abdulagatov and Alkhasov 1998; Gorban' et al. 1990; Tkachev 1981; Beschastnov et al. 1973; Nozdrenko 1968; Holman and Boggs 1960; Griffith and Sabersky 1960).

A limited number of studies were devoted to heat transfer and pressure drop in annuli, rectangular-shaped channels and bundles.

Chapter 2

PHYSICAL PROPERTIES OF FLUIDS IN CRITICAL AND PSEUDOCRITICAL REGIONS

2.1 GENERAL

Heat transfer at supercritical pressures is influenced by the significant changes in thermophysical properties at these conditions. For many working fluids, which are used at critical and supercritical conditions, their physical and thermophysical properties are well established (Kirillov 2003). This is especially important for the creation of generalized correlations in non-dimensional form, which allows the experimental data for several working fluids to be combined into one set, as well as the use of numerical solutions (Polyakov 1991). The most significant thermophysical property variations occur near the critical and pseudocritical points (Supercritical Fluids 2002).

In general, at the critical point (Kaye and Laby 1973)

$$\left(\frac{\partial p}{\partial V}\right)_{T=T_{cr}} = 0 \quad \text{and} \quad \left(\frac{\partial^2 p}{\partial^2 V}\right)_{T=T_{cr}} = 0. \qquad \text{(2.1) and (2.2)}$$

At temperatures above the critical temperature, a fluid cannot be liquefied. The supercritical fluid is neither a gas or a liquid, and approximates the behavior of a perfect gas. To account for the actual intermolecular effects at high densities, it is necessary to correct the equation of state and physical properties for non-ideal departure from perfect gas values. The equation of state of a fluid can be accurately expressed by a semi-empirical compressibility factor equation of the form:

$$z = \frac{p\,V_m}{R\,T} = 1 + \frac{B}{V_m} + \frac{C}{V_m^2} + \cdots \qquad (2.3)$$

where R is the molar gas constant and B, C, ..., are the second and third, virial coefficients, which are functions of temperature only, and higher order correction terms also exist.

Table 2.1. Critical parameters of fluids (NIST 2002; Kalinin et al. 1998; Hewitt et al. 1994; Kaye and Laby 1973; Griffith and Sabersky 1960).

Fluid	p_{cr}, MPa	t_{cr}, °C	ρ_{cr}, kg/m³	$\frac{p_{cr} V_{m\,cr}}{R T_{cr}}$
Air	3.8	−140.5	333.3	–
Ammonia (NH_3)	11.333	132.25	225.0	0.243
Argon (Ar)	4.863	−122.46	535.6	0.292
Benzene (C_6H_6)	4.89	288.9	309.0	0.266
iso-Butane (2-Methyl-propane, C_4H_{10})	3.64	134.7	224.4	0.283
Carbon dioxide (CO_2)	7.3773	30.978	467.6	0.274
Di-iso-propyl-cyclo-hexane	1.96	376.9	–	–
Ethanol (C_2H_6O)	6.15	240.8	276.0	0.240
Freon-12 (Di-chloro-di-fluoro-methane, CCl_2F_2)	4.1361	111.97	565.0	0.280
Freon-13B1 (Bromo-tri-fluoro-methane, $CBrF_3$)	3.95	67.0	770.0	–
Freon-22 (Chloro-di-fluoro-methane, $CHClF_2$)	4.99	96.145	523.84	0.264
Freon-114a (1,1-Di-chloro-tetra-fluoro-ethane, $C_2Cl_2F_4$)	3.257	145.68	579.97	–
Freon-134a (1,1,1,2-tetra-fluoro-ethane, CH_2FCF_3)	4.0593	101.06	511.9	–
Helium (He)	0.22746	−267.95	69.641	0.307
n-Heptane (C_7H_{16})	2.736	266.98	232.0	0.260
n-Hexane (C_6H_{14})	3.034	234.67	233.18	0.264
Hydrogen (H_2)	1.315	−239.96	30.118	0.309
Kerosene RT	2.5	392.9	–	–
Methanol (CH_4O)	8.1035	239.45	275.56	0.224
Nitrogen (N_2)	3.3958	−146.96	313.3	0.291
Nitrogen tetroxide (N_2O_4)	10.1	157.9	–	–
n-Octane (C_8H_{18})	2.497	296.17	234.9	0.258
Oxygen (O_2)	5.043	−118.57	436.14	0.308
iso-Pentane	3.396	187.2	236.0	0.268
Poly-methyl-phenyl-siloxane	0.75	502	–	–
RT (jet propulsion fuel)	2.19	395	–	–
Sulphur hexafluoride (SF_6)	3.7546	45.583	743.81	0.277
T-6 (jet propulsion fuel)	2.24	445	–	–
Toluene (C_7H_8)	4.1263	318.6	291.99	0.267
Water (H_2O)	22.064	373.95	322.39	0.243

Critical parameters of fluids listed in this monograph are shown in Table 2.1.

Critical constants of gases and many other substances can be found in NIST (2002), Kreglewski (1984), and Kaye and Laby (1973).

The thermophysical properties of water at different pressures and temperatures, including the supercritical region, can be calculated using the NIST software (2002, 1997, 1996). In this software, the fundamental equation for the Helmholtz energy per unit mass (kg) as a function of temperature and density is used. This equation was combined with a function for the ideal gas Helmholtz energy to define a complete Helmholtz energy surface. All other thermodynamic properties are stated to be obtained by differentiation of this surface.

Also, the latest NIST software (2002) with supplemental fluids (as per March 2006) calculates the thermophysical properties of the following 82 fluids: acetone, ammonia, argon, benzene, butane, 1-butene (CH_3-CH_2-CH=CH_2), C_4F_{10}, C_5F_{12} (FC-87), carbon dioxide, carbon monoxide, carbonyl sulphide, cis-butene (C_4H_8), cyclo-hexane, cyclo-propane, decane, deuterium, di-methyl-ether

(C_2H_6O), do-decane, ethane, ethanol, ethylene, fluorine, heavy water, helium, heptane, hexane, hydrogen, hydrogen sulphide, iso-butane, iso-butene, iso-hexane, iso-pentane, krypton, methane, methanol, neon, neo-pentane ($C(CH_3)_4$), nitrogen, nitrogen tri-fluoride (NF_3), nitrous oxide (N_2O), nonane, octane, oxygen, para-hydrogen, pentane, propane, propylene, propyne (CH_2-C-CH_2), refrigerants R-11–14, 21, 22, 23, 32, 41, 113–116, 123–125, 134a, 141b, 142b, 143a, 152a, 218, 227ea, 236ea, 236fa, 245ca, 245fa and RC318, sulphur dioxide, sulphur hexa-fluoride (SF_6), toluene, trans-butene (CH_3-CH=CH-CH_3), water and xenon within a wide range of pressures and temperatures including supercritical pressures.

2.2 PARAMETRIC TRENDS

General trends of various properties at near-critical and pseudocritical points can be illustrated on the basis of those of water (see Figures 2.1[2] – 2.3). Water as a working fluid is used in supercritical "steam" generators/turbines and is considered as the primary coolant for SCWR concepts (see Chapter 4). For comparison purposes, trends in thermophysical properties of carbon dioxide, R-134a and helium are shown in Appendix B.

Figure 2.1 shows basic thermophysical properties of water near critical (p = 22.064 MPa) and pseudocritical (p = 25.0 MPa) points calculated according to NIST (2002). In general, all thermophysical properties undergo significant changes near the critical and pseudocritical points. Near the critical point, these changes are dramatic (see Figure 2.1). In the vicinity of pseudocritical points, with an increase in pressure, these changes become less pronounced (see Figure 2.1). It can also be seen from Figure 2.1 that properties such as density and dynamic viscosity undergo a significant drop (near the critical point, this drop is almost vertical) within a very narrow temperature range, while specific enthalpy and kinematic viscosity undergo a sharp increase. Volume expansivity, specific heat, thermal conductivity and Prandtl number have a peak near the critical and pseudocritical points. The magnitudes of these peaks decrease very quickly with an increase in pressure.

Figure 2.2 shows a comparison between different thermophysical property sources (US and UK tables). For critical and supercritical pressures, it is very important to use original correlations for thermophysical properties (e.g., NIST (2002) or NIST/ASME Steam Properties (1997, 1996)), rather than the primary table data. This is because significant changes in the thermophysical properties are confined to very narrow temperature or pressure ranges and the primary data are usually tabulated with relatively large temperature or pressure increments (for details, see ASME International Steam Tables for Industrial Use, 2000).

Figure 2.3 shows a comparison between calculated values of thermal conductivity and experimentally measured values near critical and pseudocritical points.

The specific heat of water (as well as of other fluids) has a maximum value at the critical point. The exact temperature that corresponds to the specific heat peak

[2] Data in Figure 2.1 were obtained with temperature increments of 0.01°C. It should be noted that height of the peaks in specific heat, thermal conductivity, volume expansivity, and Prandtl number in the critical point and pseudocritical points near the critical point may vary with a temperature increment value.

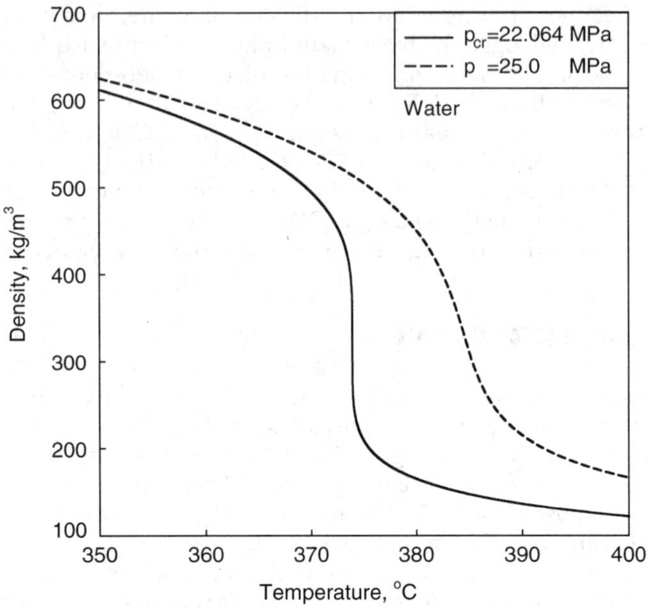

Figure 2.1a. Density vs. temperature.

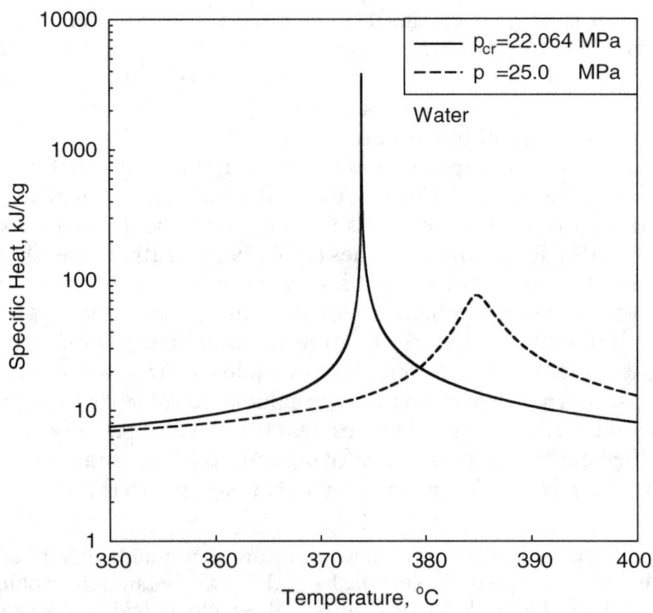

Figure 2.1b. Specific heat vs. temperature.

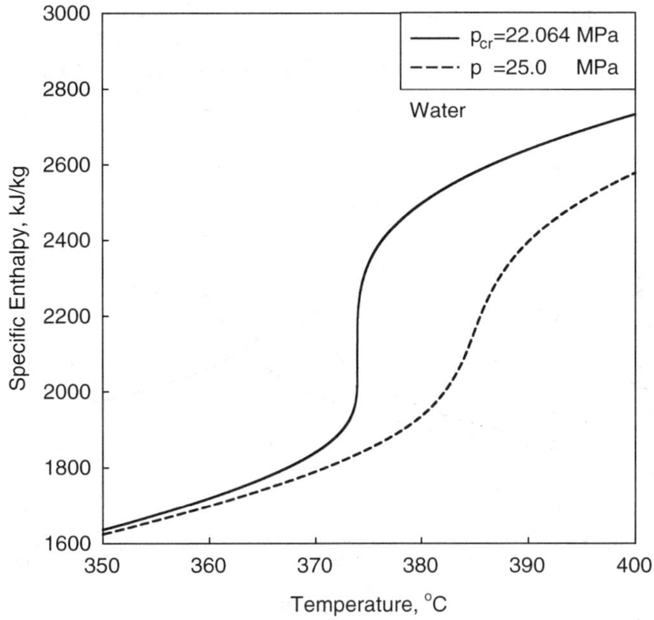

Figure 2.1c. Specific enthalpy vs. temperature.

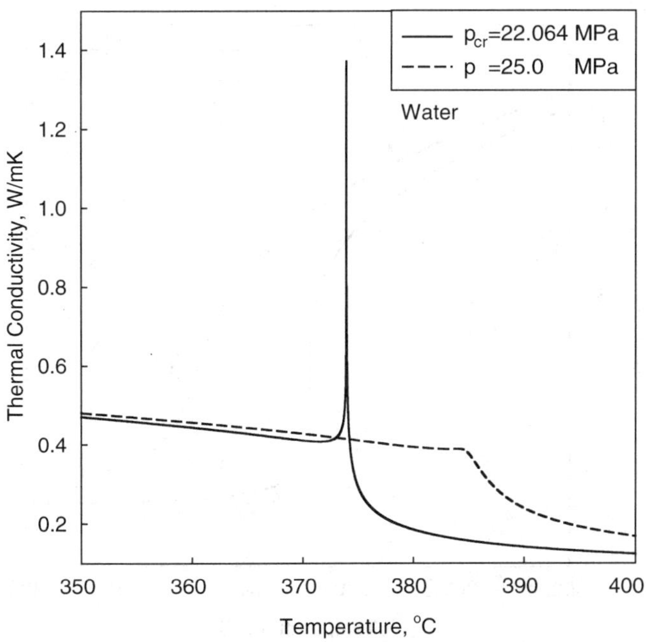

Figure 2.1d. Thermal conductivity vs. temperature.

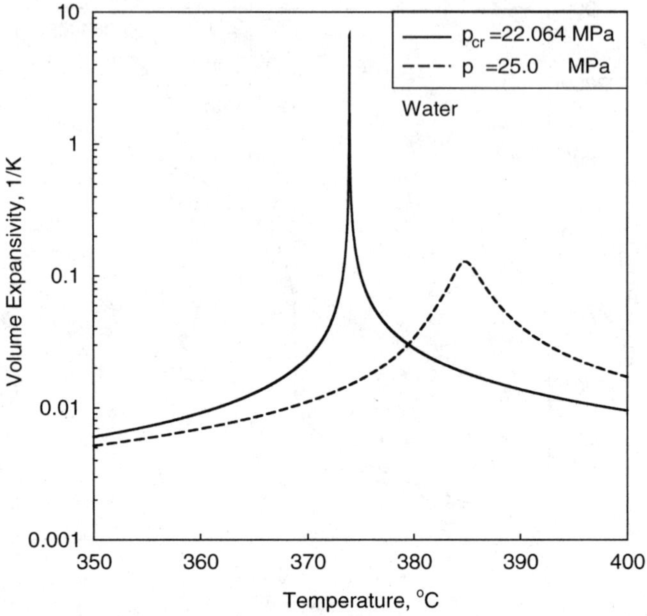

Figure 2.1e. Volume expansivity vs. temperature.

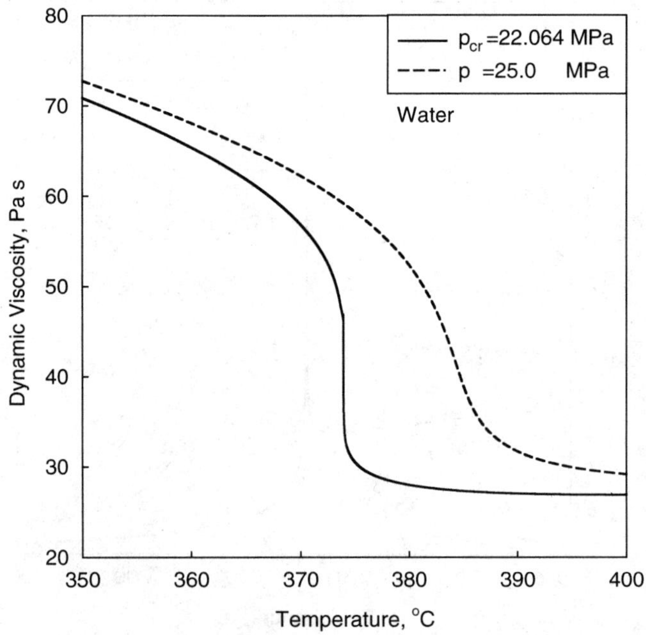

Figure 2.1f. Dynamic viscosity vs. temperature.

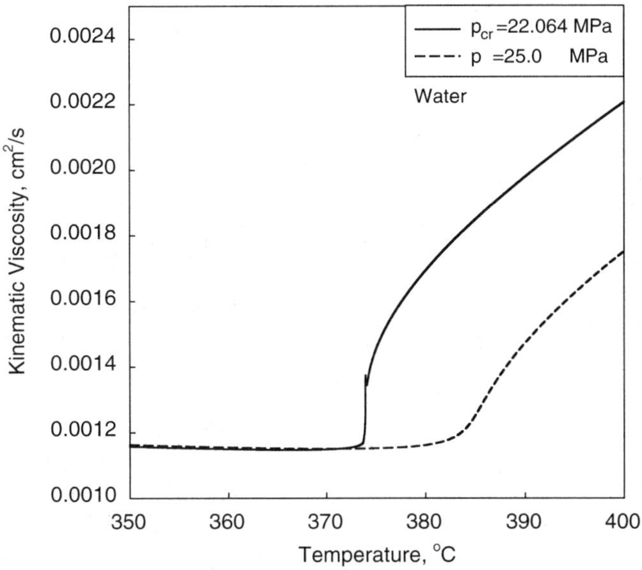

Figure 2.1g. Kinematic visosity vs. temperature.

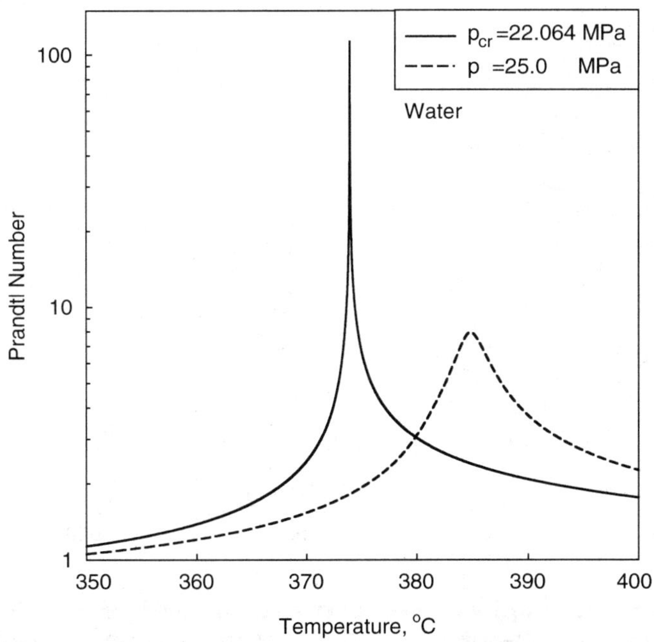

Figure 2.1h. Prandtl number vs. temperature.

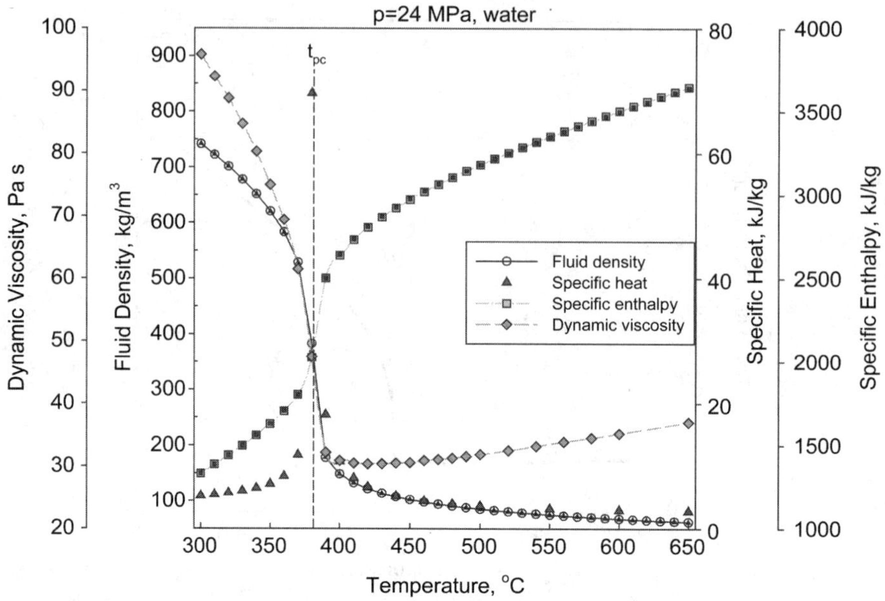

Figure 2.2a. Thermophysical properties of water near pseudocritical point: Open symbols—ASME Steam Tables (2000); and closed symbols—UK Steam Tables (1970).

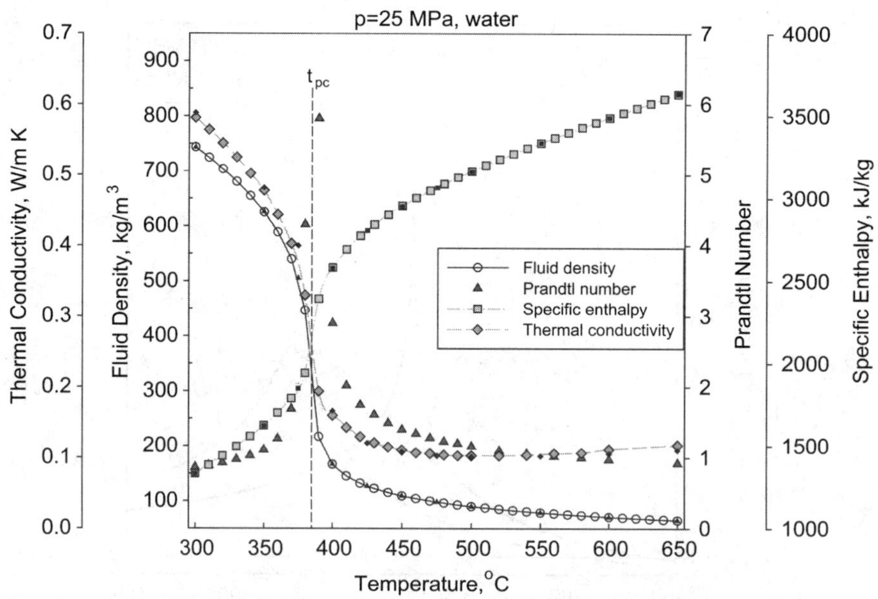

Figure 2.2b. Thermophysical properties of water near pseudocritical point: Open symbols—UK Steam Tables (1970) and closed symbols—Vargaftik et al. (1996).

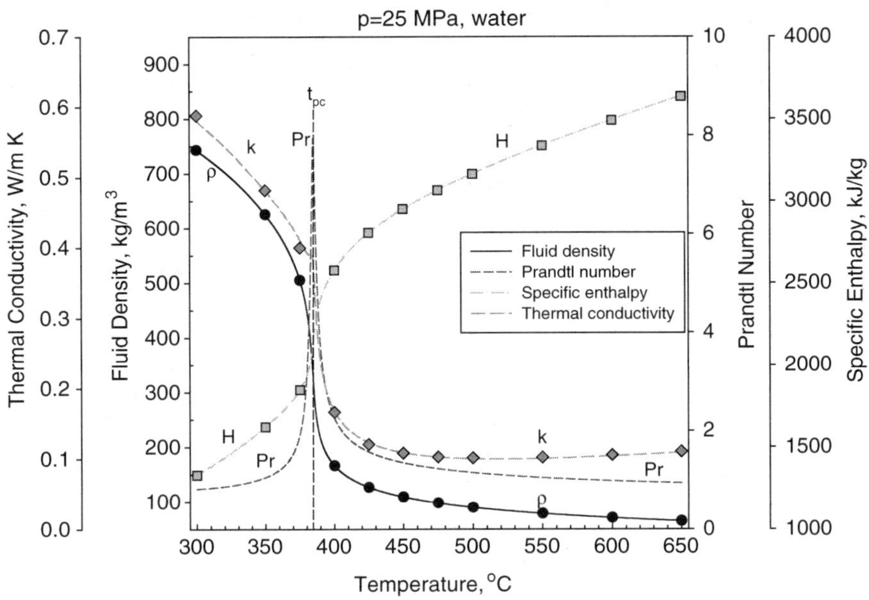

Figure 2.2c. Thermophysical properties of water near pseudocritical point: Lines—NIST (2002); and closed symbols—Vargaftik et al. (1996).

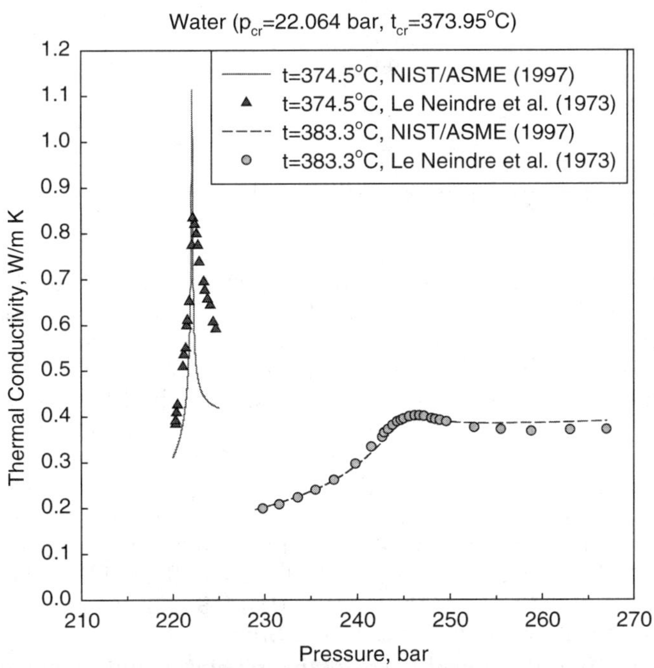

Figure 2.3. Thermal conductivity of water near critical; and pseudo-critical points: Lines—calculations and symbols—experiment.

Table 2.2. Values of pseudocritical temperature and corresponding peak values of specific heat within wide range of pressures (NIST 2002).

Pressure, MPa	Pseudocritical Temperature, °C	Peak Value of Specific Heat, kJ/kg·K
23	377.5	284.3
24	381.2	121.9
25	384.9	76.4
26	388.5	55.7
27	392.0	43.9
28	395.4	36.3
29	398.7	30.9
30	401.9	27.0
31	405.0	24.1
32	408.1	21.7
33	411.0	19.9
34	413.9	18.4
35	416.7	17.2
36	419.5	16.1
37	422.2	15.2
38	425.0	14.5
39	427.7	13.8
40	430.3	13.2
41	433.0	12.7
42	435.6	12.2
43	438.1	11.8
44	440.6	11.4
45	443.1	11.0
46	445.5	10.7
47	447.9	10.4
48	450.2	10.1
49	452.5	9.9
50	454.8	9.6

above the critical pressure is known as the pseudocritical point (International Encyclopedia of Heat & Mass Transfer 1998) (see Table 2.2.). It should be noted that peaks in thermal conductivity and volume expansivity may not correspond to the pseudocritical point (see Table 2.3 and Figure 2.4).

Thermophysical properties of carbon dioxide are listed in the handbook by Vargaftik et al. (1996), papers by Rivkin and Gukov (1971) and Altunin et al. (1973) or can be calculated with the NIST (2002) software and ChemicaLogic Corporation (1999) software.

In early studies (e.g., Ornatskiy et al. 1980; Petukhov 1970; Shitsman 1959; Bringer and Smith 1957), the peak in thermal conductivity was not taken into account. Later, this peak (see Figure 2.1d) was well established (NIST 2002; Harvey 2001; Levelt Sengers 2000; NIST/ASME Steam Properties 1997, 1996; Neumann and Hahne 1980; Alekseev and Smirnov 1976; Altunin 1975; Sirota et al. 1974; Le Neindre et al. 1973; Vukalovich and Altunin 1968) and was accounted for in developing generalized correlations. The peak in thermal conductivity diminishes at about 25.5 MPa for water (see Figure 2.5). Therefore,

Table 2.3. Peak values of specific heat, volume expansivity and thermal conductivity in critical and near pseudocritical points (NIST 2002).

Pressure, MPa	Pseudocritical Temperature, °C	Temperature, °C	Specific Heat, kJ/kg·K	Volume Expansivity, 1/K	Thermal Conductivity, W/m·K
p_{cr}=22.064	t_{cr}=374.1	–	∞	∞	∞
22.5	375.6	–	690.6	1.252	0.711
23.0	–	377.4	–	–	0.538
	377.5	–	284.3	0.508	–
23.5	–	379.2	–	–	0.468
	–	379.3	–	0.304	–
	379.4	–	171.9	– –	
24.0	–	381.0	–	–	0.429
	381.2	–	121.9	0.212	–
24.5	–	382.6	–	–	0.405
	–	383.0	–	0.161	–
	383.1	–	93.98	–	–
25.0	–	384.0	–	–	0.389
	384.9	–	76.44	–	–
	–	385.0	–	0.128	–
25.5	386.7	–	64.44	0.107	no peak

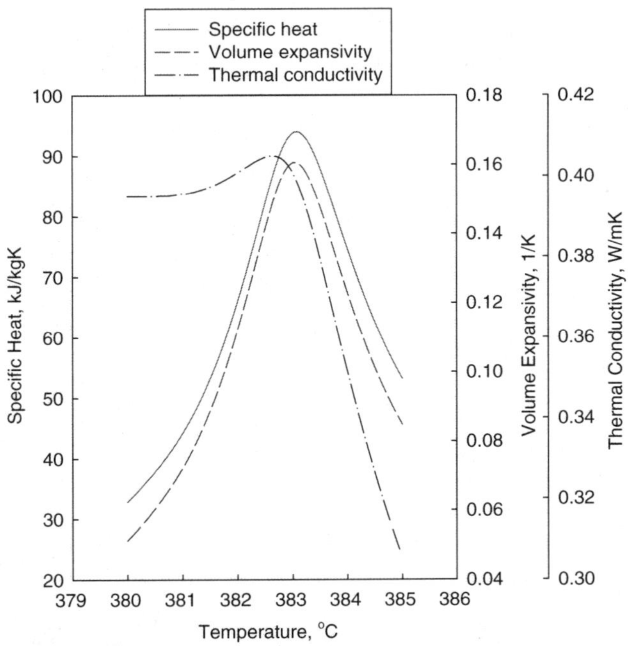

Figure 2.4. Specific heat, volumetric thermal expansion coefficient; and thermal conductivity vs. temperature (NIST 2002): p = 24.5 MPa.

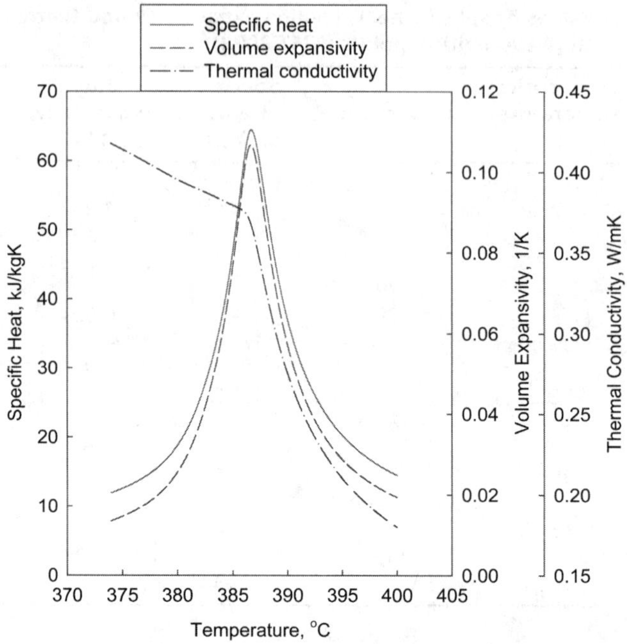

Figure 2.5. Specific heat, volumetric thermal expansion coefficient; and thermal conductivity vs. temperature (NIST 2002): p = 25.5 MPa.

some early correlations were affected by this finding near the critical and pseudocritical points. However, it is not always possible to determine which correlations were affected.

Chapter 3

POWER-PLANT "STEAM" GENERATORS WORKING AT SUPERCRITICAL PRESSURES: REVIEW AND STATUS

3.1 RUSSIAN SUPERCRITICAL UNITS

Early studies in Russia related to the heat transfer at supercritical pressures started in the late 1940s. In the 1950s, Podol'sk Machine-Building plant "ЗиО" ("ZiO") (Plant by the name of S. Ordzhonikidze) manufactured several small experimental supercritical "steam" generators for research institutions such as: *(i)* "ВТИ" ("VTI")—All-Union Heat Engineering Institute (Moscow) with "steam" parameters of 29.4 MPa and 600°C (Shvarts et al., 1963); *(ii)* "ЦКТИ" ("TsKTI")—Central Boiler-Turbine Institute by Polzunov (St.-Petersburg); and *(iii)* Kiev Polytechnic Institute with "steam" parameters of 39 MPa and 700°C (Kirillov, 2001).

The implementation of supercritical power-plant "steam" generators in Russia (the former USSR) started with units having thermal powers of 300 MW (Ornatskiy et al. 1980). Two leading Russian manufacturing plants: "ТКЗ" ("TKZ")—Taganrog Power-Plant Steam Generator's Manufacturing plant (Taganrog, Ukraine) and "ЗиО" (Podol'sk, Russia) developed and manufactured the first units, with the assistance of research institutions such as "ЦКТИ" and "ВТИ." Supercritical "steam" generators are usually the once-through type boilers (Belyakov 1995).

Power-plant "steam" generator ТПП-110 (TPP-110) manufactured at "ТКЗ" in 1961 was the first industrial unit operating at supercritical conditions in the former USSR, and was used at coal-fired power plant. Its design included a liquid slug drain. A total of six units were put into operation.

Also, a power-plant "steam" generator (model ПК-39 (PK-39)) was built at "ЗиО" in 1961. Next year, "ЗиО" designed a new unit, ПК-41, to work with residual fuel oil and natural gas. Later, in 1964 and 1967, upgraded designs of ТПП-110 (units ТПП-210 and ТПП-210A) were developed and manufactured. In these units, it was decided to decrease the temperature of the primary "steam" from 585°C to 565°C.

Based on industrial experience, several upgraded designs were manufactured at "ТКЗ" (units ТГМП-144 (TGMP-144) for residual fuel oil and natural gas, ТПП-312 (1970) for coal, ТПП-314 (1970) for residual fuel oil and natural gas, and ТГМП-144 (1971) for residual fuel oil and natural gas with pressurized combustion chamber) and "ЗиО" (units П-50 (1963) for coal, ПК-59 (1972) for brown coal (lignite), and П-64 (P-64) (1977) for Yugoslavian lignites).

The 300-MW power-"steam"-generating units have the following characteristics:

• "steam" capacity, t/h	950–1000
• Pressure (primary "steam"), MPa	25
• Temperature (primary "steam"), °C	545–585
• Pressure (secondary steam), MPa	3.5–3.9
• Feed-water temperature, °C	260–265
• Thermal efficiency, %	88–93

The next stage in further development of supercritical "steam" generators involved an increase in their thermal capacity to 500 MW (units manufactured at "ЗиО": П-49 (1965) and П-57 (1972)) and 800 MW (units manufactured at "ТКЗ": ТПП-200 (1964), ТГМП-204 (1973) and ТГМП-324; unit manufactured at "ЗиО": П-67 (1976)).

The 500-MW power-"steam"-generating units have the following characteristics:

• "steam" capacity, t/h	1650
• Pressure (primary "steam"), MPa	25
• Temperature (primary "steam"), °C	545
• Pressure (secondary steam), MPa	3.95
• Temperature (secondary steam), °C	545
• Feed-water temperature, °C	277
• Thermal efficiency, %	92

The 800-MW power-"steam"-generating units have the following characteristics:

• "steam" capacity, t/h	2650
• Pressure (primary "steam"), MPa	25
• Temperature (primary "steam"), °C	545
• Pressure (secondary steam), MPa	3.44
• Temperature (secondary steam), °C	545
• Feed-water temperature, °C	275
• Thermal efficiency, %	92–95

In 1966, the first 1000-MW ultra-supercritical plant started its operation in Kashira with a primary "steam" pressure of 30.6 MPa, and primary and reheat temperatures of 650°C and 565°C, respectively (Smith 1999).

In modern designs of supercritical units, the thermal capacity was upgraded to 1200 MW (unit manufactured at "ТКЗ": ТГМП-1204 (1978), "steam" generating capacity of 3950 t/h and thermal efficiency of 95%) (Ornatskiy et al.

1980). This, one of the largest in the world, supercritical power-generating unit operates with a single-shaft turbine at the Kostroma district power plant and is a gas-oil-fired "steam" generator (Belyakov 1995).

Over the last 25 years, more than 200 supercritical units were manufactured and put into operation in Russia (Smith 1999). Supercritical "steam" generators manufactured in Russia are listed in Table 3.1.

Table 3.1. Supercritical "steam" generators manufactured in Russia (Belyakov 1995).

Capacity, MW	Manufacturer				Total
	"ТКЗ" (Taganrog)		"ЗиО" (Podol'sk)		
	gas-oil	coal	gas-oil	coal	
300	91	49	19	36	195
500	–	–	–	16	16
800	17	2	–	1	20
1200	1	–	–	–	1
In all	109	51	19	53	–
	160		72		232

3.2 SUPERCRITICAL UNITS DESIGNED IN THE USA

In the early 1950s, the development work on supercritical "steam" generators started in the USA (Lee and Haller 1974). The first supercritical "steam" generator was put into operation at the Philo Plant of American Electric Power in 1957. The capacity of this unit was 120 MW with "steam" parameters of 31 MPa and 620°C/566°C/538°C (main/reheat/reheat) (Retzlaff and Ruegger 1996).

Later on, supercritical power-plant "steam" generators in the USA were developed, manufactured and put into operation with a "steam" generating capacity of 500 MW (1961) (Ornatskiy et al. 1980).

In the early 1960s, another plant was built with ultra-supercritical "steam" parameters (pressure of 30 MPa, temperatures (primary and reheat) of 650°C) (Smith 1999).

Major USA manufacturers of power-plant "steam" generators such as Babcock & Wilcox, Combustion Engineering, Inc., Foster Wheeler, and others were involved in the development and manufacturing of the supercritical units. The supercritical units found their application at thermal capacities from 400 to 1380 MW. Often the subcritical units for 1000 MW and higher were replaced with supercritical "steam" generators in the USA (Ornatskiy et al. 1980).

United States power "steam"-generating units have the following averaged characteristics (Ornatskiy et al. 1980):

- "steam" capacity, t/h 1110–4440
- Pressure (primary "steam"), MPa 23–26
- Temperature (primary "steam"), °C 538–543
- Temperature (secondary steam), °C 537–551

The characteristics of two supercritical units built by "Babcock & Wilcox" are listed below (Ornatskiy et al. 1980).

Power-plant "steam" generator put into operation at the "Paradise" power plant (USA) in 1969 (for 1130 MW unit):

- "Steam" capacity, t/h 3630
- Pressure (primary "steam"), MPa 24.2
- Temperature (primary "steam"), °C 537
- Steam capacity (secondary steam), t/h 2430
- Pressure (secondary steam), MPa 3.65
- Temperature (secondary steam), °C 537
- Feed-water temperature, °C 288
- Thermal efficiency, % 89

Power-plant "steam" generators put into operation at the "Emos" (1973) and "Gevin" (1974 – 1975) power plants (USA) (for 1130 MW units):

- "Steam" capacity, t/h 4438
- Pressure (primary "steam"), MPa 27.3
- Temperature (primary "steam"), °C 543
- Steam capacity (secondary steam), t/h 3612
- Pressure (secondary steam), MPa 4.7
- Temperature (secondary steam), °C 538
- Feed-water temperature, °C 291
- Thermal efficiency, % 93

The largest supercritical units are rated up to 1300 MW with "steam" parameters of 25.2 MPa and 538°C (Lee and Haller 1974).

3.3 RECENT DEVELOPMENTS IN SUPERCRITICAL "STEAM" GENERATORS AROUND THE WORLD

Recently, supercritical power-plant "steam" generators are working around the world with a wide range of "steam" parameters (see Table 3.2).

On average, the usage of supercritical "steam" generators instead of subcritical ones increased overall power plant efficiency from 45% to about 50% (Smith 1999).

In Japan, the first supercritical "steam" generator (600 MW) was commissioned in 1967 at the Anegasaki plant (Oka and Koshizuka 2002; Tsao and Gorzegno 1981). Nowadays, many power plants are equipped with supercritical "steam" generators and turbines. Hitachi operating supercritical pressure "steam" turbines have the following average parameters: output—350 (1 unit), 450 (2 units), 500 (3 units), 600 (11 units), 700 (4 units), and 1000 MW (4 units); "steam" pressure about 24.1 MPa (one unit 24.5 MPa); "steam" temperature (main/reheat)—538°C/566°C (the latest units 600°C/600°C (610°C)).

In Germany, at the end of the 1990s, construction was started on Unit "K" of RWE Energie's Niederaußem lignite-burning power station near Cologne (Heitmüller et al. 1999). This power plant is described as the most advanced lignite-fired power

Table 3.2. Characteristics of modern supercritical "steam" generators (Smith 1999).

	"Steam" Parameters					
	Capacity	**Primary**		**Reheat**		**Feed Water**
Country	**t/h**	***p*, MPa**	***t*, °C**	***p*, MPa**	***t*, °C**	***t*, °C**
China	–	25	538	–	566	–
Denmark	–	30	580	7.5	600	320
Germany	2420	26.8	547	5.2	562	270
Japan*	350–1000	24.1	538	–	566	–
		25	600	–	610	300
		31.1	566	–	566	–

*Updated with recent data.

plant in the world with 45.2% planned thermal efficiency. At a later date, with new dry lignite technology introduced, a further increase in efficiency of 3% – 5% is expected. The new Unit "K" will have the following parameters: output of about 1000 MW and "steam" conditions of 27.5 MPa and 580°C/600°C (main/reheat).

In Denmark (Noer and Kjaer 1998), the first supercritical power plant started operation in 1984, and today a total of seven supercritical units are in operation. Main parameters of these units are: output—2 units 250 MW, the rest 350 - 390 MW; "steam" pressure 24.5–25 MPa; "steam" temperature 545–560°C, reheat temperature 540 – 560°C; feed-water temperature 260 – 280°C; and net efficiency 42%–43.5%. Main parameters of ultra-supercritical units: "steam" pressure 29–30 MPa; "steam" temperature 580°C; steam reheat temperature 580°C–600°C, feed-water temperature 300°C–310°C and net efficiency 49%–53%.

So-called "Ultra-supercritical boilers" are now being researched and deployed world wide, particularly in Japan, Korea and China. Using double-steam reheat and advanced high-temperature blade materials, the turbine inlet temperature is being extended to 625° C at pressures of up to 34 MPa, with overall efficiencies then approaching 51%–53%.

Chapter 4

SUPERCRITICAL WATER-COOLED NUCLEAR REACTOR CONCEPTS: REVIEW AND STATUS

4.1 GENERAL CONSIDERATIONS

Concepts of nuclear reactors cooled with water at supercritical pressure were studied as early as the 1950s and 1960s in the USA and former USSR (Oka 2000; Wright and Patterson 1966; Bishop et al. 1962; Skvortsov and Feinberg 1961; Marchaterre and Petrick 1960; Supercritical pressure power reactor 1959). The main characteristics of the first concepts of SCWRs are listed in Table 4.1.

After a 30-year interval, the idea of developing nuclear reactors cooled with supercritical water became attractive as the ultimate development path for water cooling. Several countries (Canada, Germany, Japan, Korea, Russia, USA, and others) have started R&D work in that direction. However, none of these concepts is expected to be implemented in practice before 2015–2020.

The main objectives of using supercritical water in nuclear reactors are:

1. Increase the efficiency of modern nuclear power plants (NPPs) from 33%–35% to about 40%–45%; and
2. Decrease capital and operational costs and hence decrease electrical energy costs (~$1000 US/kW).

Currently, the latest designs of subcritical pressure nuclear reactors (so-called Generation III+ nuclear reactors), which will be prototyped over next 10 years or so, are expected to have specific overnight capital cost of about $1500 US/kW.

Supercritical water NPPs will have much higher operating parameters (see Figure 4.1: pressure about 25 MPa and outlet temperature up to 625°C) compared to modern NPPs, and a simplified flow circuit (see Figure 4.2), in which steam generators, steam dryers, steam separators, etc., can be reduced or eliminated. Also, higher supercritical water temperatures allow direct thermo-chemical or indirect electrolysis production of hydrogen at low cost, due to increased process efficiencies. According to the IAEA (1999), the optimum required temperature is about 850°C and the minimum required temperature is around 650°C to 700°C, well within modern materials capability.

Table 4.1. First concepts of nuclear reactors cooled with supercritical water (Oka 2002; 2000).

	Company/Reactor Acronym (year)			
	Westinghouse		GE, Hanford	B & W
Parameters	**SCR (1957)**	**SCOTT-R (1962)**	**SCR (1959)**	**SCFBR (1967)**
Reactor type	Thermal	Thermal	Thermal	Fast
Pressure, MPa	27.6	24.1	37.9	25.3
Power, MW (thermal/electrical)	70/21.2	2300/1010	300/–	2326/980
Thermal efficiency, %	30.3	43.5	~40	42.2
Coolant temperature at outlet, °C	538	566	621	538
Primary coolant flow rate, kg/s	195	979	850	538
Core height/diameter, m/m	1.52/1.06	6.1/9.0	3.97/4.58	–
Fuel material	UO_2	UO_2	UO_2	MOX
Cladding material	SS	SS	Inconel-X	SS
Rod Diameter/Pitch, mm/mm	7.62/8.38	–	–	–
Moderator	H_2O	Graphite	D_2O	–

Explanations to the Table:
Acronyms: GE—General Electric; B & W—Babcock & Wilcox; SCR—SuperCritical Reactor; SCOTT-R—SuperCritical Once-Through Tube Reactor; and SCFBR—SuperCritical Fast Breeder Reactor.

Also, future nuclear reactors will have high indexes of fuel usage in terms of thermal output per mass of fuel (Kirillov 2000; Alekseev et al. 1989). Therefore, changing over to supercritical pressures increases the thermal output coefficient and decreases the consumption of natural resources of uranium. Due to the considerable reduction in water density in the reactor core, it might be possible to develop fast SCWRs with a breeding factor of more than one for converting fertile (non-fissionable fuel) to fissionable isotopes in a self-sustaining fuel cycle.

4.2 DESIGN CONSIDERATIONS

The design of SCWRs is seen as the natural and ultimate evolution of today's conventional nuclear reactors:

First, some designs of the modern Pressurized Water Reactors (PWRs) work at pressures about 16 MPa.

Second, some Boiling Water Reactors (BWRs) are a once-through or a direct-cycle design, i.e., steam from a nuclear reactor is forwarded directly into a turbine.

Third, some experimental reactors used nuclear steam superheaters with outlet steam temperatures well beyond the critical temperature but

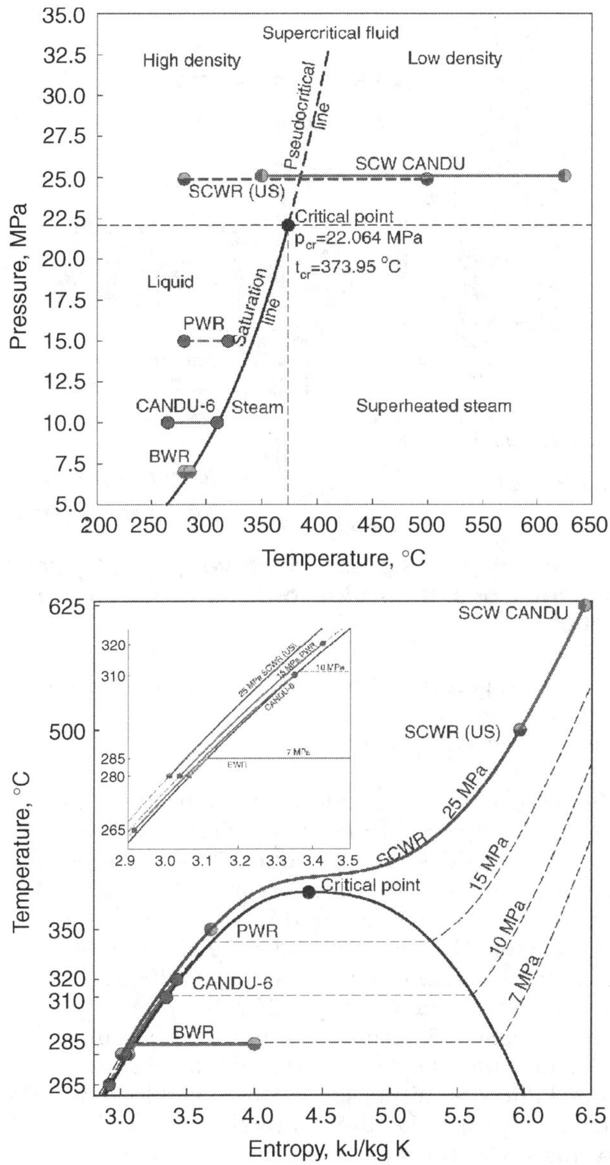

Figure 4.1. Pressure-temperature and temperature-entropy diagrams of water with typical operating conditions of SCWRs, PWR, CANDU-6 and BWR (figures partially based on data from Buongiorno and MacDonald (2003)).

at pressures below the critical pressure (DOE USA 2005; Grigor'yants et al. 1979; Baturov et al. 1978; Samoilov et al. 1976; Aleshchenkov et al. 1971; Dollezhal' et al. 1974, 1971, 1958; Kornbichler 1964; Margen 1961; Spalaris et al. 1961; Wallin et al. 1961).

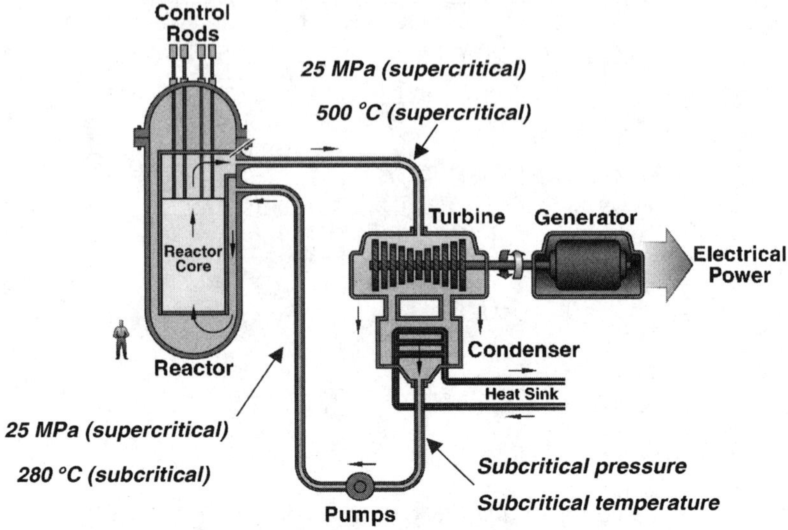

Figure 4.2. Schematic of US pressurized-vessel SCW nuclear reactor (courtesy of Professor J. Buongiorno (MIT) (Buongiorno and MacDonald 2003)).

Fourth, modern supercritical turbines, at pressures about 25 MPa and inlet temperatures of about 600°C, operate successfully at thermal power plants for many years.

The SCWR concepts therefore follow two main types, the use of either (a) a large reactor pressure vessel (Figures 4.2 and 4.3) (Buongiorno and MacDonald 2003) with wall thickness of about 0.5 m to contain the reactor core (fuelled) heat source, analogous to conventional PWRs and BWRs, or (b) distributed pressure tubes or channels analogous to conventional CANDU®[3] and RBMK nuclear reactors (Duffey et al. 2006, 2005a,b, 2004, 2003; Duffey and Pioro 2006; Khartabil et al. 2005; Torgerson et al. 2005; Gabaraev et al. 2005, 2004, 2003a,b; Kuznetsov and Gabaraev 2004).

The pressure-vessel SCWR design is developed largely in the USA, EU, Japan, Korea and China and allows using a traditional high-pressure circuit layout.

The pressure-channel SCWR design is developed largely in Canada (Figure 4.4) and in Russia (Figure 4.5) to avoid a thick wall vessel, and allows, in principle, the following key features for safety and performance:

a) Passive accident and decay heat removal by radiation and convection from the distributed channels even with no active cooling and no fuel melting. Thus, the system is potentially inherently safe.
b) Use of multi-pass reactor flows, so that reheat and superheat are possible while still keeping the pressure tube cool. Thus, the system can

[3] CANDU® (CANada Deuterium Uranium) is a registered trademark of Atomic Energy of Canada Limited (AECL).

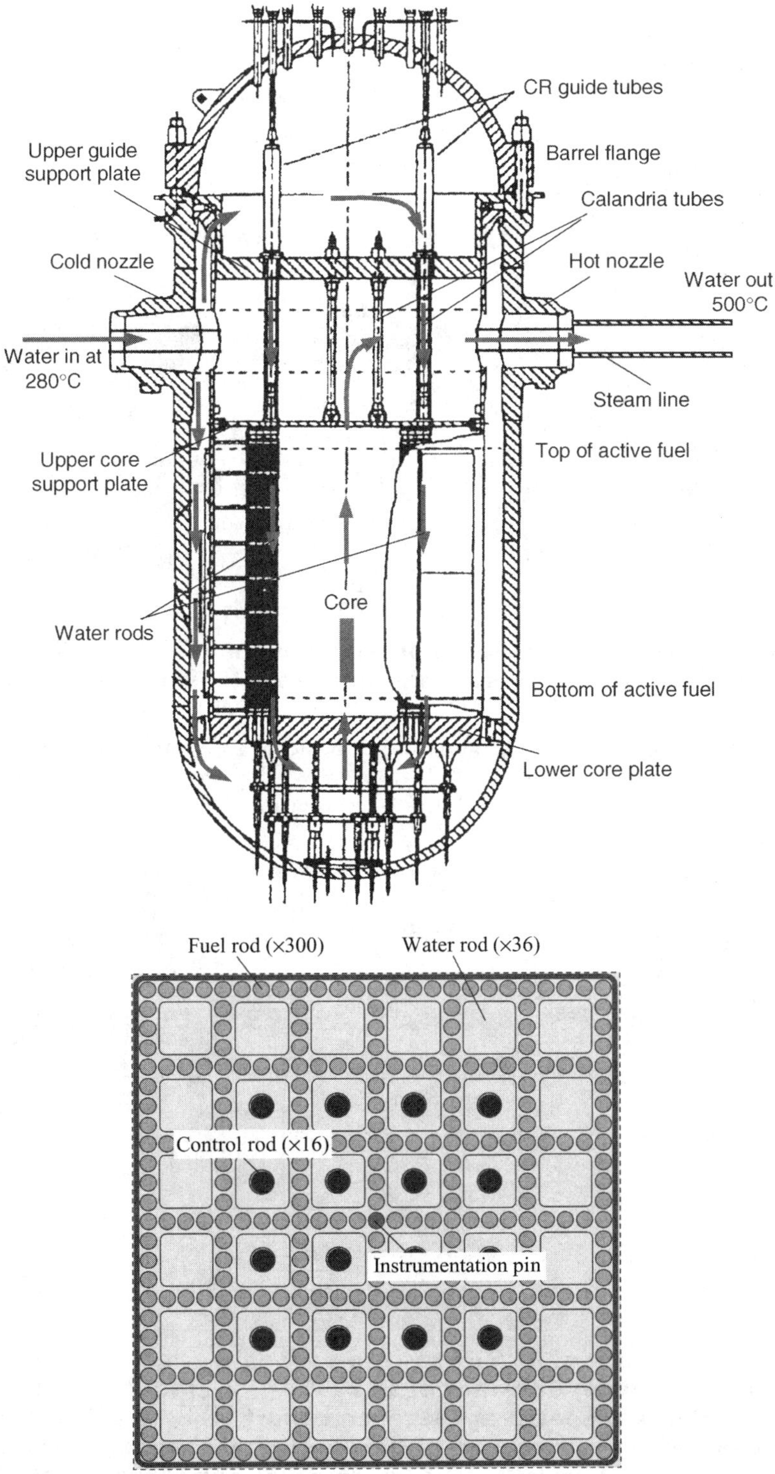

Figure 4.3. Schematics of US SCWR pressure vessel and reference SCWR fuel assembly with water rods (courtesy of Professor J. Buongiorno (MIT) (Buongiorno and MacDonald 2003)).

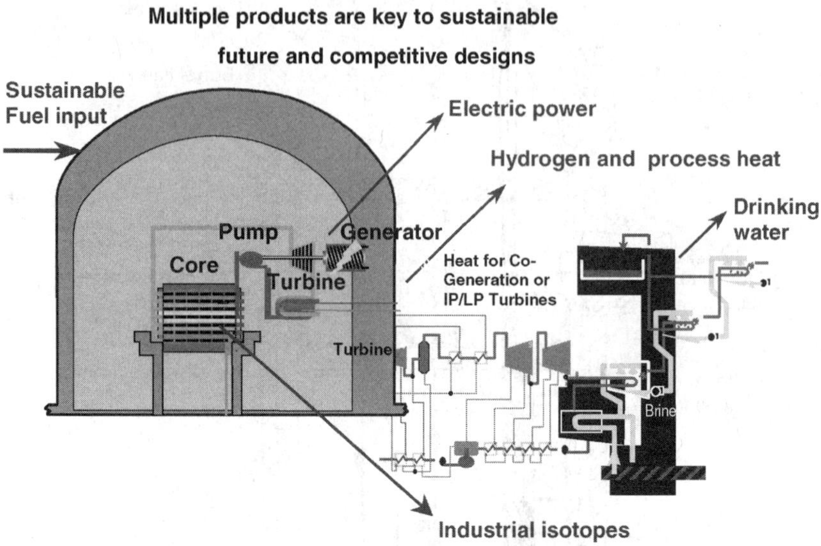

Figure 4.4. General concept of pressurized-channel SCW CANDU reactor: IP—intermediate-pressure turbine, and LP—low-pressure turbine.

produce process heat on demand. Reactor size (and thermal power) can be adjusted from 300 to 1400 MW_e depending on the customer site, financing and product mix application.

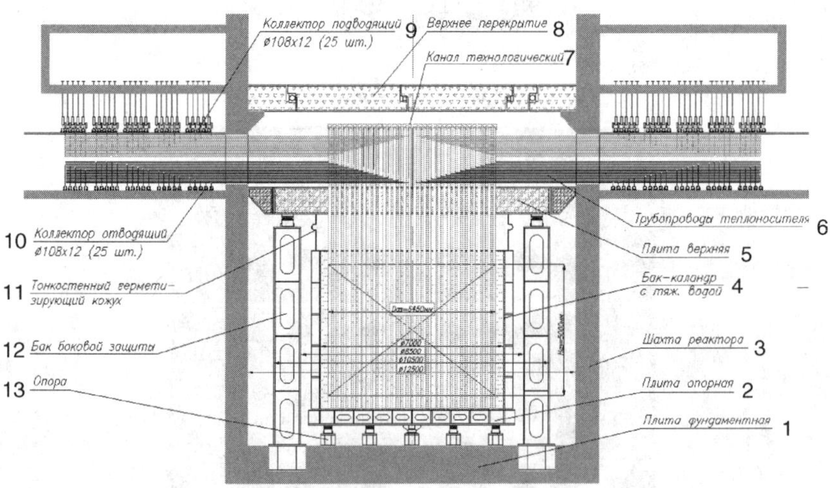

Figure 4.5. Layout of RDIPE pressure-channel nuclear reactor (Duffey et al. 2006; Gabaraev et al. 2005): 1) & 2) foundation and bearing plates, respectively; 3) reactor shaft; 4) calandria tank; 5) top plate; 6) coolant pipes; 7) technological channel; 8) top cover; 9) & 10) inlet and outlet collectors, respectively; 11) thin-wall sealing casing; 12) lateral shielding tank; and 13) supports.

The system features together set the fuel design, channel power, core lattice pitch, enrichment, and flow circuit parameters, where the coolant is usually water. A thermal neutron spectrum is used with either a solid moderator using graphite or zirconium hydride, or a liquid using heavy water moderator.

To reduce the severe axial flux tilt due to the large density decrease as the coolant is heated, the core flow path can be a re-entrant in the vessel option, coming down unheated first and then turning into an upflow; or interlaced or re-entrant in channels with flow in opposite directions. Both options allow the chance to reduce pressure boundary temperatures, by partly insulating the pressure-retaining vessel of the channel wall using the first pass of the unheated flow. Typical outlet temperatures are expected to be up to near 600°C to match supercritical turbine inlet needs. There is also the option of a superheat pass (return flow) to further raise outlet temperatures if needed (for example, for hydrogen production).

The limit on supercritical water outlet temperature is effectively set by the fuel cladding, since the peak clad temperature will be some 20% higher than the average, and the corrosion rates much higher. Hence, the thermal limits depend on the wall heat transfer, and estimates of the peak values have been made to establish the margins and clad lifetime expected before refuelling.

Moreover, one of the unique features of the SCWRs is the very low coolant mass-flow rates that are required through the reactor core because of the high thermal capacity. Preliminary calculations showed that the rate could be about *five to eight times less* than in modern PWRs, significantly reducing the pumping power and costs. This improvement is due to the considerable increase in the specific enthalpy at supercritical conditions, of about 2000 kJ/kg. Therefore, tightly packed cylindrical fuel bundles, which are more acceptable in SCWRs than in other types of reactors, can be used. These tight bundles have a significant pressure drop, which in turn can enhance the hydraulic stability of the flow. Since the supercritical water is a single-phase "gas," then the cladding surfaces can and should be finned or ridged to enhance turbulence levels to increase the HTC. This is already done for Advanced Gas-cooled Reactors (AGRs) today, and hence will increase the heat transfer and reduce peak cladding temperatures in normal operation.

To optimize thermal efficiency and capital cost, there are also options for the thermal cycles (Spinks et al. 2002; Bushby et al. 2000a,b; Oka et al. 1996), with either a direct cycle into a supercritical turbine or an indirect using a heat exchanger. Analyses by Spinks et al. (2002) show a cost reduction of 15% or more for the direct-cycle option, without use of the process heat.

One advantage of separating the moderator and coolant in the pressure-channel design is that the moderator can act as a backup heat sink in the event when emergency core cooling is not available. The advanced fuel-channel design (Khartabil et al. 2005) combined with a passive-moderator cooling system (Khartabil 1998) results in a design where severe core damage is practically eliminated. In this design, the passive-moderator loop operates continuously to remove heat deposited in the moderator during normal operation. The moderator heat during normal operation is comparable to decay heat following reactor shutdown; therefore, the moderator can also be used to remove decay heat following postulated accidents.

In summary, the use of supercritical water in nuclear reactors will, according to the US DOE (Roadmap) Generation IV[4] Nuclear Energy Systems Report (2001):

- significantly increase thermal efficiency from 33%–35% up to 40%–45%;
- eliminate steam generators, steam separators, steam dryers and recirculation pumps;
- allow the production of hydrogen due to high coolant outlet temperatures;
- decrease reactor coolant pumping power;
- reduce frictional losses;
- lower containment loadings; and
- eliminate dryout.

The latest concepts of SCWRs are summarized in Table 4.2. Figure 4.2 shows the schematic of the US pressurized-vessel SCWR, Figure 4.4—the general concept of the pressurized-channel SCW CANDU nuclear reactor, and Figure 4.5—the general scheme of the RDIPE pressurized-channel SCWR.

Specific features of SCWRs (see Table 4.2) being developed:

- in Canada are listed in Duffey et al. 2006, 2005a,b, 2004, 2003; Duffey and Pioro 2006; Khartabil et al. 2005; Torgerson et al. 2005; Spinks et al. 2002; Bushby et al. 2000a,b; and Khartabil 1998;
- in Europe are described in Hofmeister et al. (2005), Mori et al. (2005), Waata et al. (2005), Marsault et al. (2004); Starflinger et al. (2004), Aksan et al. (2003), Bittermann et al. (2003a,b), Cheng et al. (2003), Rimpault et al. (2003) and Squarer et al. (2003a,b);
- in Japan can be found in Kamei et al. (2005), Kitou and Ishii (2005); Ookawa et al. (2005), Yamada and Oka (2005), Yang et al., (2005), Yamaji et al. (2005a-d, 2004, 2003a-d), Yi et al. (2005a,b; 2004a-c, 2003), Yoo et al. (2005), Ishiwatari et al. (2004, 2003a-e, 2002, 2001), Oka and Yamada (2004), Tanabe et al. (2004), Kataoka et al. (2003), Koshizuka et al. (2003, 1993), Oka et al. (2003a-d, 2002, 2000, 1996, 1995a,b, 1994a,b, 1993, 1992a,b), Shioiri et al. (2003), Cheng et al. (2002), Oka (2003, 2002, 2000), Kitoh et al. (2001, 1999), Nakatsuka et al. (2001, 2000, 1998), Koshizuka and Oka (2000, 1998), Mukohara et al. (2000a,b, 1999), Oka and Koshizuka (2000, 1998, 1993), Lee et al. (1999, 1998), Kitoh et al. (1998), Dobashi et al. (1998a,b, 1997), Jevremovich et al. (1996, 1994, 1993a,b), Okano et al. (1996a,b, 1994), Oka and Kataoka (1992) and Kataoka and Oka (1991);
- in Korea—in Joo et al. (2005) and Bae et al. (2004);
- in Russia—in Gabaraev et al. 2005, 2004, 2003a,b; Kuznetsov and Gabaraev 2004; and
- in the—USA Buongiorno et al. (2006, 2003), MacDonald et al. (2005), Modro (2005), Yang and Zavaljevski (2005), Zhao et al. (2005), Fischer et al. (2004), Buongiorno (2004, 2003), Buongiorno and MacDonald (2003a,b,c), and Davis et al. (2003).

[4] On progress of the Generation IV nuclear energy systems, see Sagayama (2005).

Table 4.2. Modern concepts of nuclear reactors cooled with supercritical water.

Parameters	Unit	SCW CANDU	HPLWR	SCLWR-H	SCFBR-H	SCWR	B-500 SKDI	ChUWR	ChUWFR	KP-SKD
Reference	–	Bushby et al. 2000; Khartabil et al. 2005	Squarer et al. 2003	Yamaji et al. 2004	Oka, Koshizuka 2000	Bae et al. 2004; Bae 2004	Silin et al. 1993	Kuznetsov 2004 (project from 80s)	Gabaraev et al. 2003a,b; 2004	Gabaraev et al. 2005
Country	–	Canada	EU/Japan	Japan		Korea	Russia	Russia	Russia	Russia
Organization	–	AECL	EU / U of Tokyo	University of Tokyo		KAERI / Seoul NU	Kurchatov Institute	RDIPE (НИКИЭТ)		
Reactor type	–	PT	RPV	RPV	RPV	RPV	RPV	PT	PT	PT
Spectrum	–	Thermal	Thermal	Thermal	Fast	Thermal	Thermal	Thermal	Fast	Thermal
Power thermal	**MW**	2540	2188	2740	3893	3846	1350	2730	2800	1960
electrical	**MW**	1140	1000	1217	1728	1700	515	1200	1200	850
linear max/ave	**kW/m**		39/24	39/18	39	39/19		38/27		69/34.5
Thermal eff.	**%**	45	44	44.4	44.4	44	38.1	44	43 (48)	42
Pressure	**MPa**	25	25	25	25	25	23.5	24.5	25	25
T_{in} coolant	**°C**	350	280	280	280	280	355	270	400	270
T_{out} coolant	**°C**	625	500	530	526	508	380	545	550	545
Flow rate	**kg/s**	1320	1160	1342	1694	1862	2675	1020		922
Core height	**m**		4.2	4.2	3.2	3.6	4.2	6	3.5	5
Diameter	**m**	~4		3.68	3.28	3.8	2.61	11.8	11.4	6.45
Fuel	–	UO_2/Th	UO_2 or MOX	UO_2	MOX	UO_2	UO_2	UCG	MOX	UO_2
Enrichment	**% wt.**	4	<6%	~6.1		5.8	3.5	4.4		6

(*continued*)

Table 5.1. (*continued*)

Parameters	Unit	SCW CANDU	HPLWR	SCLWR-H	SCFBR-H	SCWR	B-500 SKDI	ChUWR	ChUWFR	KP-SKD
Cladding material	–	Ni alloy	SS	Ni alloy	Ni alloy	SS	Zr alloy/ SS	SS	SS	SS
# of FA		300	121	121	419	157	121	1693	1585	653
# of FR in FA		43	216/252	300		284	252	10	18	18
D_{rod}/δ_w	**mm/mm**	11.5 and	8	10.2/0.63	12.8	9.5/0.635	9.1 (Zr),	12/1	12.8	10/1
Pitch	**mm**	13.5	9.5			11.5	8.5 (SS)			
T_{max} cladding	**°C**	<850	620	650	620	620	425	630	650	700
Moderator	–	D_2O	H_2O	H_2O	H_2O	ZrH_2	H_2O	Graphite	–	D_2O

Explanations to the Table:

Concepts appear according to the alphabetical order of the country of origin.

ChUWR – Channel-type Uranium-graphite Water Reactor with annular-type elements cooled from inside; ChUWFR – Channel-type Uranium-graphite Water Fast Reactor; FA – fuel assembly; FR – fuel rod; HPLWR – High Performance Light Water Reactor; KP-SKD – Channel Reactor of Supercritical Pressure (in Russian abbreviations); PT – Pressure Tube (reactor); PVWR – Pressure-Vessel Water Reactor; RPV – Reactor Pressure Vessel; SCFBR-H – SuperCritical Fast Breeder Reactor with High temperature; SCLWR-H – SuperCritical Light Water Reactor with High temperature; Seoul NU – Seoul National University; SKDI – SuperCritical Pressure Integral (reactor) (in Russian abbreviations); TBD – To Be Determined; U of Tokyo – University of Tokyo; VNIIAM – All-Union Scientific-Research Institute of Atomic Machine Building (in Russian abbreviations); WWPR-SCP – Water-Water Power Reactor ("VVER" in Russian abbreviations) of SuperCritical Pressure.

Continuation of Table 1

Parameters	Unit	PVWR	WWPR-SCP	SCWR-US
Reference	–	Filippov et al. 2003	Baranaev et al. 2004	Buongiorno, MacDonald 2003
Country	–	Russia	Russia	USA
Organization	–	VNIIAM / Kurchatov Institute	IPPE (ФЭИ)	INEEL
Reactor type	–	RPV	RPV	RPV
spectrum	–	Thermal	Fast	Thermal
Power thermal	MW	3500	3830	3575
electrical	MW	1500	1700	1600
linear max/ave	kW/m		35/15.8	39/19.2
Thermal eff.	%	43	44	44.8
Pressure	MPa	25	25	25
T_{in} coolant	°C	280	280	280
T_{out} coolant	°C	550–610	530	500
Flow rate	kg/s	1600	1860	1843
Core height	m	3.5	4.05	4.87
diameter	m	2.92	3.38	3.91
Fuel	–	UO_2	MOX	UO_2 95%
Enrichment	% wt.			5
Cladding material	–		Ni alloy	TBD
# of FA		37	241	145
# of FR in FA			252	300
D_{rod}/δ_w	mm/mm	Sphere 1.8	10.7/0.55	10.2/0.63
Pitch	mm		12	11.2
T_{max} cladding	°C	630–730	630	
Moderator	–	H_2O	$ZrH_{1.7}$	H_2O

Typical values of the HTC and wall temperatures at SCWR operating conditions are presented in Section 11.1.4.

4.3 SUPERCRITICAL WATER-COOLED CANDU NUCLEAR REACTOR CONCEPT

The SCW CANDU nuclear reactor is a pressurized-channel type reactor (Torgerson et al. 2005; Spinks et al. 2002). The general concept of a SCW CANDU reactor is shown in Figure 4.4. Supercritical water (dense fluid) at a temperature of about 350°C (inlet temperature is below the pseudocritical temperature of 384.9°C) from a circulation pump enters the reactor core and heats up caused by the heat of fission to 625°C (outlet temperature is above the pseudocritical temperature of 384.9°C) at a pressure of about 25 MPa, which is above the critical pressure of 22.1 MPa. After that, supercritical water is directed to a turbine to perform useful work and returns back through the cooler to the circulation pump. Due to high operating parameters, the coolant in the second circuit may be used for a heat supply or be directed to intermediate or low-pressure turbines.

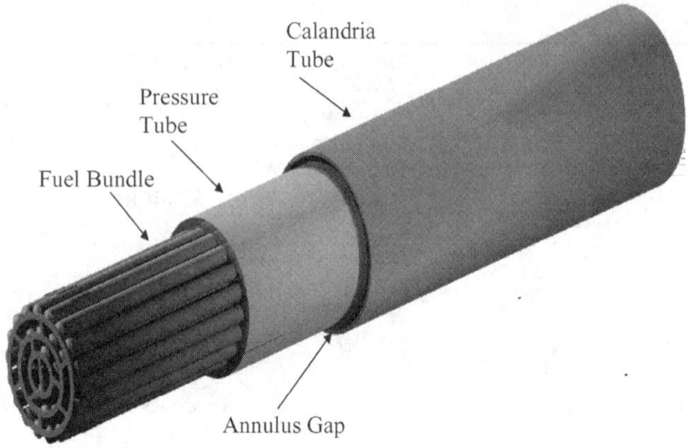

Figure 4.6. Current CANDU reactor fuel channel design (Duffey et al. 2003).

High pressures and temperatures inside the reactor core require a new design of the fuel channel (Duffey et al. 2003). Current CANDU reactors use a channel design that consists of a pressure tube that is insulated from the cool heavy-water moderator by an annulus gap and a calandria tube (see Figure 4.6).

The insulated pressure tube (Figure 4.7) is a key technology that is needed to make use of supercritical water in channel-type reactors feasible (however, other options such as a solid moderator using graphite or zirconium hydride are possible). In this design, the pressure tube is insulated from the coolant

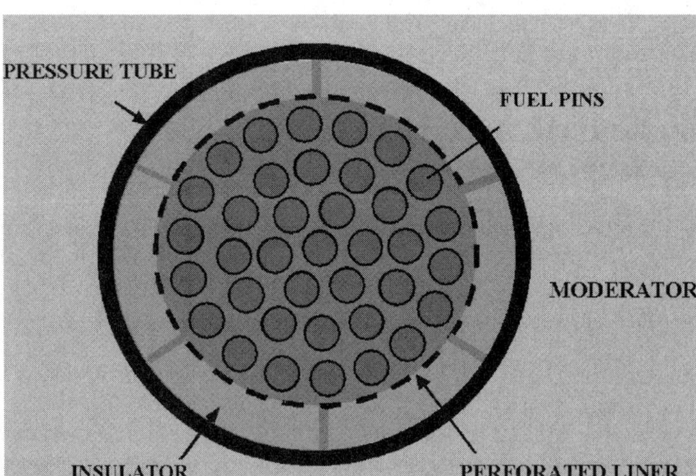

Figure 4.7. Schematic of CANTHERM fuel channel design (Duffey et al. 2006).

Table 4.3. Typical values of HTC (normal heat transfer) for reactor coolants within operating ranges.

Reference	Reactor Coolant	Heat-Transfer Cooling Conditions	Typical Geometry	HTC Range (kW/m^2K)
Hewitt and Collier 2000	Subcritical water	Forced convection	Fuel bundles	30
		Flow boiling		60
	Subcritical CO_2	Forced convection		1
Yamagata et al. 1972	SCW	Forced convection: G = 1,120 kg/m^2s	Inside circular tube (10 mm ID)	10–15
Dyadyakin and Popov 1977 (correlation)	SCW	Forced convection: G = 860 kg/m^2s	7-element helically finned bundle model (correlation used for $D_{hy}\approx 8$ mm)	4
Pioro and Khartabil 2005	Supercritical CO_2	Forced convection: G = 900 kg/m^2s G = 2000 kg/m^2s	Inside circular tube (8 mm ID)	2–3 3–4

by using an internal layer of low-neutron absorbing material. Furthermore, an internally insulated pressure tube operates at much lower temperatures (close to the moderator temperature) than in current reactors, which means that any increase in pressure-tube thickness, if any, is negligible.

Comparison of thermophysical properties of water for operating conditions of subcritical and supercritical pressure reactors is presented in Table 4.3.

Like the commercialization of High-Temperature Reactors (HTRs), Very High-Temperature Reactors (VHTRs) and SCWRs themselves, direct application of heat from HTRs to produce hydrogen is not an immediate prospect. In the near term, electrolysis can gradually supplement first-generation production by Steam-Methane-Reforming (SMR) process ($CH_4+2{\cdot}H_2O = 4{\cdot}H_2+CO_2$) with the electricity produced in low-cost Generation III+ reactors such as the ACR™ (Advanced CANDU Reactor) at other than periods of peak electrical demand. Economic assessments show this is competitive with the SMR process for large-scale, industrial production of hydrogen as well as for dispersed, smaller-scale production.

Since so-called high-temperature electrolysis using solid oxide fuel cells (SOFCs) seems to have advantages in efficiency at higher temperatures, at say around 750°C and above, a possibility was examined of adding steam superheat channels to the SCWR concept to give even higher outlet temperatures.

This is relatively simple in principle, and has been demonstrated in practice at the Russian Beloyarsk NPP, named by I.V. Kurchatov (Grigor'yants et al. 1979; Baturov et al. 1978; Samoilov et al. 1976; Aleshchenkov et al. 1971; Dollezhal' et al. 1974, 1971, 1958). In these 170–500 MW_e power reactors of the Beloyarsk NPP, the superheat channels were supplied as a second pass with the exit steam from the first pass through the reactor core, to give an average outlet temperature of about 500°C at a pressure of about 7 to 8 MPa. Operation of these reactors with superheat channels was reported as entirely satisfactory, once some initial manufacturing issues had been resolved.

The chosen reactor exit temperature can be increased by either extending the channel length, or having one or more additional passes through the core. Superheat channels are then located at the periphery of the reactor core and have about 1.5 times lower heat flux compared to the average heat flux. For the SCWR superheat version of the SCW CANDU reactor, the following option was examined: the use of steam discharged from the HP turbine outline, reheated and then introduced at 7 MPa and 350°C into the superheat channels. The entire cycle is shown in outline in Figure 4.8.

Since these superheat channels are now at a much lower (highly subcritical) pressure near standard steam conditions, re-entrant flow or stainless steel can be used for the pressure tubes for these superheat channels. In addition, a degree of the superheat can be chosen and varied together with the mass-flow fraction that is superheated, as the steam demand may vary.

To estimate the core exit temperatures, the flow and heat transfer were calculated. Values of the HTC, cladding and bulk steam temperatures at steam superheat conditions in SCW CANDU reactor are shown in Figure 4.9. The HTC was calculated using the conventional Dittus-Boelter correlation (1930), as introduced by McAdams (1942) (for details, see Winterton (1998)):

$$\mathbf{Nu}_D = \mathbf{0.023\,Re}_D^{0.8}\,\mathbf{Pr}^{0.4}. \tag{4.1}$$

In general, the Dittus-Boelter correlation (Incropera and DeWitt 2002) is valid for single-phase heat transfer in channels within the following range: $0.7 \le \mathbf{Pr} \le 160$; $\mathbf{Re}_D \ge 10{,}000$; and $L/D_{hy} \ge 10$.

By simply tuning the channel flow (which can be done using orificing), the exit temperatures can be attained up to 850°C, as set by the channel thermal limits.

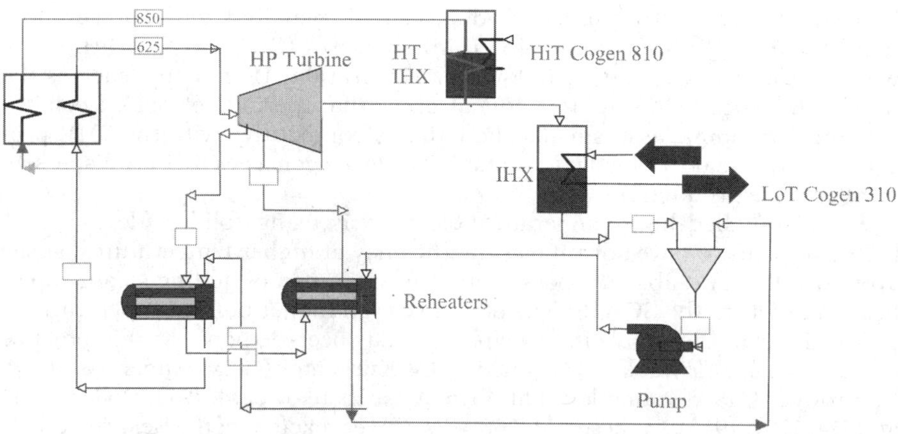

Figure 4.8. SCWR with superheat cycle and hydrogen co-generation (Duffey and Pioro 2006).

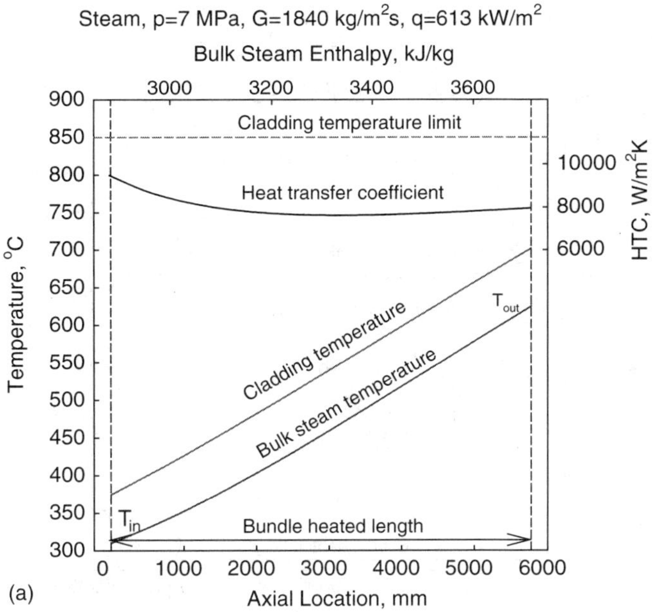

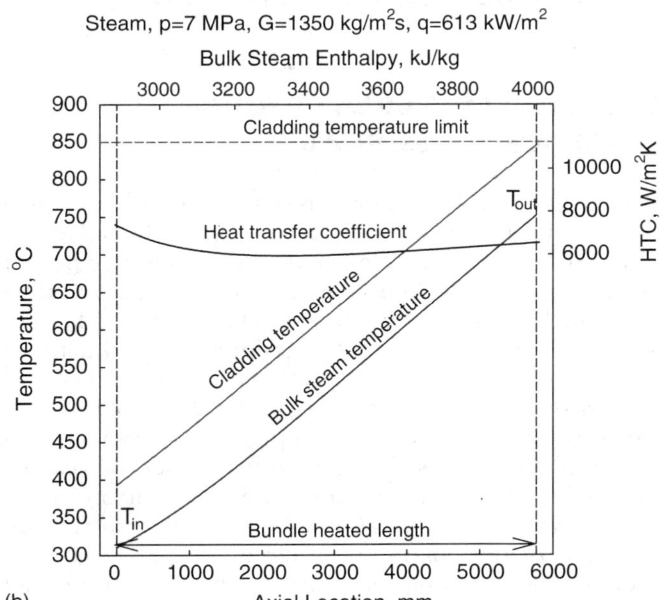

Figure 4.9. Continued

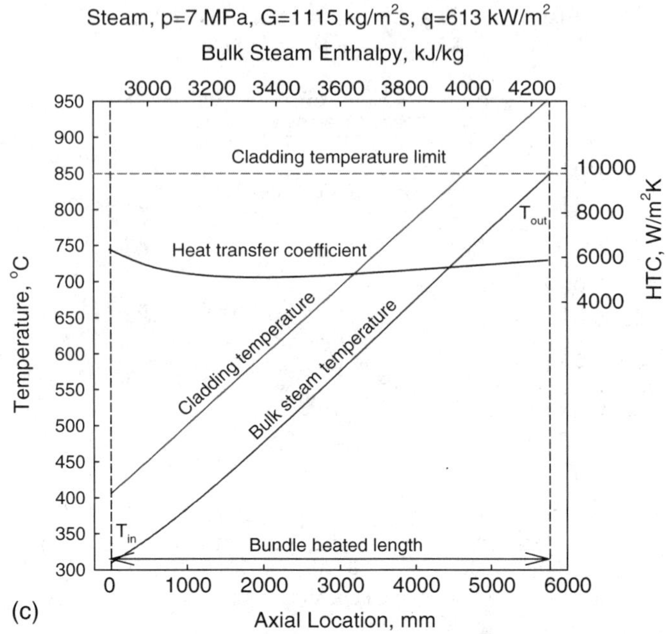

(c)

Figure 4.9. Temperature and HTC variations along 43-element bundle (D_{hy}≈8 mm) cooled with steam at various operating conditions (steam inlet temperature 310°C) (Duffey and Pioro 2006): (a) steam outlet temperature 625°C; (b) 750°C; and (c) 850°C.

4.4 SOME DESIGN FEATURES OF RDIPE PRESSURE-CHANNEL SCWR

The current concept of a direct-flow, pressure-channel reactor operating at supercritical pressures (KP-SKD) with vertical channels (Gabaraev et al. 2005) (Figure 4.5 and Table 4.1) possesses all advantages of current pressure-channel reactors. This reactor is a thermal-spectrum type, because of the special features of the channel design eliminating the closely spaced lattice of fuel elements in the core. Therefore, it cannot be tailored to a fast neutron spectrum. It has been shown (Gabaraev et al. 2003) that a fast pressure-channel reactor with a gaseous medium instead of the moderator does not possess an efficient fuel cycle because of the extremely high neutron leakage and low fuel-breeding ratio. Consequently, the proposed concept aims at thermal reactors also with a heavy-water moderator, which gives an acceptable neutron balance and deep fuel burn-up. The required temperature 80°C–90°C is easily achieved for such moderator because of the possibility of circulation and external cooling. A graphite moderator is not considered due to the difficulty of achieving the required cooling of the structure at supercritical operating conditions.

Neutron-physical characteristics and economic indicators of KP-SKD have been investigated for two types of fuel elements and channels:

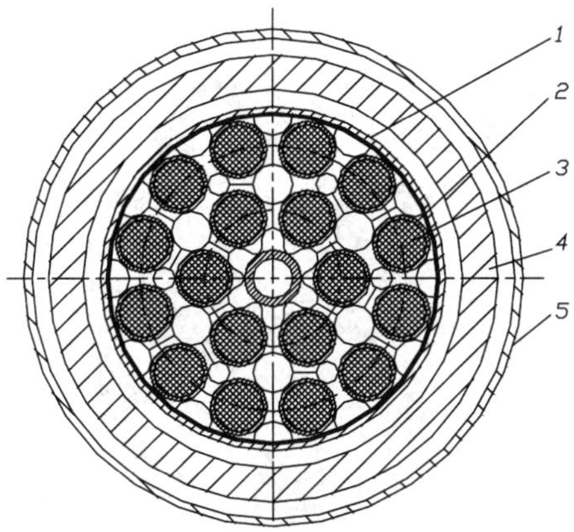

Figure 4.10. Cross-section of fuel channel with 18-rod fuel elements (Duffey et al. 2006; Gabaraev et al. 2005): 1) thin-wall 60 × 1 mm separating casing; 2) 10 × 1 mm fuel element cladding; 3) fuel element; 4) 78 × 6 mm channel tube; and 5) 90 × 3 mm calandria tube.

- A channel containing a pressure-carrying zirconium channel tube (placed in a calandria zirconium tube), a steel screen casing, and RBMK-type fuel assemblies; fuel—uranium dioxide pellets or cermet—micro-particles of uranium dioxide in a metallic matrix; fuel element coating–heat-resistant steel (Figure 4.10);
- A channel containing an assembly of tubular (annular) AMB-type fuel elements, likewise placed in a calandria tube; fuel—cermet; the outer coating and a central-pressure carrying fuel-element tube are made of heat-resistant steel (Figure 4.11).

The first type of channel design was developed with nuclear superheating of steam (Dollezhal' and Emel'yanov 1980). Here, the channel tube is protected from heating by a casing (in contrast to the Canadian concept, which uses a layer of thermal insulation). This design has been successfully tested as a steam-superheating channel in the No. 2 unit at the Beloyarskaya NPP.

In the first design (Figure 4.10), coolant with density of 0.78 g/cm^3 and temperature of 275°C is fed into the channel from below into a gap between the steel casing and the channel tube and rises upward, cooling the tube. In the top of the channel, the coolant flow with a temperature of 360°C and a density of 0.61 g/cm^3 turns into the space between the fuel elements, and is heated, on average, up to 550°C at the core exit. The maximum wall temperature is about 620°C.

In the second design (Figure 4.11), the coolant flows into the channel from above into the downflow central tube of each fuel element and the downflow (interior) tube of the central tube of the channel. At the bottom of the channel, coolant with temperature 380°C and density 0.43 g/cm^3 turns and enters the ring-shaped gap of the rising branch, cooling directly the interior wall of the fuel element and having the same water parameters at the

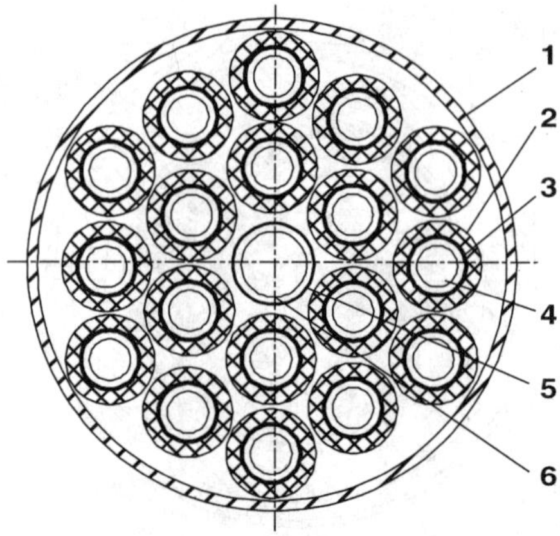

Figure 4.11. Cross-section of fuel channel with 18-annular fuel elements (Duffey et al. 2006; Gabaraev et al. 2005): l) 107 × 3 mm calandria tube; 2) 19 × 0.3 mm outer cladding of a fuel element; 3) 12 × 1.2 mm pressure-carrying fuel-element tube; 4) 70 × 0.2 mm separating tube; 5) 20 × 2 mm pressure-carrying central tube of a fuel assembly; and 6) 13 × 0.3 mm separating tube.

exit – 550°C and density of 0.08 g/cm^3. And the maximum wall temperature is about 600°C.

To implement this technology, the fuel elements differ from those of the conventional ceramic fuel as follows.

The prototype of the second type of channel is a nominal steam-superheating channel used in the reactors in the first phase of the Beloyarskaya NPP (Samoilov et al. 1982). This design was found to be reliable for prolonged operation with temperatures close to those of KP-SKD. It was possible to increase the steam temperature at the exit from a group of such channels up to 560°C–565°C and to operate a channel successfully up to average burn-up of 43–44 MW·days/kg.

4.5 HEAT-TRANSFER OPTIMIZATION

It is known that the HTC from a fuel element to a gaseous coolant (supercritical water is considered physically as a dense gas) is lower than in subcritical water-cooled nuclear reactors (Hewitt and Collier 2000) (for details, see Table 4.3) for the same velocity. Hence the fuel centreline temperature will be higher in a SCWR than in a subcritical water-cooled nuclear reactor.

Using simple logic, the wall temperature would be too high after allowance for hot spots and flow distribution uncertainties. However, if high temperature cladding and enhanced heat-transfer surfaces are used in bundles, then the wall temperature is well within safety limits. The objective is simply to increase the supercritical water turbulence level and inter-channel mixing in the bundles. Evidence for the efficiency of bundles is available. It is important

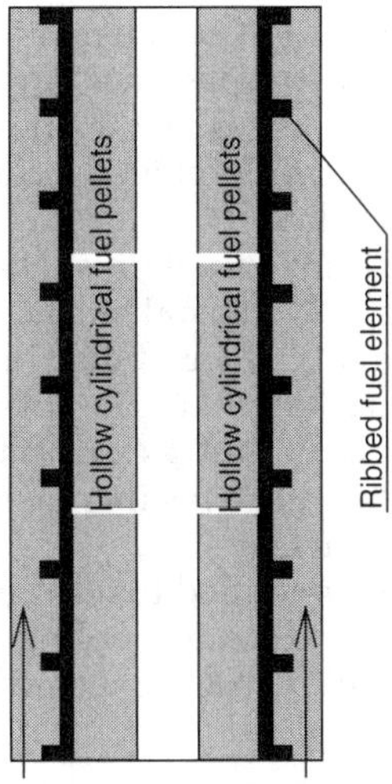

Figure 4.12. AGR ribbed fuel element (Hewitt and Collier 2000).

to implement these special design features directed to decreasing the fuel centreline temperature and hence the centreline fuel temperature within values allowed by safety limits. These features can be:

- Manufacturing fins on the external surface of fuel-element cladding (Hewitt and Collier 2000, Dyadyakin and Popov 1977). In Magnox

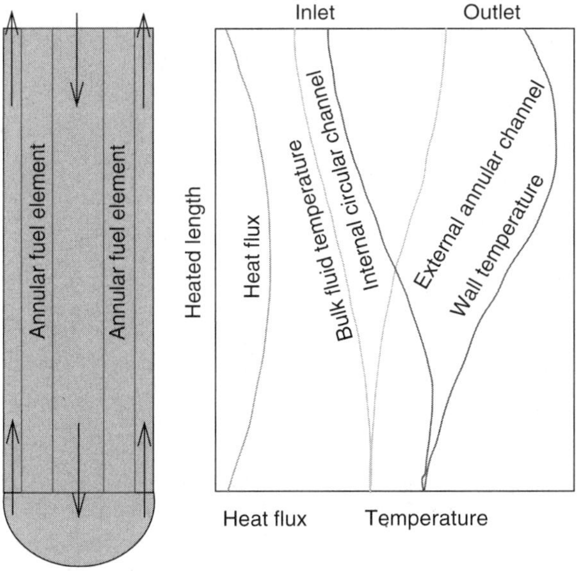

Figure 4.13. Closed-end annular-type fuel element with internal and external cooling (Dement'ev 1990): Scheme and heat flux/temperature profiles along heated length.

gas-cooled nuclear reactors, fuel elements equipped with the herringbone pattern of fins with splitters (Hewitt and Collier 2000) were used. This design feature works as a heat-transfer enhancing device by mixing the gas and as a developed heat-transfer surface (finned surface); hence, the total heat transfer rate was increased by up to 5 to 6 times compared to that of a plain surface.
- Manufacturing ribs on the external surface of fuel element cladding (Figure 4.12) (Hewitt and Collier 2000). In AGRs, fuel elements are machined to produce rectangular ribs of a relatively small height. This design feature works mainly as a heat-transfer enhancing device by mixing of the gas. The HTC was increased by up to 2.5 times compared to that of a plain surface.
- Using hollow fuel pellets installed inside annular-type elements with internal or internal and external cooling (Figure 4.13) (Dement'ev 1990; Dollezhal' et al. 1971; Aleshchenkov et al. 1971; Kornbichler 1964; Spalaris et al. 1961).

Therefore, in spite of some technical difficulties, there are definitely proven ways to overcome them. The supercritical water and other supercritical fluid literature show that enhancements of 2 to 5 times are possible using ribs, grids and "turbulizers" (for details on the heat-transfer enhancement at supercritical pressures, see Chapter 9). Therefore the supercritical water bundle HTC is expected to be between ~ 4–20 kW/m^{2}K (see Table 4.4).

Table 4.4. Comparison of values of thermophysical properties of water* and values of HTC for conditions of SCW CANDU reactor, CANDU-6 reactor and PWR.

Parameter	Unit	SCW CANDU		Typical CANDU-6		PWR	
Pressure	MPa	25**		10.5		15	
Location	–	Inlet	Outlet	Inlet	Outlet	Inlet	Outlet
Temperature	°C	350	625	265	310	290	325
Increase in temperature from inlet to outlet	°C	275		45		35	
Density	kg/m³	625.5	67.58	782.9	692.4	745.4	664.9
Enthalpy	kJ/kg	1624	3567	1159	1401	1285	1486
Increase in enthalpy from inlet to outlet	kJ/kg	1943		242		201	
	kJ/kg·K	7.06		5.38		5.74	
Specific heat	J/kg·K	6978	2880	4956	6038	5257	6460
Expansivity	1/K	$5.17 \cdot 10^{-3}$	$1.74 \cdot 10^{-3}$	$2.09 \cdot 10^{-3}$	$3.71 \cdot 10^{-3}$	$2.54 \cdot 10^{-3}$	$4.36 \cdot 10^{-3}$
Thermal conductivity	W/m·K	0.481	0.107	0.611	0.530	0.580	0.508
Dynamic viscosity	Pa·s	$7.28 \cdot 10^{-5}$	$3.55 \cdot 10^{-5}$	$10.12 \cdot 10^{-5}$	$8.24 \cdot 10^{-5}$	$9.23 \cdot 10^{-5}$	$7.81 \cdot 10^{-5}$
Kinematic viscosity	m²/s	$11.63 \cdot 10^{-8}$	$52.47 \cdot 10^{-8}$	$12.93 \cdot 10^{-8}$	$11.90 \cdot 10^{-8}$	$12.38 \cdot 10^{-8}$	$11.75 \cdot 10^{-8}$
Diffusivity	m²/s	$11.02 \cdot 10^{-8}$	$54.72 \cdot 10^{-8}$	$15.75 \cdot 10^{-8}$	$12.68 \cdot 10^{-8}$	$14.80 \cdot 10^{-8}$	$11.83 \cdot 10^{-8}$
Surface tension	N/m	–	–	$22.5 \cdot 10^{-3}$	$12.1 \cdot 10^{-3}$	$16.7 \cdot 10^{-3}$	$8.77 \cdot 10^{-3}$
Prandtl number	–	1.06	0.96	0.82	0.94	0.84	0.99
Reynolds number ($\times 10^6$) at G***=860 kg/m²s and D_{hy}=8 mm	–	0.946	1.940	0.680	0.835	0.745	0.881
Nusselt number**** ($=0.023 \cdot \mathbf{Re}^{0.8} \cdot \mathbf{Pr}^{0.4}$)	–	1418	2425	985	1225	1068	1308
HTC	W/m²K	8527	3228	7522	8114	7744	8303

* All thermophysical properties of water were calculated according to NIST (2002).

** Pseudocritical temperature at pressure of 25 MPa is 384.9°C.

*** This value of mass flux corresponds to SCW CANDU reactor operating conditions. Mass flux values in subcritical pressure nuclear reactors are much higher; therefore, values of Reynolds number, Nusselt number and HTC will be also much higher in subcritical pressure reactors.

**** Nusselt number is calculated according to Equation (4.1) (Dittus and Boelter 1930) for forced convective heat transfer in a circular tube as a first estimate only.

Chapter 5

EXPERIMENTAL HEAT TRANSFER TO WATER AT SUPERCRITICAL PRESSURES

5.1 HEAT TRANSFER IN VERTICAL CIRCULAR TUBES AND COILS

Water is the most investigated fluid in near-critical and supercritical regions. All[5] primary sources of heat-transfer experimental data of water flowing inside vertical circular tubes are listed in Table 5.1. Ranges of investigated parameters for selected experiments with water in circular tubes at supercritical pressures that are relevant to SCWR operating range are shown in Figure 5.1.

Swenson et al. (1965) found that HTC has a peak when the film temperature is within the pseudocritical temperature range (Figures 5.2 and 5.3). This peak in HTC decreases with increase in pressure and heat flux.

Vikhrev et al. (1971, 1967) conducted experiments in supercritical water flowing in a vertical tube (Figure 5.4). To be able to cover a wide range of bulk-fluid enthalpies (see Figure 5.4c,d), experiments were conducted in the same flow geometry, same mass flux and heat flux, but at various inlet temperatures (enthalpies). Later, these data were combined into one curve. However, this method is not the perfect one, because entrance and/or exit (i.e., deteriorated heat transfer at the exit) effects do not allow matching properly two or several series in one plot (for details, see original figures in Vikhrev et al. (1967)). Therefore, it is important to perform supercritical heat-transfer experiments in one sufficiently long-heated test section.

Vikhrev et al. (1971, 1967) found that at a mass flux of 495 kg/m²s two types of deteriorated heat transfer existed (Figure 5.4a): the first type appeared in the entrance region of the tube $\left(L/D \leq 40-60\right)$ and the second type appeared at any section of the tube, but only within a certain enthalpy range. In general, the deteriorated heat transfer occurred at high heat fluxes.

The first type of deteriorated heat transfer observed was due to the flow structure within the entrance region of the tube. However, this type of deteriorated heat transfer occurred mainly at low mass fluxes and at high heat fluxes (Figure 5.4a,b) and eventually disappeared at high mass fluxes (Figure 5.4c,d).

[5] "All" means all sources found by the authors from a total of 650 references dated mainly from 1950 till the beginning of 2006.

Table 5.1. Range of investigated parameters for experiments with water flowing in circular tubes and coils at supercritical pressures.

Reference	*p*, MPa	*t*, °C (*H* in kJ/kg)	*q*, MW/m²	*G*, kg/m²s	Flow Geometry
			Tubes (vertical)		
Randall 1956	27.6–55.2	t_b=204–538; t_w=204–760	0.31–9.44	2034–5425	Hastelloy C vertical tube (*D*=1.27; 1.57; 1.9 mm, *L*=203.2 mm)
Miropol'skiy and Shitsman 1957, 1958a	0.4–27.4	t_b=2.5–420; Δt_w=2.5–420	0.42–8.4	170–3000	SS tube (*D*=7.8; 8.2 mm, *L*=160 mm), upward flow
Armand et al. 1959	23–26.3	t_b=300–380	0.17–0.35	450–650	SS and nickel tubes (*D*=6; 8 mm, *L*=250; 350 mm), upward flow
Doroshchuk et al. 1959	24.3	t_b=100–250	3.06–3.9	3535–8760	Silver tube (*D*=3 mm, *L*=246 mm), downward flow
Goldmann 1961, Chalfant 1954, Randall 1956	34.5	t_b=204–538; t_w=204–760	0.31–9.4	2034–5424	Tubes (*D*=1.27–1.9 mm, *L*=0.203 m), upward flow
Shitsman 1962	22.8–26.3	t_b=300–425; t_w=260–380	0.291–5.82	100–2500	Vertical and horizontal copper and carbon steel tubes (*D*=8 mm, D_{ext}=46 mm, *L*=170 mm), upward and horizontal flows
Shitsman 1963	22.6–24.5	t_b=280–580	0.28–1.1	300–1500	SS tube (*D*=8 mm, *L*=1.5 m)
Swenson et al. 1965	23–41	t_b=75–576; t_w=93–649	0.2–1.8	542–2150	SS tube (*D*=9.42mm, *L*=1.83 m), flow (selected data are shown in Figures 5.2 and 5.3)
Smolin and Polyakov 1965	25.4; 27.4; 30.4	t_b=250–440	0.7–1.75	1500–3000	SS tube (*D*=10; 8 mm, *L*=2.6 m), upward flow
Vikhrev et al. 1967	24.5; 26.5	H_b=230–2750	0.23–1.25	485–1900	SS tube (*D*=7.85; 20.4 mm, *L*=1.515; 6 m) (selected data are shown in Figure 5.4)
Shitsman 1967	24.3–25.3	t_{in}=300–320	0.73–0.52	600–690	Vertical and horizontal tubes (*D*=8; 16 mm, *L*=1.5; 1.6 m), upward and horizontal flows
Bourke and Denton 1967	23.0–25.4	t_b=310–380	1.2–2.2	1207; 2712	Tube (*D*=4.06 mm, *L*=1.2 m)

Reference	p, MPa	t, °C (H in kJ/kg)	q, MW/m²	G, kg/m²s	Flow Geometry
Styrikovich et al. 1967	24	H_b=1260–2500	0.35–0.87	700	Tube (*D*=22 mm, *L* was not provided in the original paper) (selected data are shown in Figure 5.6)
Krasyakova et al. 1967	23	H_{in}=837–2721	0.23–0.7	300–1500	Vertical and horizontal tubes (*D*=20 mm, *L*=2.8 m), upward and horizontal flows
Shitsman 1968	10–35	t_b=100–250	0.27–0.7	400	Vertical and horizontal SS tubes (*D*/*L*=3/0.7; 8/0.8; 8/3.2; 16/1.6 mm/m), upward, downward, and horizontal flows
Krasyakova et al. 1968	15; 18.8; 23	H_{in}=840–1890	0.23–0.7	300–2000	Vertical and horizontal SS tube (*D*=20 mm, *L*=2.2 m), upward, downward, and horizontal flows
Alferov et al. 1969	14.7–29.4	t_b=160–365	0.17–0.6	250–1000	SS tubes (*D*/*L*=14/1.4; 20/3.7 mm/m)
Kamenetsky and Shitsman 1970	24.5	H_b=80–2300	0.19–1.33	50–1750	Vertical and horizontal SS tube (*D*=22 mm, *L*=3 m), non-uniform circumferential heat flux, upward and horizontal flows
Ackerman 1970	22.8–41.3	t_b=77–482	0.126–1.73	136–2170	Smooth (*D*=9.4; 11.9 and 24.4 mm, *L*=1.83 m; *D*=18.5 mm, *L*=2.74 m) and ribbed (*D*=18 mm(from rib valley to rib valley), *L*=1.83 m, six helical ribs, pitch 21.8 mm) tubes
Ornatsky et al. 1970	22.6; 25.5; 29.4	H_{in}=420–1400	0.28–1.2	450–3000	Five SS parallel tubes (*D*=3 mm, *L*=0.75 m), upward, stable, and pulsating flows
Barulin et al. 1971	22.5–26.5	t_b=50–500; t_w=60–750	0.2–6.5	480–5000	Vertical and horizontal tubes (*D*=3; 8; 20 mm, *L*/*D* < 300), upward, downward, and horizontal flows

(continued)

Table 5.1. (*continued*)

Reference	*p*, MPa	*t*, °C (*H* in kJ/kg)	*q*, MW/m²	*G*, kg/m²s	Flow Geometry
Belyakov et al. 1971	24.5	H_b=420–3140	0.23–1.4	300–3000	Vertical and horizontal SS tube (*D*=20 mm, *L*=4–7.5 m), upward and horizontal flows
Ornatskii et al. 1971	22.6, 25.5, 29.7	H_b=100–3000	0.4–1.8	500–3000	SS tube (*D*=3 mm, *L*=0.75 m), upward and downward flows
Yamagata et al. 1972	22.6–29.4	t_b=230–540	0.12–0.93	310–1830	Vertical and horizontal SS tubes (*D*/*L*=7.5/1.5; 10/2 mm/m), upward, downward, and horizontal flows (selected data are shown in Figure 5.7)
Glushchenko et al. 1972	22.6; 25.5; 29.5	H_b=85–2400	1.15–3	500–3000	Tubes (*D*=3; 4; 6; 8 mm, *L*=0.75–1m), upward flow; *D*=3 mm, downward flow
Malkina et al. 1972	24.5–31.4	t_b=20–80	0.47–2.3	*u*=7–10 m/s	SS tubes (*D*=2; 3 mm, *L*=0.15 m)
Chakrygin et al. 1974	26.5	t_{in}=220	*q* was not provided	445–1270	SS tube (*D*=10 mm, *L*=0.6 m), upward and downward flows
Lee and Haller 1974	24.1	t_b=260–383	0.25–1.57	542–2441	SS tubes (*D*=38.1; 37.7 mm, *L*=4.57 m), tube with ribs (selected data are shown in Figure 5.8)
Alferov et al. 1975	26.5	t_b=80–250	0.48	447	Tube (*D*=20 mm, *L*=3.7 m), upward and downward flows
Kamenetskii 1975	23.5; 24.5	H_{in}=100–2300	1.2	50–1700	Steel tubes (*D*=21; 22 mm, *L*=3 m), non-uniform circumferential heat flux
Alekseev et al. 1976	24.5	t_{in}=100–350	0.1–0.9	380, 490, 650, 820	SS tube (*D*=10.4 mm, *L*=0.5; 0.7 m), upward flow (selected data are shown in Figure 5.9)

Reference	p, MPa	t, °C (H in kJ/kg)	q, MW/m²	G, kg/m²s	Flow Geometry
Ishigai et al. 1976	24.5; 29.5; 39.2	H_b=220–800	0.14–1.4	500; 1000; 1500	Vertical and horizontal SS polished tubes (*D*=3.92 mm, *L*=0.63 m–vertical; *D*=4.44 mm, *L*=0.87 m–horizontal)
Harrison and Watson 1976a,b	24.5	t_b=50–350	1.3, 2.3	940, 1560	Vertical and horizontal SS tubes (*D*=1.64; 3.1 mm, *L*=0.4, 0.12 m)
Treshchev and Sukhov 1977	23; 25	H_{in}=1331	0.69–1.16	740–770	Tubes (*L*=0.5–1 m), stable and pulsating upward flows
Krasyakova et al. 1977	24.5	t_b=90–340	0.11–1.4	90–2000	Tube (*D*=20 mm, D_{ext}=28 mm, *L*=3.5 m), downward flow (selected data are shown in Figure 5.5)
Smirnov and Krasnov 1978–1980	25; 28; 30	t_w=250–700	0.25–1	500–1200	SS tube (*D*=4.08 mm, *L*=1.09 m), upward and downward flow
Kamenetskii 1980	2	H_b=100–2200	0.37–1.3	300–1700	Vertical and horizontal SS tubes with and without flow spoiler (*D*=22 mm, *L*=3 m)
Watts and Chou 1982	25	t_b=150–310; t_w=260–520	0.175–0.44	106–1060	Tubes (*D*=25; 32.2 mm, *L*=2 m) with upward and downward flow
Selivanov and Smirnov 1984	26	t_{in}=50–450	0.13–0.65	200–10000	SS tube (*D*=10 mm, D_{ext}=14 mm, *L*=1 m)
Kirillov et al. 1986	25	t_{in}=385	0.4; 0.6	1000	SS tube (*D*=10 mm, D_{ext}=14 mm, *L*=1 m)
Razumovskiy et al. 1990	23.5	H_{in}=1400;	0.657–3.385 1600; 1800	2190	Tube (*D*=6.28 mm, *L*=1.44 m), downward flow
Chen 2004	24	H_{in}=1350; 1600	300	400	SS vertical and inclined tubes (smooth with uniform and non-uniform radial heating and ribbed)

(continued)

Table 5.1. (*continued*)

Reference	*p*, MPa	*t*, °C (*H* in kJ/kg)	*q*, MW/m²	*G*, kg/m²s	Flow Geometry
Pis'mennyy et al. 2005	23.5	t_{in}=20–380	Up to 0.515	250; 500	Vertical SS tubes (*D*=6.28 mm, L_h=600; 360 mm; *D*=9.50 mm, L_h=600; 400 mm) (for more details, see Section 10.2) (selected data are shown in Figures 5.12 – 5.20)
Kirillov et al. 2005	24–25	t_{in}=300–380	0.09–1.050	200–1500	SS tube (*D*=10 mm, *L*=1; 4 m) (for more details, see Section 10.1) (selected data are shown in Figures 5.10 and 5.11)
			Coils		
Miropolskiy et al. 1966	0.2–29.5	H_b=600–3000	0.58; 1.16	800	Tube coils (*D*=8 mm, R_{bend}=43; 90; 260 mm), bend location vertical
Miropol'skii et al. 1970: Miropolskiy and Pikus 1967–1968	24.5	H_b=651–2394	0.018–0.27	200–5700	Tube coils (*D*=8; 16 mm, R_{bend}=43; 275 mm), bend location vertical
Kovalevskiy and Miropol'skiy 1978	23.3–25.3	t_b=20–386	0.116–2.68	200–5700	SS coils (*D*=15.8; 8.3; 8.25 mm, R_{bend}=43; 46; 135; 275; 525 mm)
Breus and Belyakov 1990	25	H_b=1200–2400	0.3–0.7	1000–1500	SS helical coils (*D*=20; 24 mm, *L*=7.5m, R_{bend}=115; 455 mm, pitch between spirals, respectively, 90 and 310 mm), upward flow

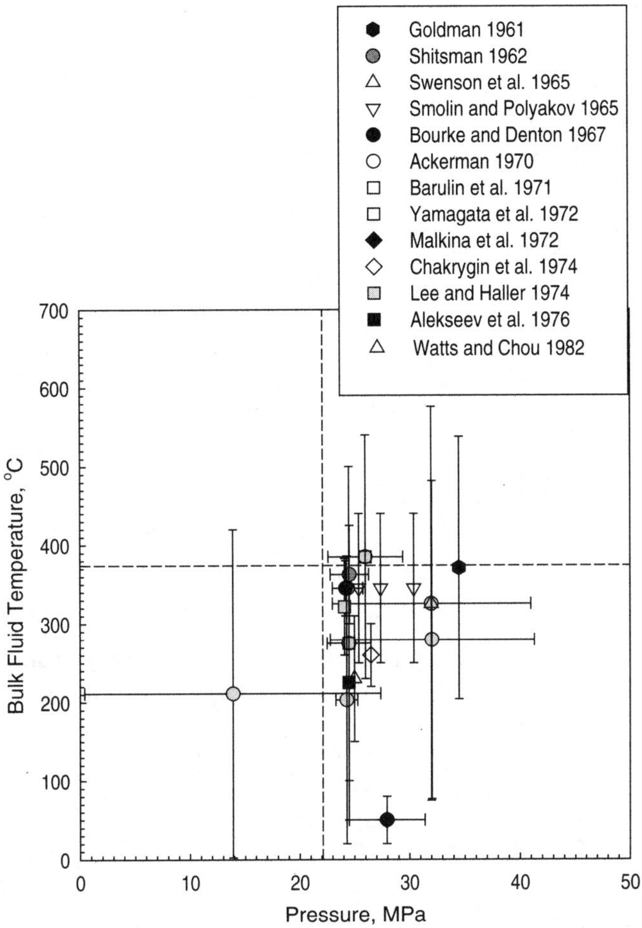

Figure 5.1. Ranges of investigated parameters for selected experiments with water in circular tubes at supercritical pressures (for details, see Table 5.1).

The second type of deteriorated heat transfer occurred when the wall temperature exceeded the pseudocritical temperature (Figure 5.4). According to Vikhrev et al. (1967), the deteriorated heat transfer appears when $q/G > 0.4$ kJ/kg (where q is in kW/m^2 and G is in kg/m^{2}s). This value is close to that suggested by Styrikovich et al. (1967) $q/G > 0.49$ kJ/kg. However, the above definitions of two types of deteriorated heat transfer are not enough for their clear identification.

Krasyakova et al. (1977) did not find the deteriorated heat transfer within the entrance region of the vertical tube with downward flow (Figure 5.5), which is typical for upward flow (see above). Also, they noticed that variations in wall temperature along the heated length of the downtake tube were smoother compared to those in the rising tube.

Results of Styrikovich et al. (1967) are shown in Figure 5.6. Improved and deteriorated heat transfer as well as a peak ("hump") in HTC near the

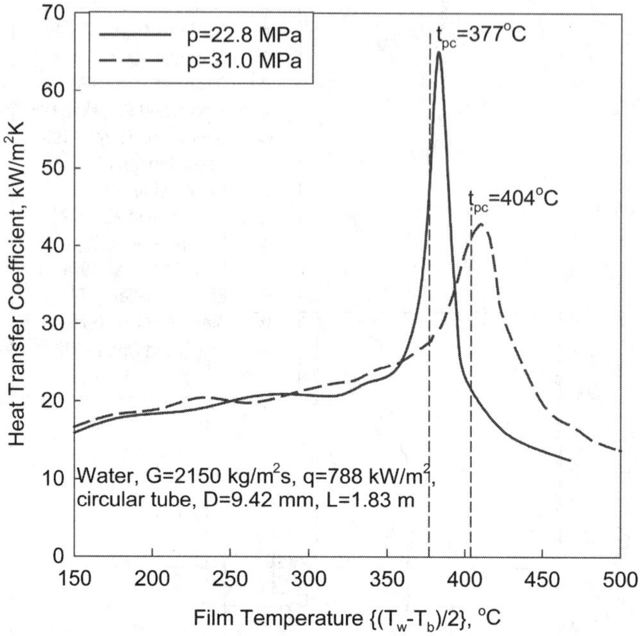

Figure 5.2. Effect of pressure on HTC (Swenson et al. 1965).

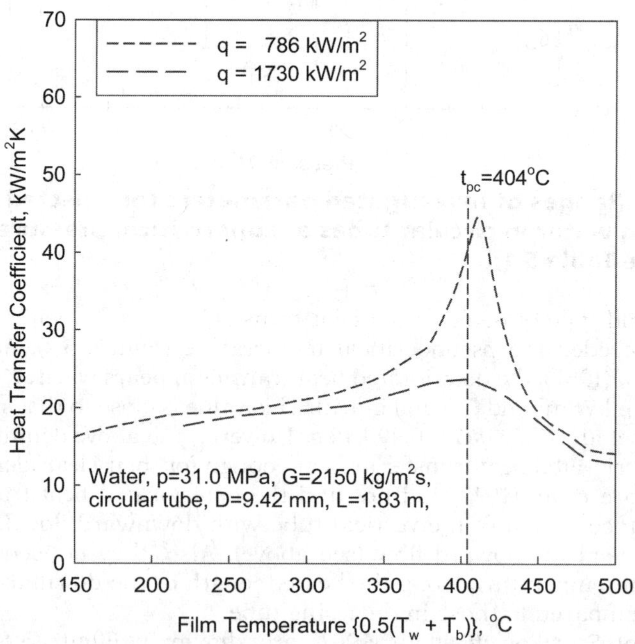

Figure 5.3. Effect of heat flux on HTC (Swenson et al. 1965).

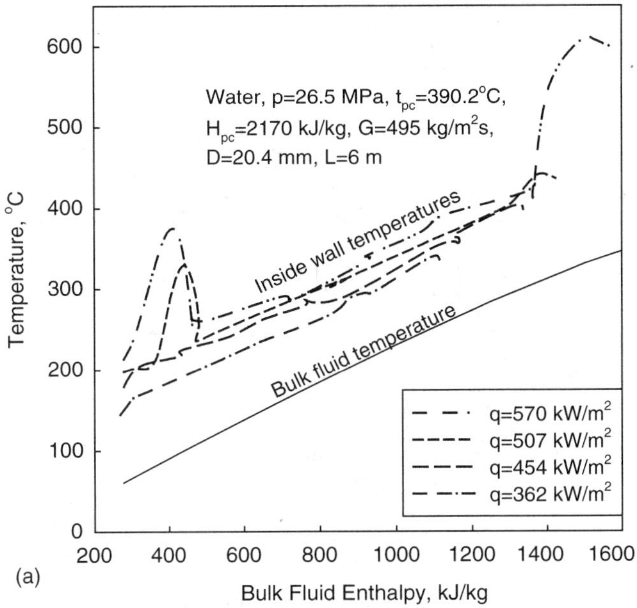

(a)

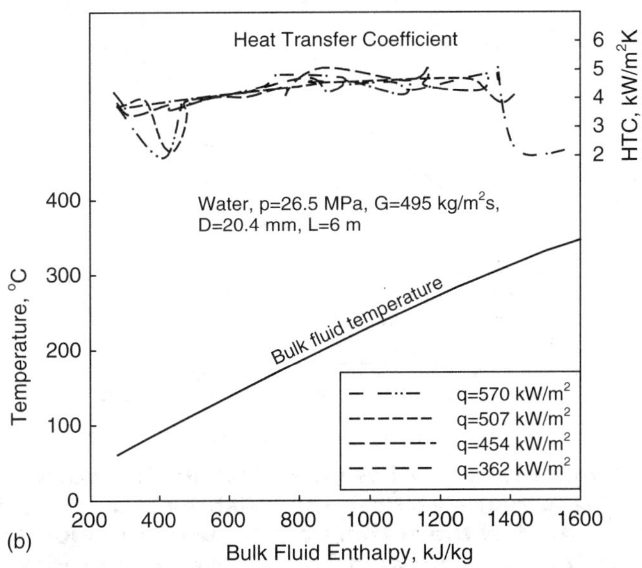

(b)

Figure 5.4. Continued

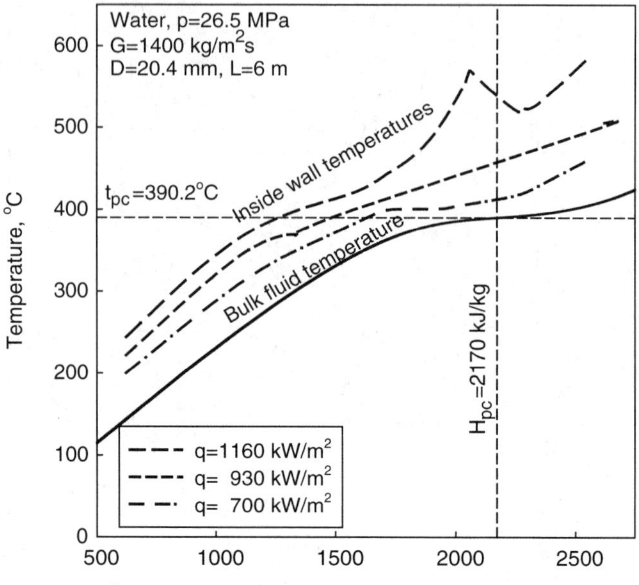

(c)

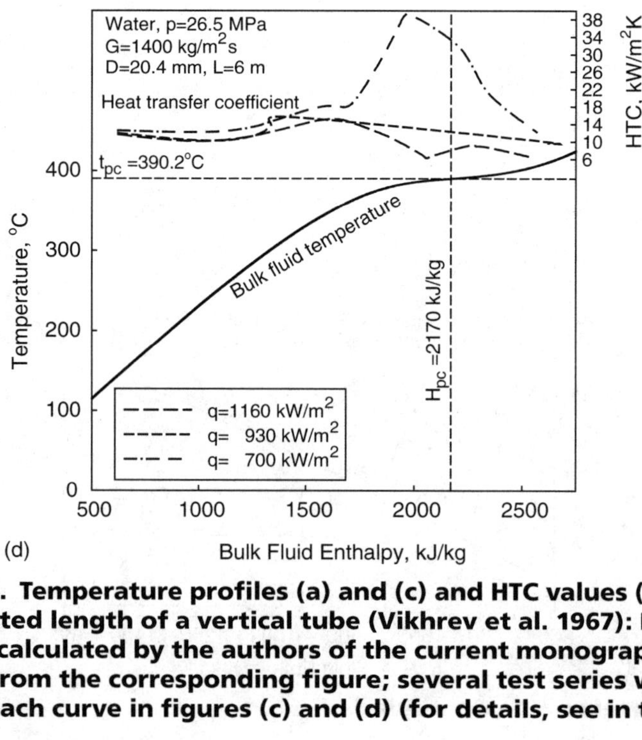

(d)

Figure 5.4. Temperature profiles (a) and (c) and HTC values (b) and (d) along heated length of a vertical tube (Vikhrev et al. 1967): HTC values were calculated by the authors of the current monograph using the data from the corresponding figure; several test series were combined in each curve in figures (c) and (d) (for details, see in the text).

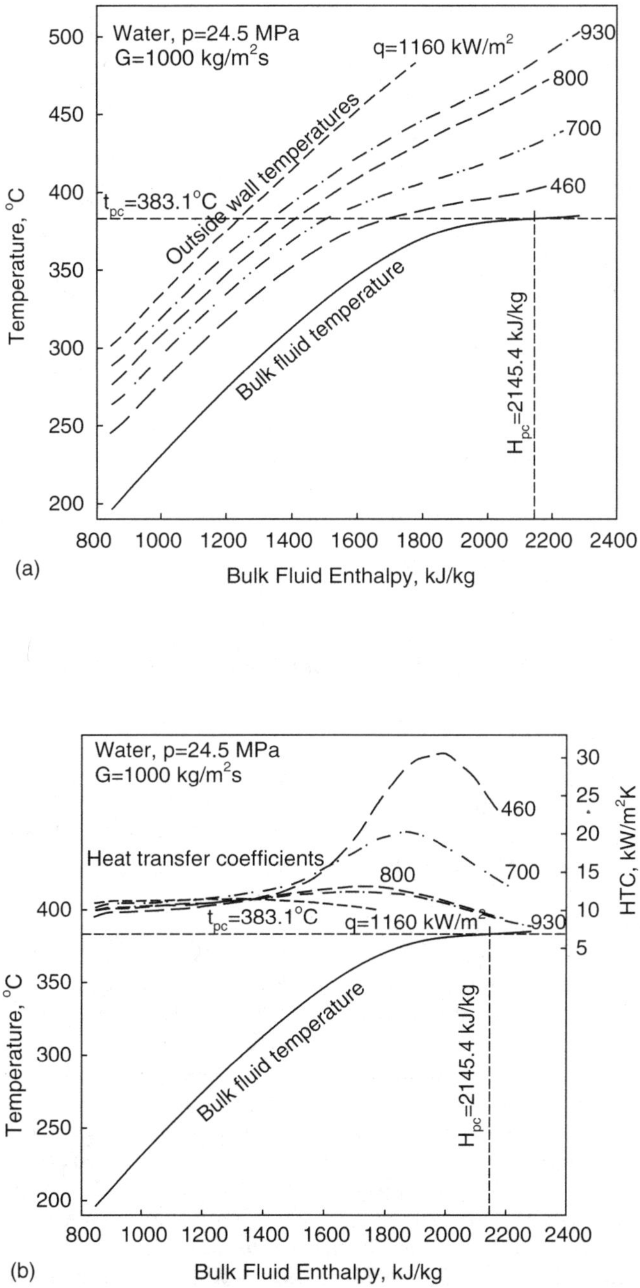

Figure 5.5. Temperature profiles (a) and HTC values (b) along heated length of a vertical tube (*D*=20 mm) with downward flow (Krasyakova et al. 1977): HTC values were calculated by the authors of the current monograph using the data from figure (a); several test series were combined in each curve (for details, see explanations to Figure 5.4c,d).

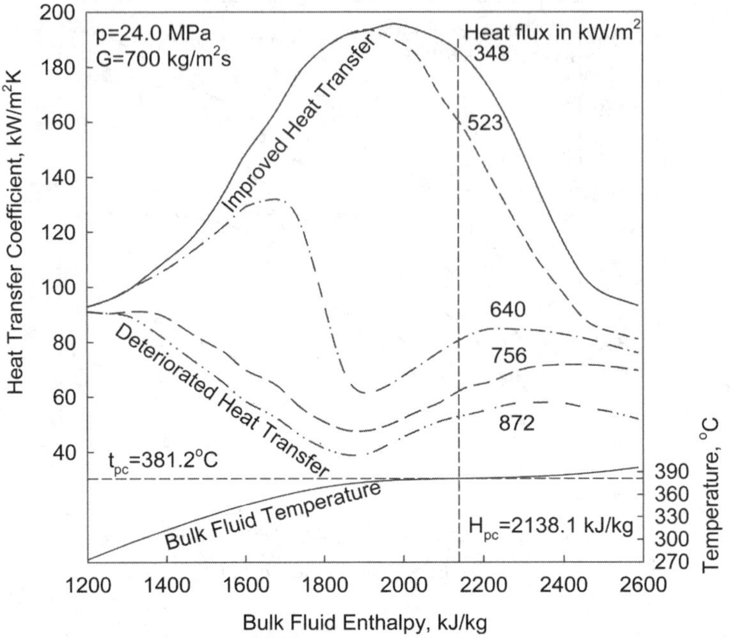

Figure 5.6. Variations in HTC values of water flowing in tube (Styrikovich et al. 1967).

pseudocritical point is clearly shown in this figure. Unfortunately, authors did not provide the tube diameter and heated length.

Shiralkar and Griffith (1970) determined both theoretically (for supercritical water) and experimentally (for supercritical carbon dioxide) the limits for safe operation, in terms of maximum heat flux for a particular mass flux. Their experiments with a twisted tape inserted inside the test section showed that heat transfer was improved by this method. Also, they found that at high heat fluxes deteriorated heat transfer occurred, when the bulk-fluid temperature was below and the wall temperature above the pseudocritical temperature.

Ackerman (1970) investigated heat transfer to water at supercritical pressures flowing in smooth vertical tubes with and without internal ribs within a wide range of pressures, mass fluxes, heat fluxes, and diameters. He found that a pseudo-boiling phenomenon could occur at supercritical pressures. The pseudo-boiling phenomenon is thought to be due to the large differences in fluid density below the pseudocritical point (high-density fluid, i.e., "liquid") and beyond (low-density fluid, i.e., "gas"). This heat transfer phenomenon was affected by pressure, bulk-fluid temperature, mass flux, heat flux, and tube diameter. The process of pseudo-film boiling (i.e., low-density fluid prevents high-density fluid from "re-wetting" a heated surface) is similar to film boiling, which occurs at subcritical pressures. Pseudo-film boiling leads to the deteriorated heat transfer. However, pseudo-film boiling phenomenon may not be the only reason for the deteriorated heat transfer. Ackerman noted that unpredictable heat transfer performance was sometimes observed when the pseudocritical temperature of the fluid was between the bulk-fluid temperature and the heated surface temperature.

Ornatsky et al. (1970) investigated the appearance of deteriorated heat transfer in five parallel tubes with stable and pulsating flow. They found that the deteriorated heat transfer in the assembly at supercritical pressures depended on the heat-flux to mass-flux ratio and flow conditions. At stable flow conditions, heat transfer deterioration occurred at values of the ratio q/G = 0.95 – 1.05 kJ/kg and at inlet bulk water enthalpies of H_{in} = 1330 – 1500 kJ/kg. In pulsating flow, deteriorated heat transfer occurred at lower ratios, i.e., $q/G \geq$ 0.68 – 0.9 kJ/kg. Flow pulsations usually occurred at regimes where the outlet water enthalpy was in the region of steep variations in thermophysical properties, i.e., critical or pseudocritical regions. The beginning of the heat transfer deterioration was usually noticed within certain zones along the tube, in which $(t_w + t_b)/2 = t_{max}$. They also established the possibility of the simultaneous existence of several local zones of deteriorated heat transfer along the tubes.

Some researchers suggested that the variations in the thermophysical properties near critical and pseudocritical points resulted in a maximum value of HTC. Thus, Yamagata et al. (1972) found that for water flowing in vertical and horizontal tubes the HTC increases significantly in the pseudocritical region (Figure 5.7). The magnitude of the peak in the HTC decreases with increasing heat flux and pressure. The maximum HTC values correspond to a bulk-fluid enthalpy, which is slightly less than the pseudocritical bulk-fluid enthalpy. Similar results were recorded by Swenson et al. (1965) (Figures 5.2 and 5.3).

Al'ferov et al. (1973) also performed forced convective heat-transfer experiments in supercritical water. They recorded the reduction in heat transfer with increasing heat flux in turbulent flow of a coolant at supercritical pressures.

Kruzhilin (1974) found considerable deterioration of heat transfer to turbulent water flow at supercritical pressure within the pseudocritical temperature range and at high heat fluxes. Within this temperature range, a drastic decrease in density occurs, with a consequent rapid expansion of this low-density

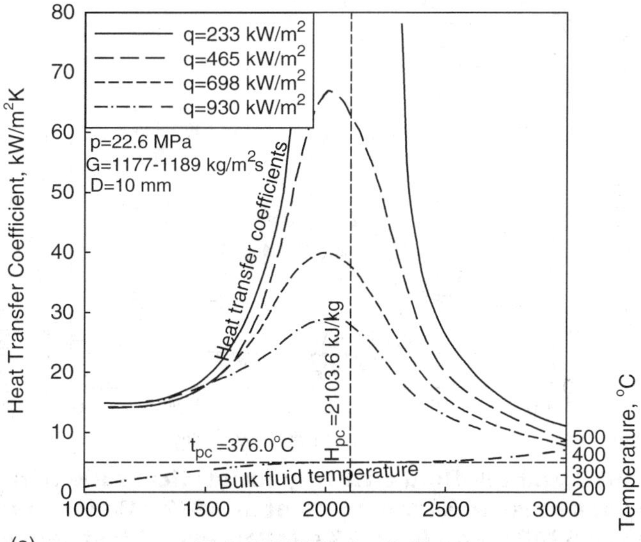

Figure 5.7. Continued

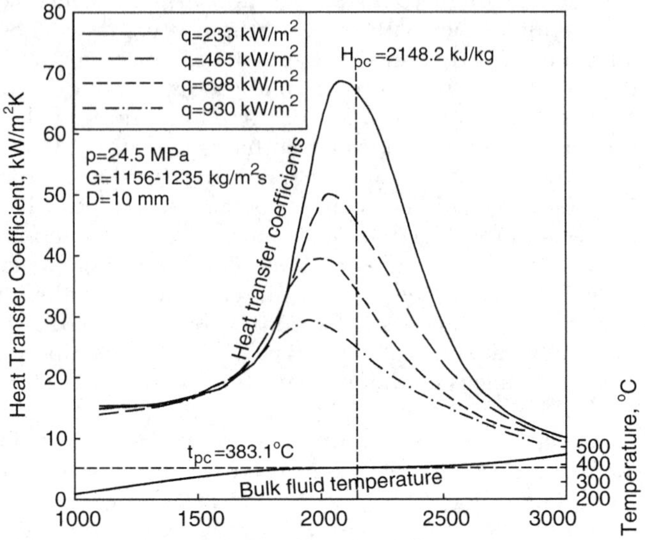

(b)

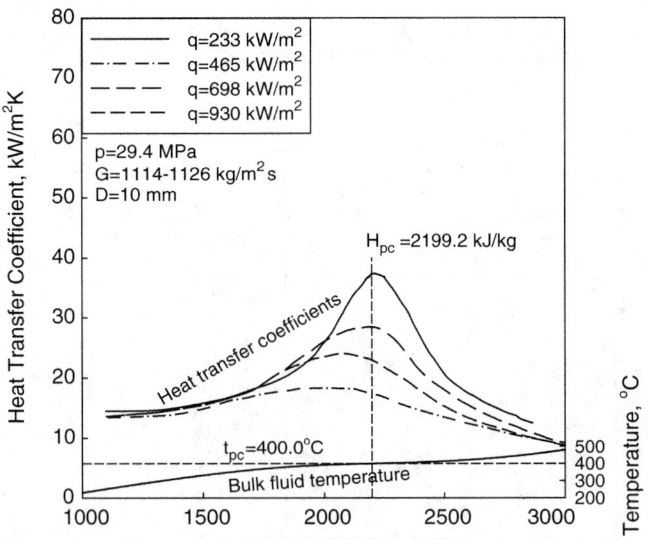

(c)

Figure 5.7. HTC vs. bulk-fluid enthalpy for vertical tube with upward flow at various pressures (Yamagata et al. 1972): Water – (a) *p*=22.6 MPa; (b) *p*=24.5 MPa; and (c) *p*=29.4 MPa; several test series were combined in each curve (for details, see explanations to Figure 5.4c,d).

layer at the wall. Both these effects, it was argued, gave rise to a flow velocity component normal to the wall. With this flow pattern, heat transfer could be considered similar to that occurring under the conditions of liquid injection into turbulent flow through a porous wall.

The findings of Lee and Haller (1974) are presented in Figure 5.8. They also combined several test series in one graph (see explanations to Figure 5.4c,d). Due to deteriorated heat-transfer region at the tube exit (one set of data) and entrance effect in another set of data, experimental curves discontinue (see Figure 5.8b,c). In general, they found heat flux and tube diameter to be the important parameters affecting minimum mass-flux limits to prevent pseudo-film boiling. Multi-lead ribbed tubes were found to be effective in preventing pseudo-film boiling.

Harrison and Watson (1976) developed conditions of similarity in convective heat transfer. These conditions were verified with water and helium data.

Kafengaus (1986, 1975), analyzing data of various fluids (water, ethyl and methyl alcohols, heptane, etc.), suggested a mechanism for "pseudo-boiling" that accompanies heat transfer to liquids flowing in small-diameter tubes at supercritical pressures. The onset of pseudo-boiling was assumed to be associated with the breakdown of a low-density wall layer that was present at an above-critical temperature, and with the entrainment of individual volumes of the low-density fluid into the cooler (below pseudocritical temperature) core of the high-density flow, where these low-density volumes collapse, with the generation of pressure pulses. At certain conditions, the frequency of these pulses can coincide with the frequency of the fluid column in the tube, resulting in resonance and in a rapid rise in the amplitude of pressure fluctuations. This theory was supported by the experimental results.

Alekseev et al. (1976) conducted experiments in a circular vertical tube cooled with supercritical water and found that at $q/G < 0.8$ kJ/kg normal heat transfer occurred. However, recalculation of their data showed that this value should be

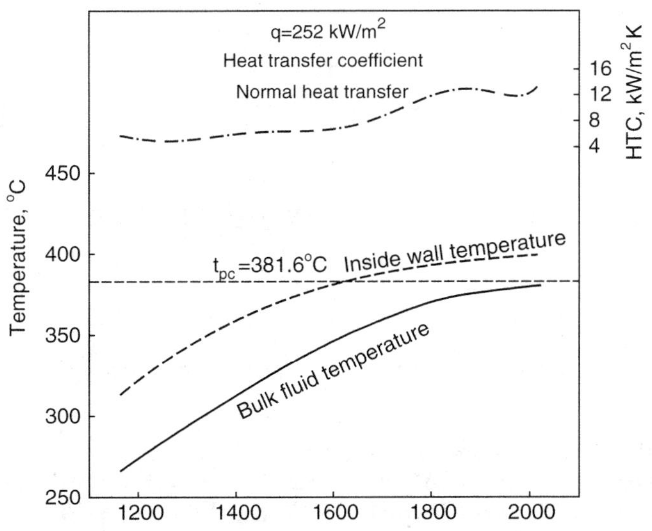

(a)

Figure 5.8. Continued

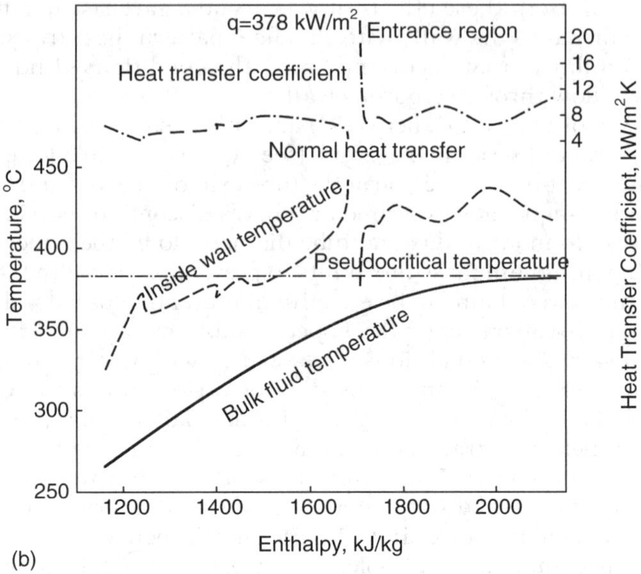

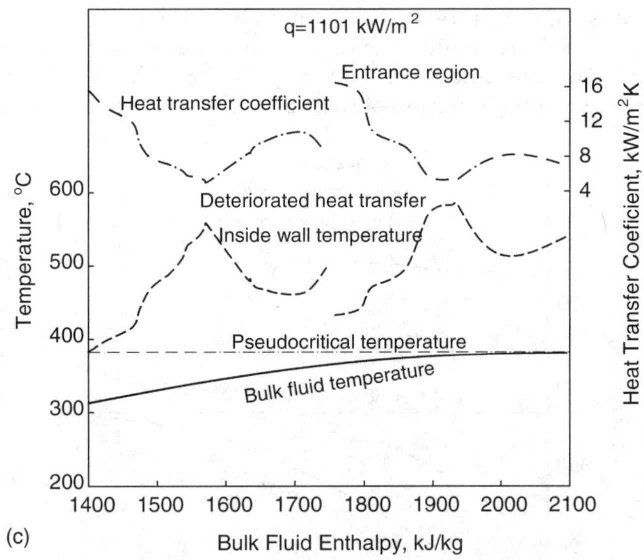

Figure 5.8. Temperature profiles and HTC values along 38.1-mm ID smooth vertical tube at different mass fluxes (Lee and Haller 1974): Water, *p*=24.1 MPa, and H_{pc}=2140 kJ/kg; (a) *G*=542 kg/m²s, (b) *G*=542 kg/m²s, and (c) *G*=1627 kg/m²s; HTC values were calculated by the authors of the current monograph using the data from the corresponding figure; several test series were combined in each curve (for details, see explanations to Figure 5.4c,d).

around $q/G < 0.92$ kJ/kg (see Figure 5.9a, $q = 0.27$ and 0.35 MW/m^2 and Figure 5.9b, $q = 0.35$ MW/m^2). At all mass fluxes and inlet bulk-fluid temperatures, the wall temperature increased smoothly along the tube. Beyond this value, the deterioration in heat transfer occurred (see Figure 5.9, the rest of the curves). With heat flux increase, a hump (Figure 5.9a, inlet temperature is 100°C) or a peak (Figure 5.9b, inlet temperature is 300°C) in the wall temperature occurs and moves towards the tube inlet as the heat flux increases.

Kirillov et al. (1986) conducted research into a radial cross-section temperature profile in water flowing in a circular tube at subcritical and supercritical pressures. They found that inside the turbulent flow core, the radial cross-section temperature profile is logarithmic at supercritical pressures and at temperatures close to the pseudocritical temperature.

Yoshida and Mori (2000) stated that supercritical heat transfer is characterized by rapid variations of physical properties with temperature change across the flow. These property variations result in a peak of HTC near the pseudocritical point at low heat flux and a peak reduction with an increase in heat flux.

Kirillov et al. (2005) performed heat transfer experiments in vertical tubes at the IPPE experimental setup (for details, see Section 10.1). The selected experimental results[6] in supercritical water are summarized in Figures 5.10 and 5.11 to illustrate key findings.

In general, the following supercritical heat-transfer cases were covered:
Within a certain heated length

1. $T_w^{in} < T_{pc}$, $T_w^{in} = T_{pc}$, $T_w^{in} > T_{pc}$ and (a) $T_b < T_{pc}$; or (b) $T_b < T_{pc}$, $T_b = T_{pc}$ and $T_b > T_{pc}$;
2. $T_w^{in} > T_{pc}$ and (a) $T_b < T_{pc}$ and $T_b = T_{pc}$; or (b) $T_b < T_{pc}$, $T_b = T_{pc}$ and $T_b > T_{pc}$.

Typically at the entrance region (i.e., $L/D \leq 30$), the wall temperature rises sharply (Figures 5.10 and 5.11). In general, this temperature profile is due to the thermal boundary layer development. At the outlet, the colder power clamp lowers heated wall temperature nearby (see Figure 5.10). Similar power clamp effect may contribute to the inlet temperature profile, however, only within a short distance. Therefore, any data, i.e., affected with the colder-clamp effect, should be eliminated from the consideration.

Experimental data of supercritical water obtained at higher mass fluxes (G = 1500 kg/m^{2}s) (see Figure 5.11) show good agreement (i.e., about 1% difference) between the calculated value of the last downstream bulk-fluid temperature, which was calculated from incremental heat balances, and the measured outlet bulk-fluid temperature just downstream of the outlet mixing chamber.

However, at lower mass flux (G = 200 kg/m^{2}s), there is a noticeable difference between the measured and calculated outlet bulk-fluid temperatures (see Figure 5.10). This seems to be due to the increased measurement uncertainty at low mass-flow rates.

In general, all experimental data shown in Figures 5.10 and 5.11 are within limits of the deteriorated heat transfer, because according to various literature sources, the deteriorated heat-transfer mode can appear at $q / G > 0.4$ kJ/kg.

Figure 5.11a shows evidence of a peak in HTC within the pseudocritical region.

Experimental results of KPI (Pis'mennyy et al. 2005) for supercritical water flowing upward and downward inside vertical tubes at the outlet pressure of

[6] The data obtained at high heat fluxes were chosen for presentation.

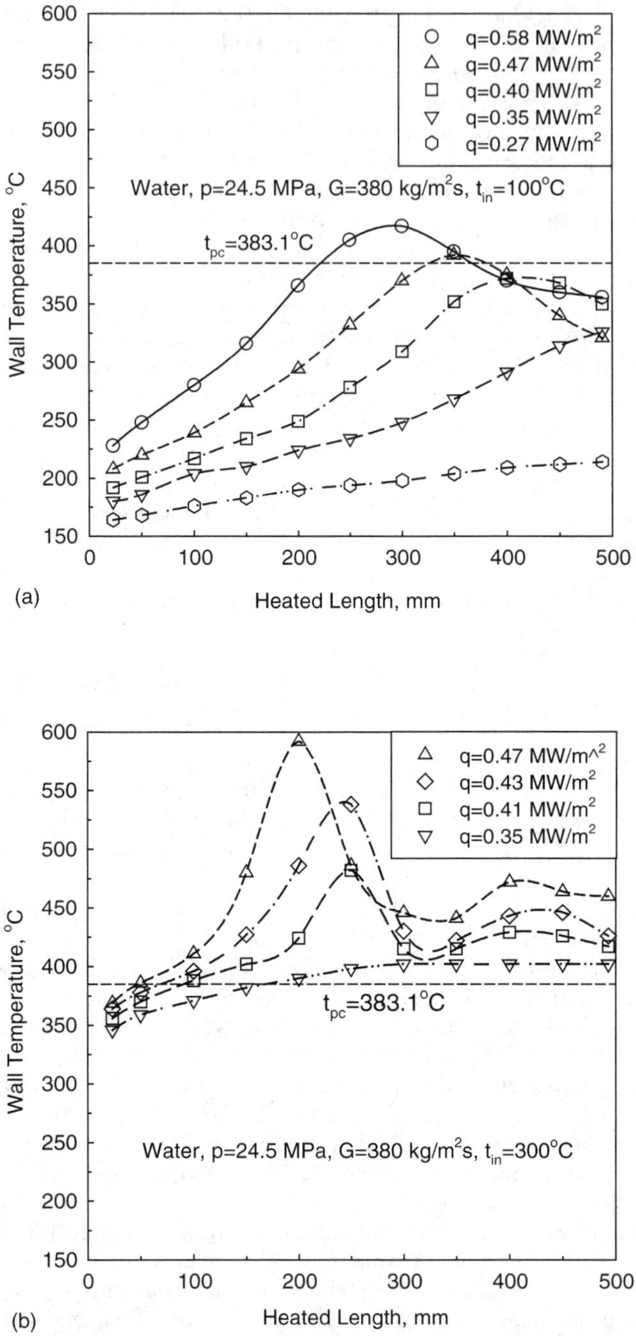

Figure 5.9. Temperature profiles along heated length of vertical circular tube (*D*=10.4 mm) with upward flow (natural circulation) (Alekseev et al. 1976).

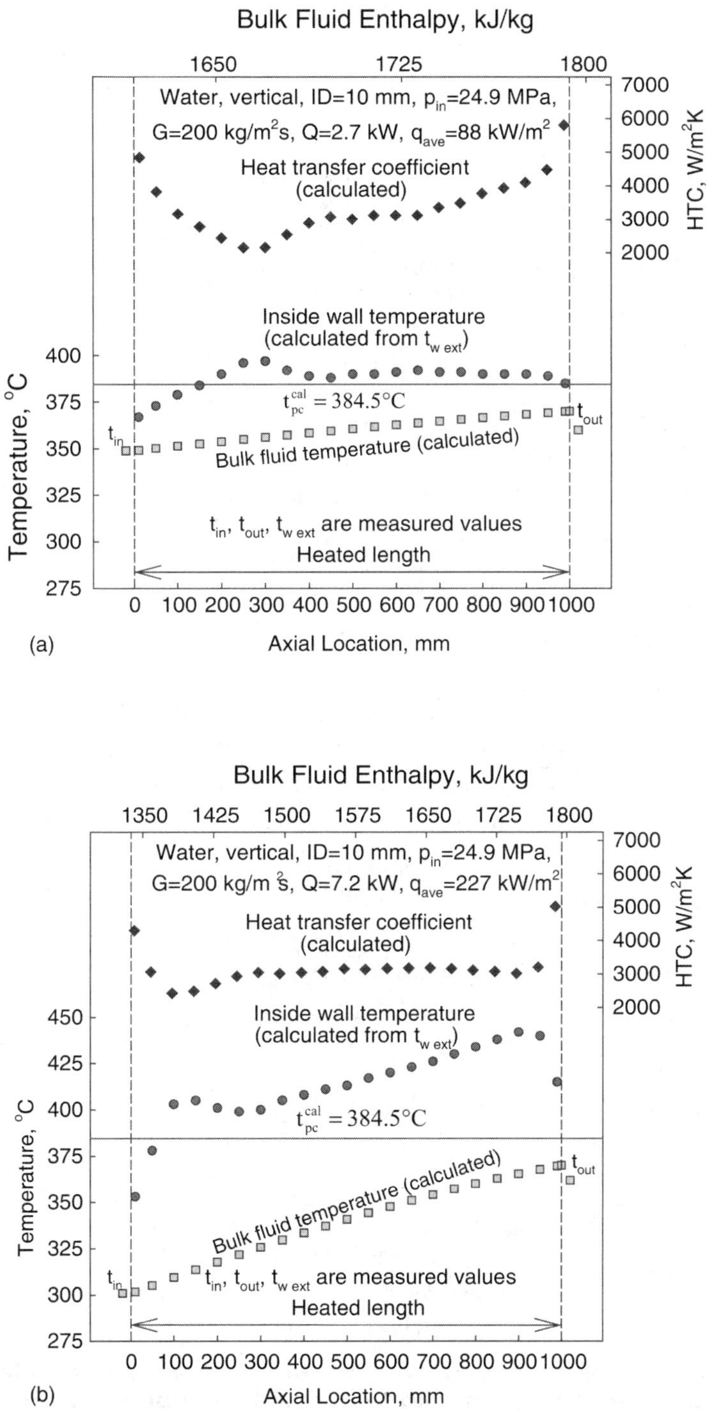

Figure 5.10. Continued

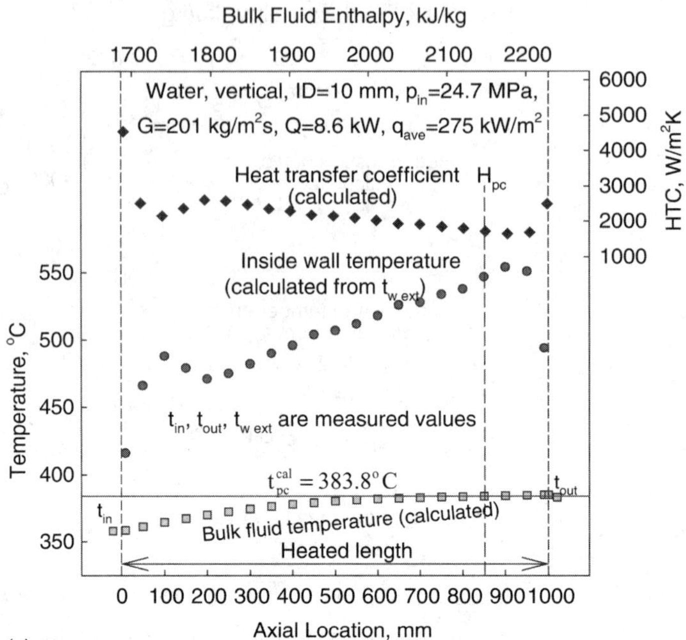

(c)

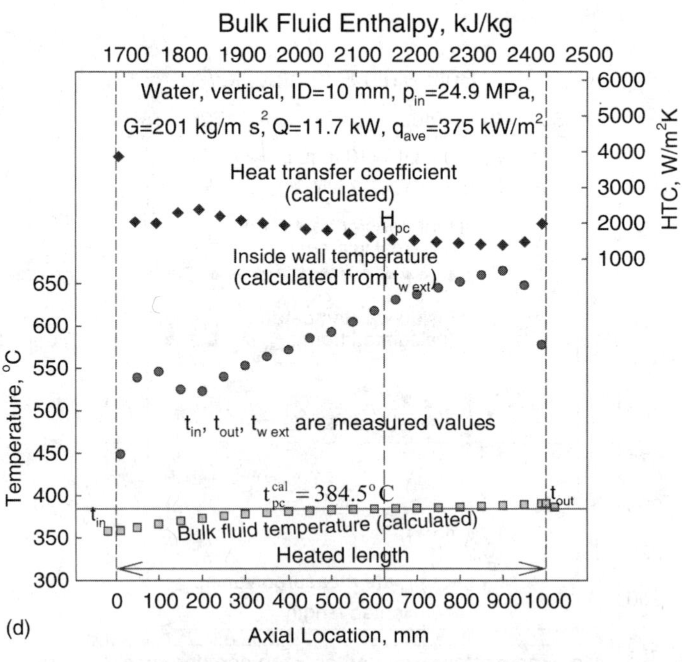

(d)

Figure 5.10. Temperature and HTC variations along 1-m circular tube at various heat fluxes and inlet temperatures (mainly deteriorated heat-transfer mode) (Kirillov et al. 2005): Nominal flow conditions–p_{in}=24.9 MPa and G=200 kg/m²s; H_{pc}=2149 kJ/kg.

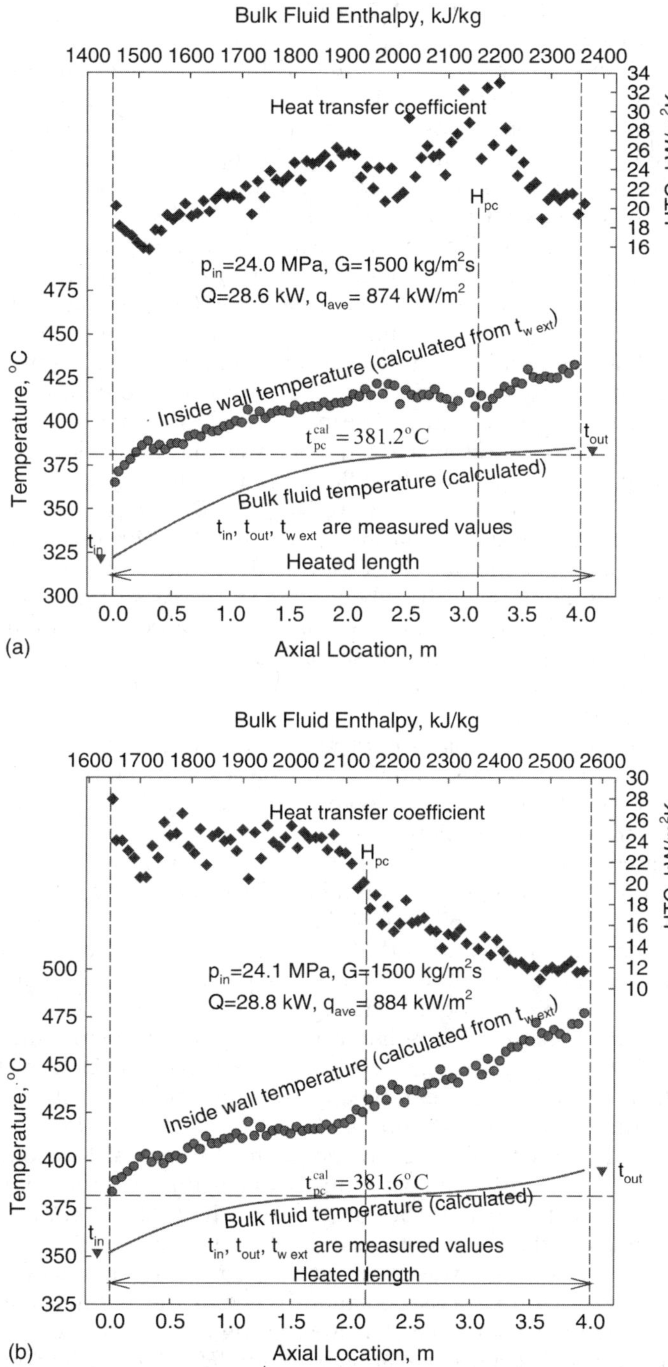

Figure 5.11. Temperature and HTC variations along 4-m circular tube (Kirillov et al. 2005): Nominal operating conditions – p_{in}=24.0 MPa, G=1500 kg/m^{2}s and q=880 kW/m^2; H_{pc}=2159 kJ/kg.

23.5 MPa (t_{pc} = 279.4°C and H_{pc} = 2127 kJ/kg) are summarized in Figures 5.12 to 5.20 (for experimental setup details, see Section 10.2).

Figures 5.12 and 5.13 show temperature profiles along the 9.50-mm and 6.28-mm ID vertical tubes with upward flow of water at $G \approx 250$ kg/m²s and at various heat fluxes. Flow conditions and variations in heat flux correspond to the following changes in the non-dimensional free-convection parameter: $\mathbf{Gr/Re^2} \approx$ 0.1 – 0.5, where the smallest value of this parameter corresponds to the lowest value of heat flux. At a $\mathbf{Gr/Re^2}$ value of about 0.5, the deteriorated heat transfer or peak ("hump") in the wall temperature appeared along the outlet section of the tube (see Figures 5.12 and 5.13). Actually, the first signs of the deteriorated heat transfer usually appeared near the tube outlet (see Figures 5.12 and 5.13). With a heat flux increase, these peaks moved towards the tube entrance.

Figure 5.14 shows variations in heat transfer along the 6.28-mm ID vertical tube with upward flow at $G \approx 250$ kg/m²s. Analysis of the data in Figure 5.14 shows that the so-called "entrance section" with developing heat transfer can vary from $L/D = 95$ (the total heated length of the channel) at lower heat fluxes to about $L/D = 50$ or less at higher heat fluxes. At the same value of an inlet bulk-fluid enthalpy, the minimum of the curve $\mathbf{Nu} = f\,(L/D)$ is located within a cross-section with the same Reynolds number, i.e., these minimums correspond to the same thermalhydraulic flow conditions. Also, values of enthalpy increase (ΔH_{in}) within the section from the tube inlet to the cross-section with the minimum value of Nusselt number (i.e., the maximum value of wall temperature) for these cases are approximately the same. Plots in Figures 5.12, 5.13, and 5.16 show that with a rise in the inletenthalpy, the value of ΔH_{in} decreases, i.e., H_{in} has a certain effect on ΔH_{in}.

Figure 5.15 shows a comparison of wall temperature profiles in a vertical tube for upward and downward flows. These data show that peaks in wall temperature were not detected in the downward flow within the same operating conditions as for the upward flow.

In the current experiments, heat-transfer improvement was detected within the exit section ($L/D = 36.6$) only at high heat fluxes ($q/G = 1.49$; 1.89 kJ/kg) (see Figure 5.16a, profiles for $q = 370$ and 471 kW/m²).

The effect of inlet bulk-fluid enthalpy on ΔH_{in}, based on averaged data values resulting from the repeated runs, is shown in Figure 5.17. There is a peak in the curves near the pseudocritical region (H_{pc} = 2128.6 kJ/kg). In general, this region is characterized with higher uncertainties in the measured parameters. Analysis of the data in Figure 5.17 for the 6.28-mm ID tube shows that the dependence of ΔH_{in} on H_{in} can be approximated with two straight lines of different inclination. The cross-point for these two lines (see Figure 5.17, lines for laminar and turbulent flow) corresponds to the critical Reynolds number $\mathbf{Re} \approx 2400$), at which laminar flow regime changes into turbulent flow regime. A similar trend is valid for the 9.50-mm ID tube. Its first left point (see Figure 5.17) corresponds to the same critical Reynolds number value.

Based on these findings, it was assumed that ΔH_{in} is an amount of heat, which should be transferred to 1 kg of the coolant in a certain thermal state for formation of the thermal boundary layer, i.e., for overcoming the thermal resistance of the entrance region. The detected correlation between $\Delta H_{in}(H_{in})$ and flow regime probably proves that the thermal boundary layer formation is a thermodynamic process and a significant additional amount of heat is required for formation of the turbulent boundary layer, when the laminar flow regime takes place at the inlet.

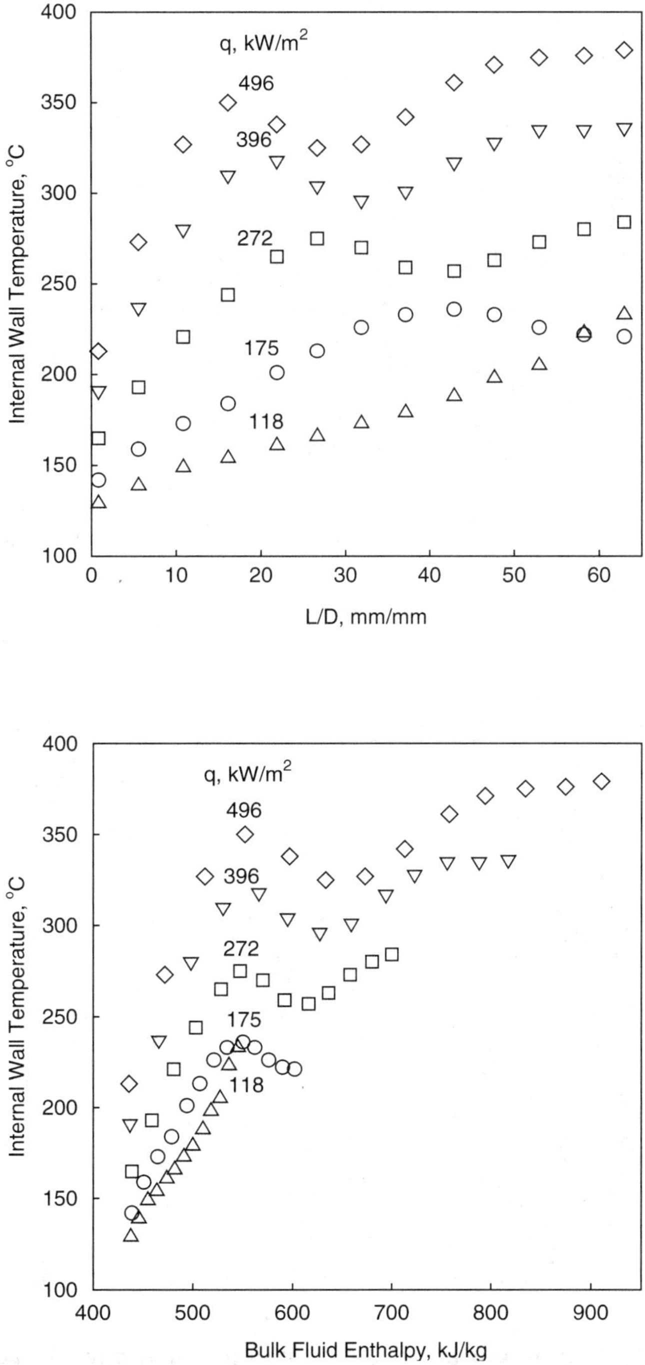

Figure 5.12. Temperature profiles at mixed convection in vertical tube (*D*=9.50 mm) with upward flow (Pis'mennyy et al. 2005): Water, *p*=23.5 MPa, *G*=248 kg/m²s and t_{in}≈100°C.

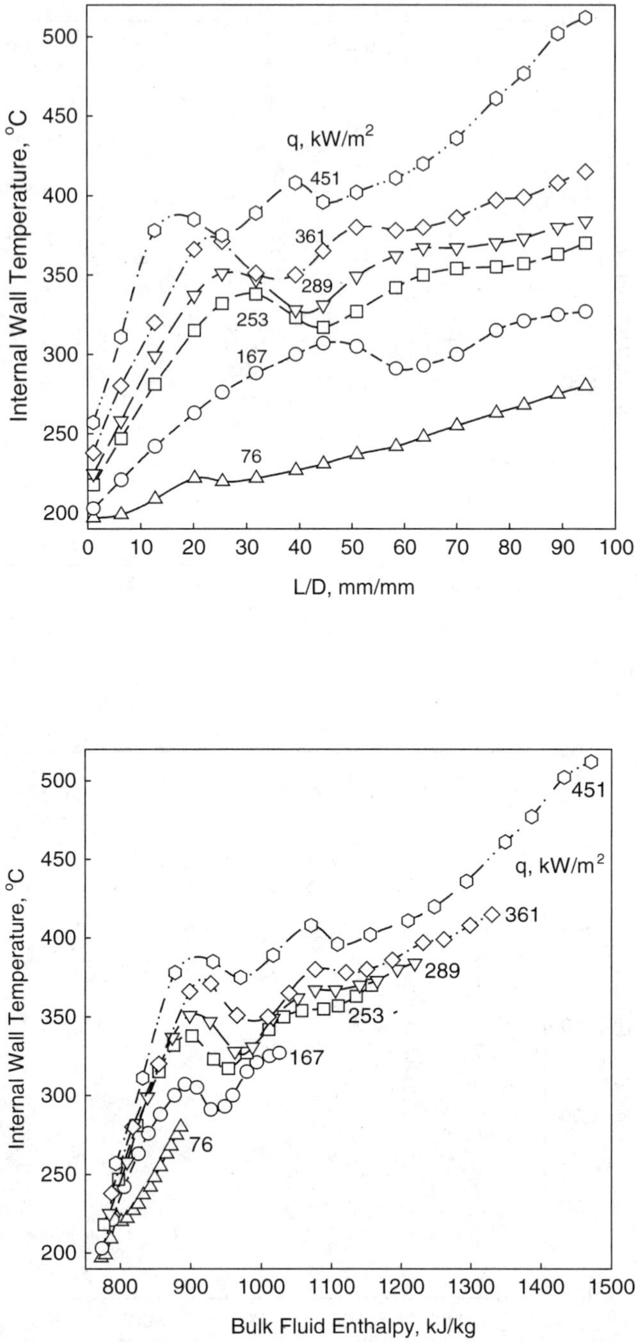

Figure 5.13. Temperature profiles at mixed convection in vertical tube (D=6.28 mm) with upward flow (Pis'mennyy et al. 2005): Water, p=23.5 MPa, G=249 kg/m²s and $t_{in} \approx$ 200°C.

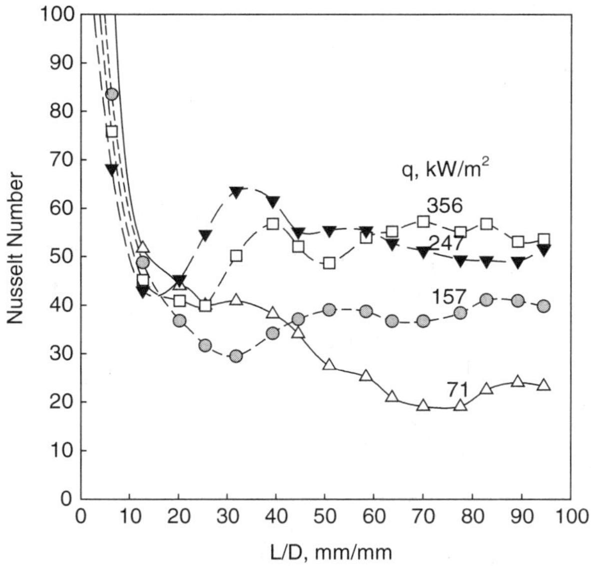

Figure 5.14. Heat-transfer variations along vertical tube (D=6.28 mm) with upward flow (Pis'mennyy et al. 2005): Water, p=23.5 MPa, G=249 kg/m^{2}s, and $t_{in} \approx$ 250°C; Re_{min}=15,500 (i.e., Reynolds number in a cross-section with Nu_{min}).

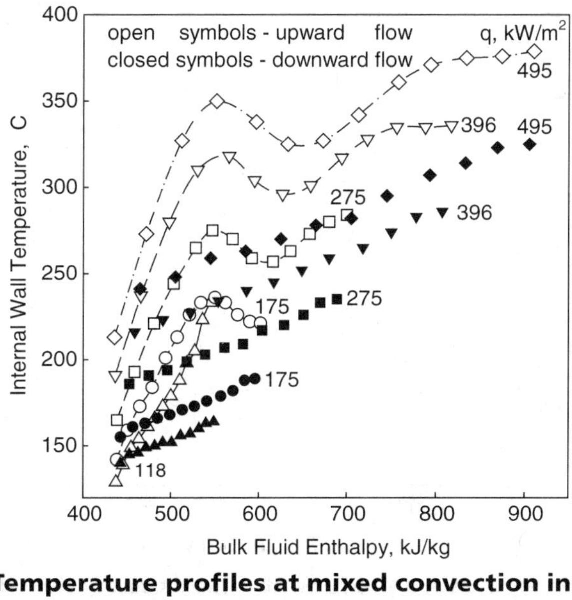

Figure 5.15. Temperature profiles at mixed convection in vertical tube (D=9.50 mm) (Pis'mennyy et al. 2005): Water, p=23.5 MPa, G=248 kg/m^{2}s, and $t_{in}\approx$100°C.

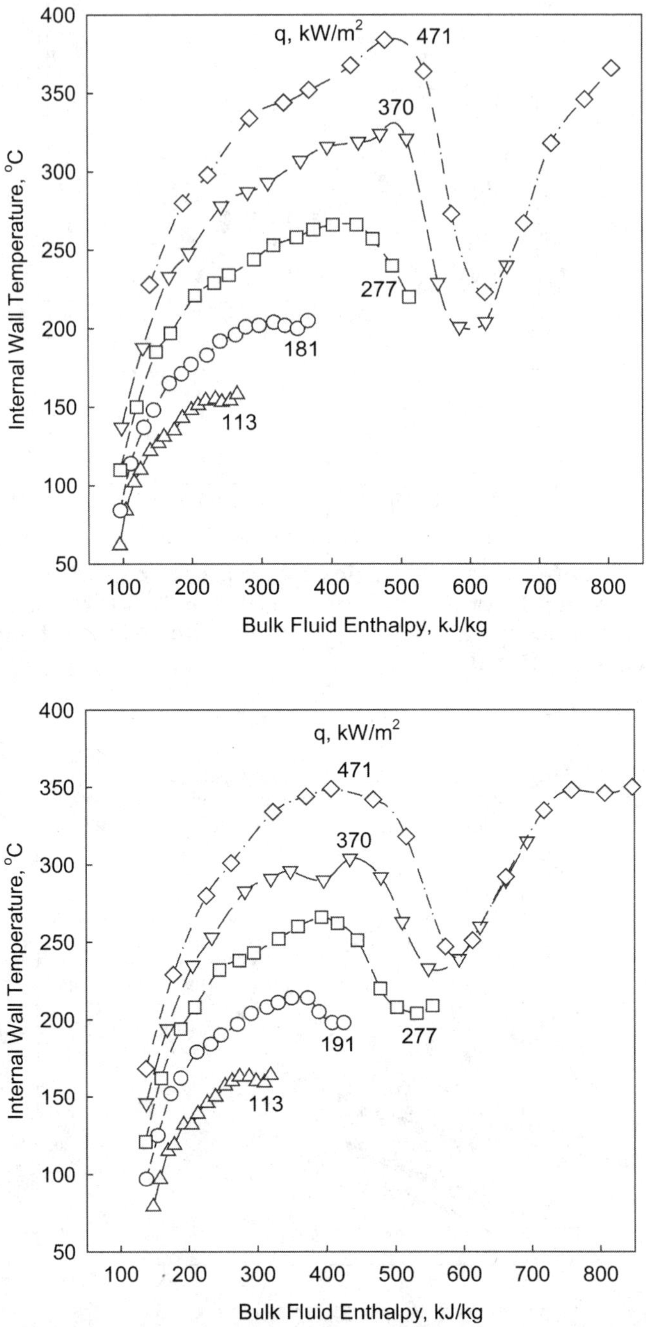

Figure 5.16. Temperature profiles at mixed convection in vertical tube (D=6.28 mm) with upward flow at different inlet enthalpies (Pis'mennyy et al. 2005): Water, p=23.5 MPa, G=248 kg/m²s; (a) t_{in}≈17°C and (b) t_{in}≈30°C.

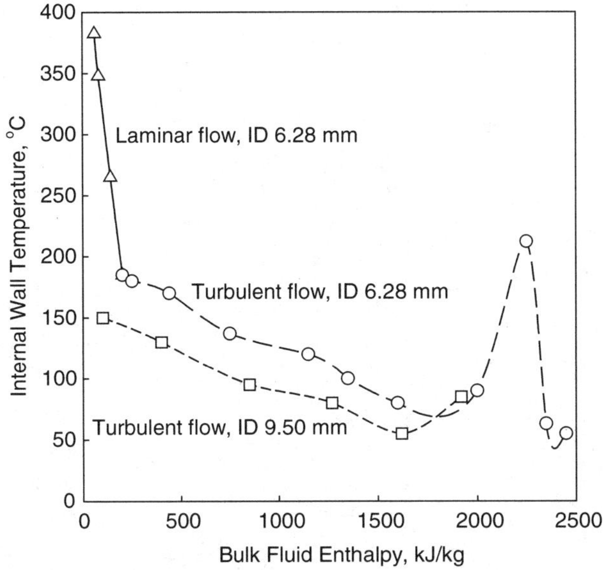

Figure 5.17. Effect of inlet bulk-enthalpy of water on size of entrance thermodynamic region of vertical tubes with upward flow expressed in H_{in} (Pis'mennyy et al. 2005): p=23.5 MPa and G=249 kg/m²s.

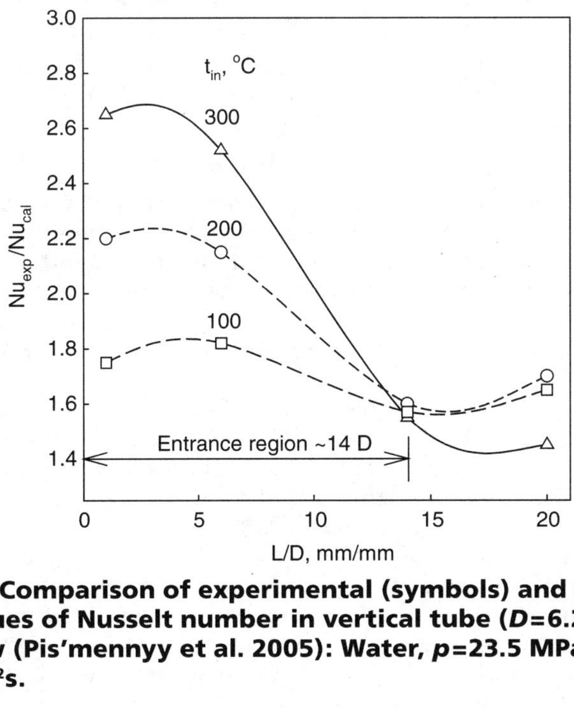

Figure 5.18. Comparison of experimental (symbols) and calculated (curves) values of Nusselt number in vertical tube (D=6.28 mm) with upward flow (Pis'mennyy et al. 2005): Water, p=23.5 MPa and G=499 kg/m²s.

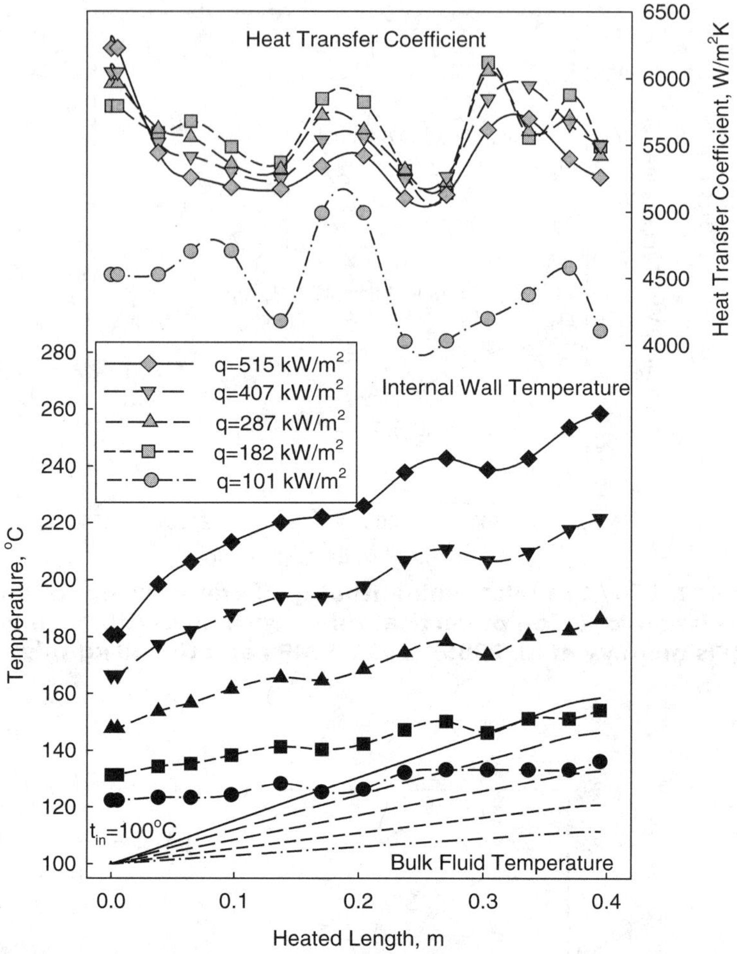

Figure 5.19. Temperature profiles at weak effect of free convection in vertical tube (*D*=9.50 mm) with upward flow (Pis'mennyy et al. 2005): Water, *p*=23.5 MPa and *G*=509 kg/m²s; HTC—grey symbols, internal wall temperature—black symbols, and bulk-fluid temperature—lines.

Figure 5.18 shows that the entrance region is about 14·*D* at various inlet temperatures.

Figure 5.19 shows temperature profiles along the heated length of the 9.50-mm ID tube at the maximum ratio of $\mathbf{Gr/Re^2}$ about 0.03. At these conditions, the inlet peaks in the wall temperature did not appear. Comparison of the data in Figure 5.19 with those in Figure 5.15 shows inlet peaks in the wall temperature appeared usually within $\mathbf{Gr/Re^2} \geq 0.06 - 0.08$.

The current experimental results show that inlet-temperature peaks are dangerous for the heating surface, because at a low mass flux the wall temperature within the peak region is subjected to significant variations even at small changes in operating parameters (t_{in}, q and G). Figure 5.20 shows such variations in wall temperature at an inlet-temperature change of 2.5°C.

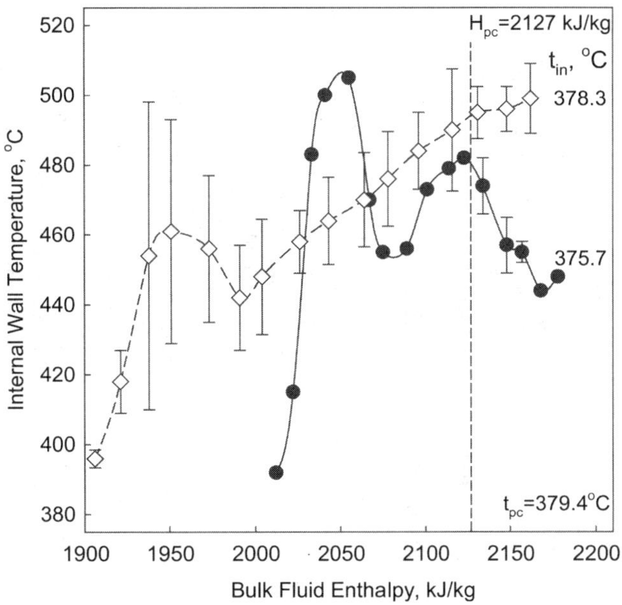

Figure 5.20. Unstable temperature regimes at mixed convection in vertical tube (*D*=6.28 mm) with upward flow (Pis'mennyy et al. 2005): Water, *p*=23.5 MPa, *G*=248 kg/m²s, and *q*=183 kW/m².

The data in this figure reflect the range of the wall temperature pulsations at a frequency of about 0.1 Hz.

5.2 HEAT TRANSFER IN HORIZONTAL TEST SECTIONS

All[7] primary sources of experimental data for heat transfer to water flowing in horizontal test sections are listed in Table 5.2.

Krasyakova et al. (1967) found that in a horizontal tube, in addition to the effects of non-isothermal flow that is relevant to a vertical tube, the effect of gravitational forces is important. The latter effect leads to the appearance of temperature differences between the lower and upper parts of the tube. These temperature differences depend on flow enthalpy, mass flux, and heat flux. A temperature difference in a tube cross-section was found at $G = 300 - 1000$ kg/m²s and within the investigated range of enthalpies ($H_b = 840 - 2520$ kJ/kg). The temperature difference was directly proportional to increases in heat-flux values. The effect of mass flux on the temperature difference is the opposite, i.e., with increase in mass flux the temperature difference decreases. Deteriorated heat transfer was also observed in a horizontal tube. However, the temperature profile for a horizontal tube at locations of deteriorated heat transfer differs from that for a vertical tube, being smoother for a horizontal tube compared to that of a vertical tube with a higher temperature increase on the upper part of the tube than on the lower part.

[7]"All" means all sources found by the authors from a total of 650 references dated mainly from 1950 till the beginning of 2006.

Table 5.2. Range of investigated parameters for experiments with water flowing in horizontal circular tubes at supercritical pressures.

Reference	p, MPa	t, °C (H in kJ/kg)	q, MW/m²	G, kg/m²s	Flow Geometry
Chakrygin and Lokshin 1957	22.5–24.5	t_{in}=190	0.2–0.44	300–800	SS tube (D=29 mm)
Dickinson and Welch 1958	24, 31	t_b=104–538	0.88–1.82	2170–3391	SS tube (D=7.62 mm, L=1.6 m)
Shitsman 1962	22.8–26.3	t_b=300–425	0.29–5.8	100–2500	Horizontal and vertical copper and carbon steel tubes (D=8 mm, L=170 mm)
Domin 1963	22.3–26.3	$t_b \leq 450$; $t_w \leq 560$	0.58–4.7	1210–5160	Inconel tubes (D=2; 4 mm, L=1.075; 1.233 m)
Vikhrev and Lokshin 1964	22.6–29.4	t_b=200–500	0.35–0.7	400, 700, 1000	SS tube (D=8 mm, L=2.5 m)
Shitsman 1966	9.8–24.5	t_b=310–410	0.33–0.45	375	SS tube (D=16 mm, L=1.6 m)
Shitsman 1967	24.3–25.3	t_b=300–400	0.73–0.52	600–690	Horizontal and vertical tubes (D_{ext}=8; 16 mm)
Krasyakova et al. 1967	23	H_{in}=837–2721	0.23–0.7	300–1500	Horizontal and vertical tubes (D=20 mm)
Shitsman 1968	10–35	t_b=100–250	0.27–0.7	400	SS tube (D/L=3/0.7; 8/0.8; 8/3.2; 16/1.6 mm/m)
Krasyakova et al. 1968	15; 18.8; 23	H_{in}=840–1890	0.23–0.7	300–2000	Horizontal and vertical SS tube (D=20 mm, L=2.2 m)
Kondrat'ev 1969	22.6; 24.5; 29.4	t_b=105–540	0.12–1.2	$Re=10^5$	Polished SS tube (D=10.5 mm, L=0.52 m)
Vikhrev et al. 1970	23–27	H_b=920–2680	232–928	500–1500	SS tubes (D=19.8; 32.1 mm, L=6 m)
Kamenetsky and Shitsman 1970	24.5	H_b=80–2300	0.19–1.33	50–1750	Horizontal and vertical SS tube (D=22 mm, L=3 m), non-uniform circumferential heat flux, horizontal and upward flows

Reference	p, MPa	t, °C (H in kJ/kg)	q, MW/m²	G, kg/m²s	Flow Geometry
Zhukovskiy et al. 1971	24.5	H_b=630–3100	0.232–1.39	300–3000	SS tube (D=20 mm, L=4 m)
Barulin et al. 1971	22.5–26.5	t_b=50–500; t_w=60–750	0.2–6.5	480–5000	Horizontal and vertical tubes (D=3; 8; 20 mm, L/D/300)
Belyakov et al. 1971	24.5	H_b=420–3140	0.23–1.4	300–3000	Horizontal and vertical SS tube (D=20 mm, L=4–7.5 m)
Yamagata et al. 1972	22.6–29.4	t_b=230–540	0.12–0.93	310–1830	Horizontal and vertical SS tubes (D/L=7.5/1.5; 10/2 mm/m)
Vikhrev et al. 1973	23.3–27.4	H_b=25.1–3056	0.23–1.16	500–1900	Horizontal and inclined (angle 15° from horizontal) tubes (D=19.8 mm, L=6 m)
Kamenetskii 1974	24.5	t_b=50–400	0.37–1.3	240–1700	SS tubes with non–uniform circumferential heating (D=21.9 mm, L=3 m)
Solomonov and Lokshin 1975	22.5–26.5	H_b=252–3066	0.23–1.16	500–1900	Horizontal and inclined SS tube (D=20 mm, L=6 m)
Ishigai et al. 1976	24.5; 29.5; 39.2	H_b=220–800	0.14–1.4	500; 1000; 1500	Horizontal and vertical SS polished tubes (D=4.44 mm, L=0.87 m)
Kamenetskii 1980	24.5	H_b=100–2200	0.37–1.3	300–1700	SS tube with and without flow spoiler (D=22 mm, L=3 m)
Robakidze et al. 1983	23.2–24	t_b=359–797	–	0.7–70	SS tube (D=16 mm, L=3.7 m)
Bazargan et al. 2005	23–27	H_b=600–3400	Up to 0.31	330–1230	Horizontal Inconel tubes (D=6.3 mm, L=1.5 m)

Table 5.3. Range of investigated parameters for experiments with water flowing in annuli at supercritical pressures.

Reference	*p*, MPa	*t*, °C (*H* in kJ/kg)	*q*, MW/m²	*G*, kg/m²s	Flow Geometry
McAdams et al. 1950	0.8–24	t_b=221–538	0.035–0.336	75–224	Gap 1.65 mm, D_{rod}= 6.4 mm, D_{tube}= 9.7 mm, L=0.312 m, upward flow, internal heating
Glushchenko and Gandzyuk 1972	23.5; 25.5; 29.5	t_b=150–375	1.15–2.96	650–3000	Gap 1 mm, D_{rod}= 6–10 mm, L=0.6 m, upward flow, external and internal heating (selected data are shown in Figure 5.21)
Glushchenko et al. 1972	23.5; 25.5; 29.5	H_b=85–2400	1.8–5.4	650–7000	Gap 0.3; 0.7; 1; 1.5 mm, D_{rod}=6–10 mm, L=0.115–0.6 m, upward flow, external and internal heating
Ornatskiy et al. 1972	23.5	H_b=400–2600	1.2–3.0	2000; 3000; 5000	Gap 0.7 mm, D_{rod}=10.6 mm, D_{tube}=12 mm, L=0.28 m, external heating (selected data are shown in Figure 5.22)

5.3 HEAT TRANSFER IN ANNULI

All[8] primary sources of heat-transfer experimental data of water flowing in annuli are listed in Table 5.3.

In general, forced convective heat transfer in an annulus is different from a circular tube even at subcritical pressures. In an annulus, three heating modes can exist: (i) outer surface heated, (ii) inner surface heated, and (iii) both surfaces heated.

McAdams et al. (1950) conducted experiments with an upward flow of water in a vertical annulus with internal heating. The objective was to estimate local HTCs for turbulent flow inside an annulus. Four chromel-alumel thermocouples, spaced at 76.2 mm intervals, were installed inside a heated rod. These measured temperatures were used to calculate local HTCs along the heated rod. The experiments showed that for given Reynolds and Prandtl numbers, a value of the local Nusselt number always decreased as a value of L/D_{hy} increased, regardless of the temperature at which the physical properties were evaluated.

[8]"All" means all sources found by the authors from a total of 650 references dated mainly from 1950 till the beginning of 2006.

The results of Glushchenko and Gandzyuk (1972) are presented in Figure 5.21. Their results showed that with heat flux increasing deteriorated heat transfer may appear at the outlet section or at inlet and outlet sections of the annulus.

Glushchenko et al. (1972) conducted experiments with an upward flow of water in annuli with external and internal one-side heating. In general, the results of the investigation showed that variations in wall temperature of a heated tube and of an annulus, when the tubes and annuli are fairly long, were similar. However, in annuli with normal and deteriorated heat transfer no decrease in temperature (past the zone of deteriorated heat transfer) was noticed in their experiments.

Ornatskiy et al. (1972) investigated normal and deteriorated heat transfer in a vertical annulus (Figure 5.22). The deteriorated heat-transfer zone was observed visually as a red-hot spot, appearing in the upper section of the test tube. The hot spot elongated in the direction of the annulus inlet with increasing heat flux.

5.4 HEAT TRANSFER IN BUNDLES

The two primary sources of heat transfer to water flowing in heated-rod bundles are listed in Table 5.4.

Dyadyakin and Popov (1977) conducted experiments with a tight-lattice 7-rod bundle (finned rods) cooled with water. They tested five bundles with different flow areas and hydraulic diameters:

Test Section #	1	2	3	4	5
A_{fl}, mm^2	112	134	113	121	102
D_{hy}, mm	2.35	2.77	2.38	2.53	2.15

The HTC was measured with a movable thermocouple installed in the central rod. However, the data reduction, in terms of the HTC, was based on heat transfer through the rod wall without taking into account internal heat generation (heating due to electrical current passing through the wall). They found that at mass fluxes greater than or equal to 2000 kg/m^{2}s with $H_b^{in} = 1000 - 1800$ kJ/kg, and at high heat fluxes, significant pressure oscillations occurred (±5 MPa with frequency of 0.04 – 0.033 Hz). Similar pressure oscillations were recorded at G = 2000 kg/m^{2}s and q = 1.2 MW/m^2 when the inlet bulk-fluid temperature was 305°C. At G= 2000 kg/m^{2}s and q = 2.3 MW/m^2 the pressure oscillations appeared at an inlet bulk-fluid temperature of 260°C and resulted in burnout of the test section. In experiments with G = 3800 kg/m^{2}s and inlet bulk temperature of 280°C, the pressure oscillations were recorded at q = 3.5 MW/m^2. At the same flow and at an inlet bulk-fluid temperature of 370°C, the pressure oscillations were recorded at q = 2.3 MW/m^2.

Silin et al. (1993) reported that a large database for water flowing in large bundles at supercritical pressures was created in the Russian Scientific Centre (RSC) "Kurchatov Institute" (Moscow, Russia). Experimental heat-transfer data were satisfactorily described by correlations obtained for water flow in tubes at supercritical pressures and for the normal heat-transfer regime. The most

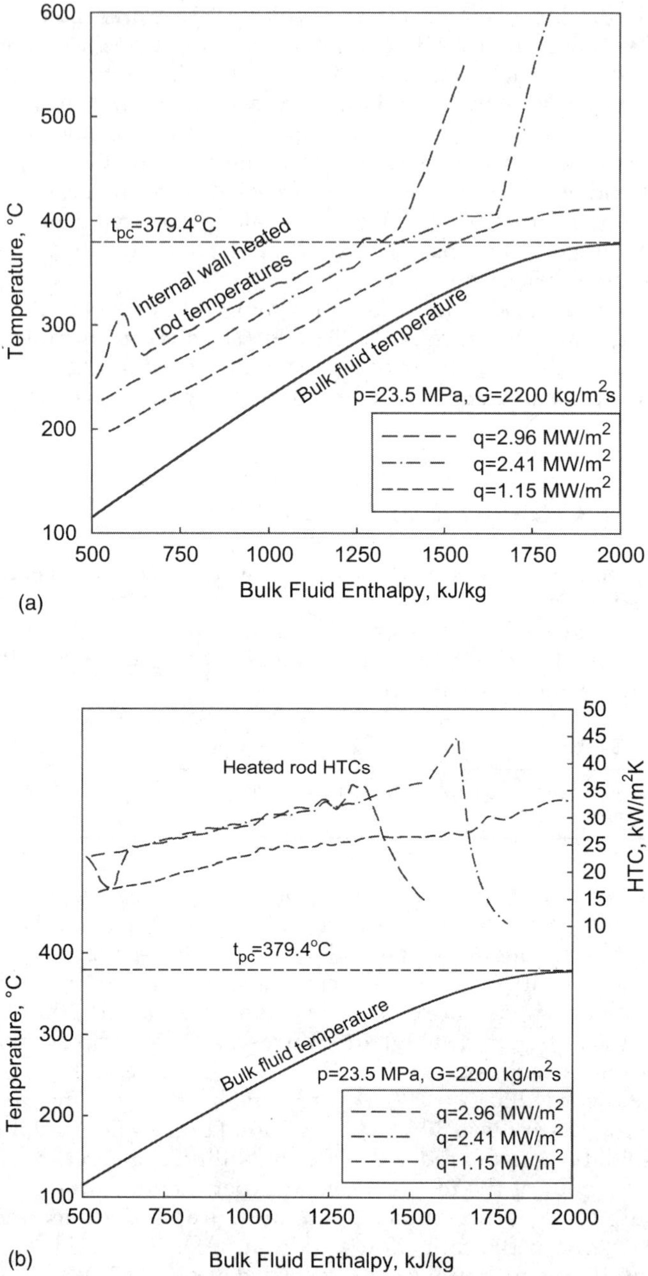

Figure 5.21. Temperature profiles (a) and HTC values (b) along heated rod in annulus (inner rod heating, gap 1 mm, L = 0.6 m) (Glushchenko and Gandzyuk 1972): Water, H_{pc} = 2126.9 kJ/kg; HTC values were calculated by the authors of the current monograph using the data from figure (a).

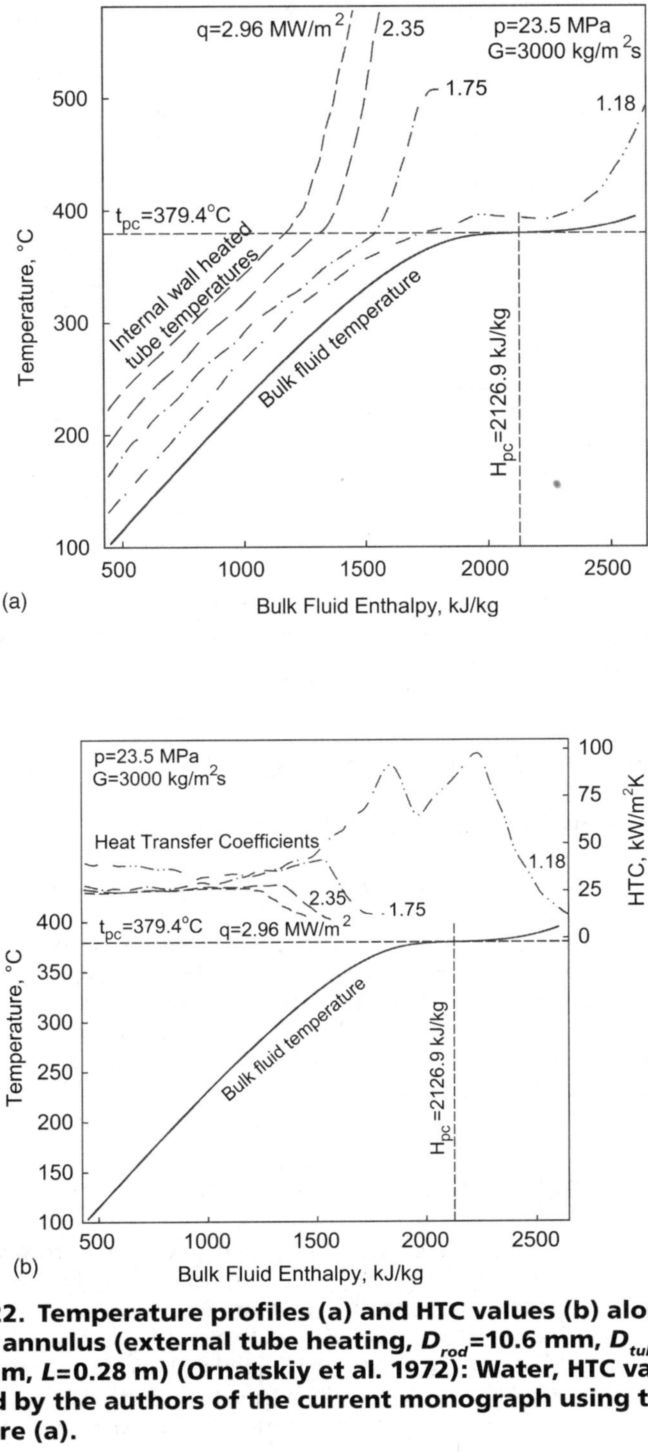

Figure 5.22. Temperature profiles (a) and HTC values (b) along heated length of annulus (external tube heating, D_{rod}=10.6 mm, D_{tube}=12 mm, gap 0.7 mm, L=0.28 m) (Ornatskiy et al. 1972): Water, HTC values were calculated by the authors of the current monograph using the data from figure (a).

Table 5.4. Range of investigated parameters for experiments with water flowing in bundles at supercritical pressures.

Reference	*p*, MPa	*t*, °C (*H* in kJ/kg)	*q*, MW/m²	*G*, kg/m²s	Flow Geometry
Dyadyakin and Popov 1977	24.5	t_b=90–570; H_b=400–3400	<4.7	500–4000	Tight bundle (7 rods (6+1),D_{rod}=5.2 mm, *L*=0.5 m), each rod has four helical fins thickness 1 mm, helical pitch 400 mm), pressure tube hexagonal in cross-section
Silin et al. 1993	23.5; 29.4	H_b=1000–3000	0.18–4.5	350–5000	Vertical full-scale bundles, D_{rod}=4 and 5.6 mm, rod's pitch 5.2 and 7 mm)

important difference between water behavior inside tubes and water behavior inside bundles was that there was *no heat transfer deterioration* in the multi-rod bundles, within the same test parameter range for which heat transfer deterioration occurred in tubes.

5.5 FREE-CONVECTION HEAT TRANSFER

Just several papers have been found related to free-convection heat transfer at supercritical pressures.

Fritsch and Grosh (1963) investigated laminar free convection heat transfer from a vertical plate to water close to its critical point. Their experiments have a systematic deviation of about 20% from their previous analytical results. In general, their analytical results were lower than the experimental data.

Larson and Schoenhals (1966) studied analytically and experimentally heat transfer from a vertical plate by turbulent free convection to water near its critical point. Their experiments indicated reasonable agreement between analytical and experimental results.

5.6 FINAL REMARKS AND CONCLUSIONS

1. The majority of the experimental studies deal with heat transfer to supercritical water in vertical and horizontal circular tubes. A few studies were devoted to heat transfer in annuli and bundles.
2. In general, experiments showed that there are three modes of heat transfer in fluids at supercritical pressures: (1) normal heat transfer, (2) deteriorated heat transfer with lower values of the HTC within some part of a test section compared to those of normal heat transfer and

(3) improved heat transfer with higher values of the HTC compared to those of normal heat transfer. Also, a peak in HTC near the critical and pseudocritical points was recorded.

3. The deteriorated heat transfer usually appears at higher heat fluxes and lower mass fluxes.
4. There are very few publications, which are devoted to heat transfer in bundles cooled with water at supercritical pressures. Therefore, more work is needed to provide reliable information for design purposes.
5. Heat transfer at supercritical pressures can be accompanied by flow oscillations and other instabilities at some operating conditions. However, experimental data on these aspects are still very limited.

Chapter 6

EXPERIMENTAL HEAT TRANSFER TO CARBON DIOXIDE AT SUPERCRITICAL PRESSURES

As mentioned previously, carbon dioxide has often been used as a supercritical working fluid in various thermodynamic cycles and heat exchangers and as a modeling fluid instead of supercritical water, because of its lower critical pressure and temperature.

6.1 FORCED–CONVECTION HEAT TRANSFER

6.1.1 Heat Transfer in Vertical Tubes

Carbon dioxide is the most investigated fluid in near-critical and supercritical regions after water. All[9] primary sources of heat-transfer experimental data of carbon dioxide flowing inside vertical circular tubes are listed in Table 6.1. Ranges of investigated parameters for selected experiments with carbon dioxide in circular tubes at supercritical pressures that are relevant to supercritical water-cooled nuclear reactor operating range are shown in Figure 6.1.

Hall et al. (1966) conducted heat-transfer experiments in turbulent flow of carbon dioxide between planes at slightly supercritical pressures. One surface was maintained at subcritical temperatures and the other was at temperatures from below to slightly above critical. The objective of these experiments was to investigate fully developed temperature and velocity profiles.

Shiralkar and Griffith (1968) conducted experiments with supercritical carbon dioxide in circular tubes over a wide range of flow conditions (Figures 6.2 – 6.4). They found that deteriorated heat transfer started at certain ratio[10] of q/G (Figure 6.2, $q/G = 0.116$) and was affected with inlet temperature (Figure 6.3) and direction of flow (Figure 6.4). In general, wall temperature excursion within the deteriorated heat-transfer region is more significant in downward flow than in upward flow at similar conditions (Figure 6.4).

[9] "All" means all sources found by the authors from a total of 650 references dated mainly from 1950 till the beginning of 2006.

[10] In supercritical water, the deteriorated heat transfer appears at $q/G > 0.4$ (Pioro and Duffey 2003a).

Table 6.1. Range of investigated parameters on heat transfer experiments with carbon dioxide at supercritical pressures in vertical tubes.

Reference	P_{out}, MPa	t_b, °C	q, kW/m²	G, kg/m²s	Flow Geometry
Bringer and Smith (1957)	8.3	t_{in}=21–49	31–310	100–1300	Inconel tube (D=4.57 mm, L=610 mm)
Petukhov et al. (1961)	8.8–10.8	Δt=4–50	–	**Re**=(50–300)·10³	Copper tube (D=6.7 mm, L=0.67 m)
Wood and Smith (1964)	7.4	t_{in}=27.5–31.4	30–120	u=3–5 m/s	Tube (D=22.9 mm, L=1435 mm)
Krasnoshchekov and Protopopov (1966)	7.9; 9.8	20–110	430–2520	1135–7520	SS tube (D=4.08 mm, L=208 mm)
Krasnoshchekov et al. (1967)	7.85; 9.81	20–110	≤2600	350	SS tube (D=4.08 mm, L=208.1 mm)
Tanaka et al. 1967	8.2	27–40	14–81	m=0.23–0.47 kg/s	Tube (D=10 mm), natural circulation
Shiralkar and Griffith (1969 and 1968)	7.6; 7.9	t_{in}=10–32	125–190	680–2710	SS tube (D=6.22 mm, L=1.52 m; D=3.175 mm), tube with twisted tape and annulus; up and down flow (selected data are shown in Figures 6.2 – 6.4)
Hall and Jackson (1969)	7.58	t_{in}=14–24	40–57	**Re**$_{in}$=113·10³	Tube, upward and downward flows
Bourke et al. (1970)	7.44–10.32	t_{in}=15–35	8–350	311–1702	SS tube (D=22.8 mm, L=4.56 m), upward and downward flows
Shiralkar and Griffith (1970)	7.6; 7.9	t_{in}=–18–31	50–453	**Re**=(267–835)·10³	SS tube (D=3.18; 6.35 mm, L=1.52 m), upward and downward flows, including tube with twisted tape
Tanaka et al. (1971)	8.1	0–170	488; 640	m=120–240 kg/h	Tubes (D=6 mm, L=1 m, smooth–tube surface (roughness 0.2) and rough tube (14 µm)
Ikryannikov et al. (1972)	7.8–9.8	t_{in}=15–50	5.8–9.3	**Re**=(30–500)·10³	SS tube (D=29 mm, L=2.3 m)
Petukhov et al. (1972)	9.8	t_{in}=12–13	85–505	960	SS tube (D=4.3 mm,L=0.33 m), upward flow
Silin (1973)	7.9; 9.8	t_w<860	<1100	200–2600	Vertical and horizontal tubes (D=2.05, 4.28 mm)
Miropol'skiy and Baigulov (1974)	7.9	t_{in}=22–30	67–224	670–770	SS tube (D=21 mm, L=1.7 m)
Baskov et al. (1974)	8; 10;	17–212	≤640	1560–4170	Copper tube (D=4.12 mm, L=375 mm)
Protopopov and Sharma (1976)	7.5; 8; 9; 10	t_{in}=14–54	3.5–110	–	SS tube (D=8.2; 19.6 mm)
Fewster (1976)	7.6	t_{in}=10–25	10–300	180–2000	Vertical SS tube (D=8; 19 mm, L=1 m) (selected data are shown in Figures 6.5–6.8)
Baskov et al. (1977)	8; 10; 12	17–212	≤640	1560–4170	Vertical tube (D=4.12 mm, L=375 mm), upward and downward flows

Reference	P_{out}, MPa	t_b, °C	q, kW/m²	G, kg/m²s	Flow Geometry
Ankudinov and Kurganov (1981)	7.7	t_{in}=20	up to 1540	2100–3200	Vertical and horizontal tubes (D=8 mm, L=1.84 m) with helical wire insert, upward, downward, and horizontal flow
Vlakhov et al. 1981	–	–	0.5–20	0.8–6.2	Vertical SS tube (D=7.85 mm, L=1520 mm)
Afonin and Smirnov (1985)	0.98–9.8	H_{in}=730	180–1250	100–1100	Steel tube (D=6 mm, L=2.5 m)
Dashevskiy et al. (1986)	7.5–7.9	29.5–54.7	0.5–85	1.9–265	Tube (D=7.85 mm, L=1520 mm), downward flow
Dashevskii et al. (1987)	7.53–8.04	300–370	1.7–23	7–90	Inclined SS tube (D=16 mm, L=3.718 m, inclination angles –10°, –21° and –40°), downward flow
Kurganov and Kaptil'ny (1992)	9	H_{in}=500–740	40–460	800–2100	Tube (D=22.7 mm, L=5.22 m)
Kurganov and Kaptilnyi (1993)	9	t_{in}=25–35	40–260	800, 1200	Tube (D=22.7 mm), upward and downward flow
Walisch et al. (1997)	8–40	50–120	Q=0.4–2 kW	m=0.8–50 g/s	Vertical, horizontal and inclined tubes (D=10 mm, L=1.5 m, Inconel 600)
Pettersen et al. (2000)	8.1–10.1	15–70	10–20	600–1200	25 parallel circular micro-channels (D=0.79 mm, L=540 mm)
Liao and Zhao (2002a,b)	7.4–12	t_{in}=20–110	(1–20)·10⁴	m=0.02–0.2 kg/min	Vertical and horizontal miniature SS tubes (D=0.70; 1.40; 2.16 mm; L=110 mm)
Jackson et al. (2003)	7.25–8.27	t_{in}=8–20	5–57	100–560	Vertical SS tube (D=19.05 mm, L=2.46 m), upward and downward flows.
Fewster and Jackson (2004)	7.6	t_{in}=10	50–460	300–3300	Vertical SS tube (D=5.08; 7.88 mm, L up to 1.2 m), upward and downward flows
Jiang et al. (2004)	9.5	t_{in}=33–51	31–108	m=1.5–4.2 kg/h	Vertical SS tube (D=0.95 mm, L=50 mm), copper tube (D=4 mm) and porous tube (D=4 mm, particle OD 0.2–0.28 mm)
Kim et al. (2005a)	7.8; 8.1; 8.9	t_{in}=27	Up to 150	400–1200	Vertical Inconel tube (D=4.4 mm, L=2 m), upflow
Kim et al. (2005b)	8	t_{in}=15–32	3–180	210–1230	Vertical channels with L=1.2 m: circular (D=7.8 mm); triangular–shaped (D_{hy}=9.8 mm); and square–shaped (D_{hy}=7.9 mm)
Pioro and Khartabil (2005); Yang and Khartabil (2005); Khartabil (2002)	7.4–8.8	t_{in}=20–40	15–615	900–3000	Vertical tube (D=8 mm, L=2.2 m, Inconel 600) (data of I.L. Pioro) (selected data are shown in Figures 6.9 – 6.13)

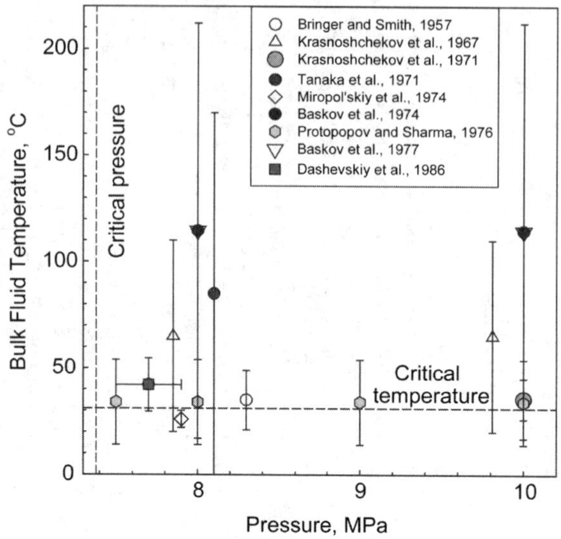

Figure 6.1. Ranges of investigated parameters for selected experiments with carbon dioxide in circular tubes at supercritical pressures (for details, see Table 6.1).

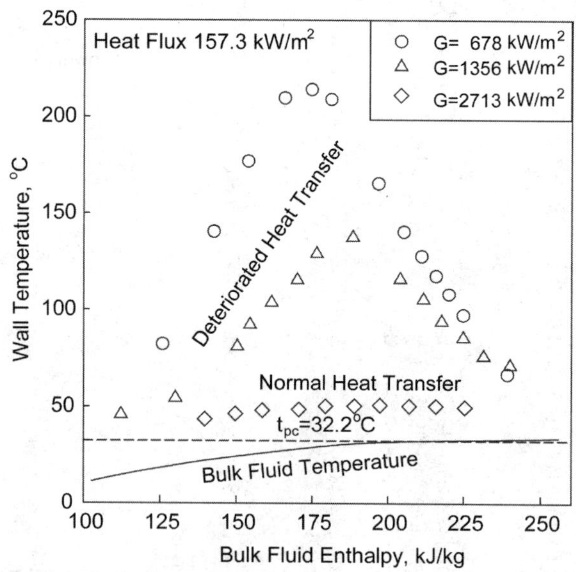

Figure 6.2. Variations in wall temperature with enthalpy at various mass fluxes (Shiralkar and Griffith 1968): CO_2, p_{out}=7.58 MPa, upward flow, D=6.35 mm, and H_{pc}=337.9 kJ/kg.

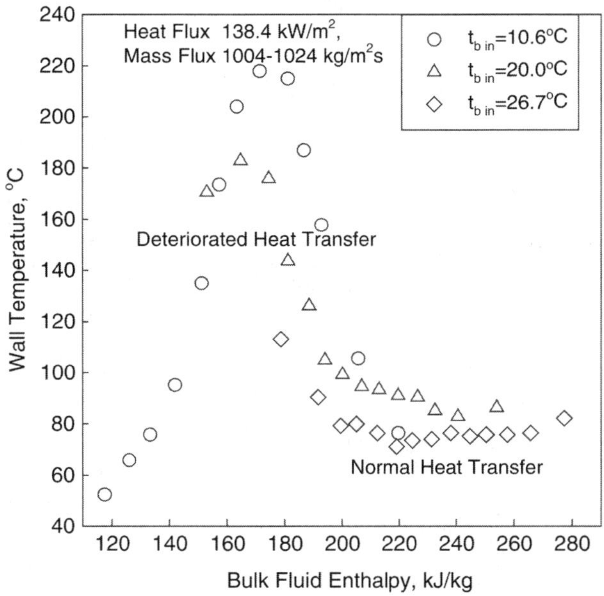

Figure 6.3. Variations in wall temperature with enthalpy at various inlet temperatures (Shiralkar and Griffith 1968): CO_2, p_{out}=7.58 MPa, upward flow, *D*=6.35 mm, t_{pc}=32.2°C, and H_{pc}=337.9 kJ/kg.

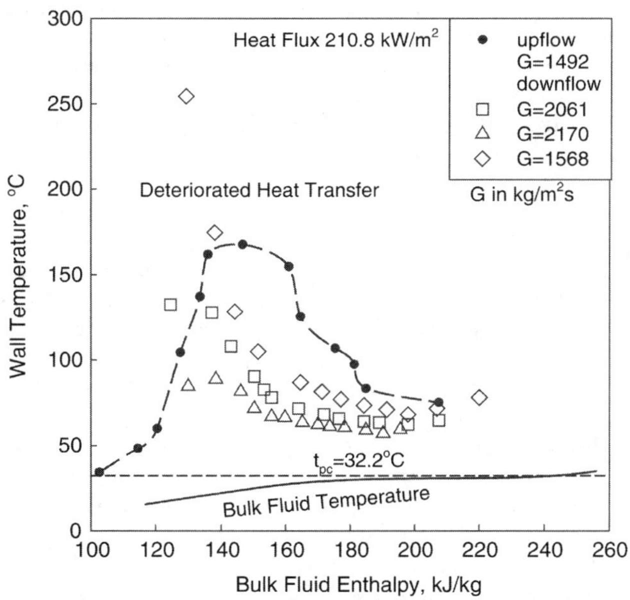

Figure 6.4. Variations in wall temperature with enthalpy at various mass fluxes (Shiralkar and Griffith 1968): CO_2, p_{out}=7.58 MPa, upward and downward flow, *D*=6.35 mm, and H_{pc}=337.9 kJ/kg.

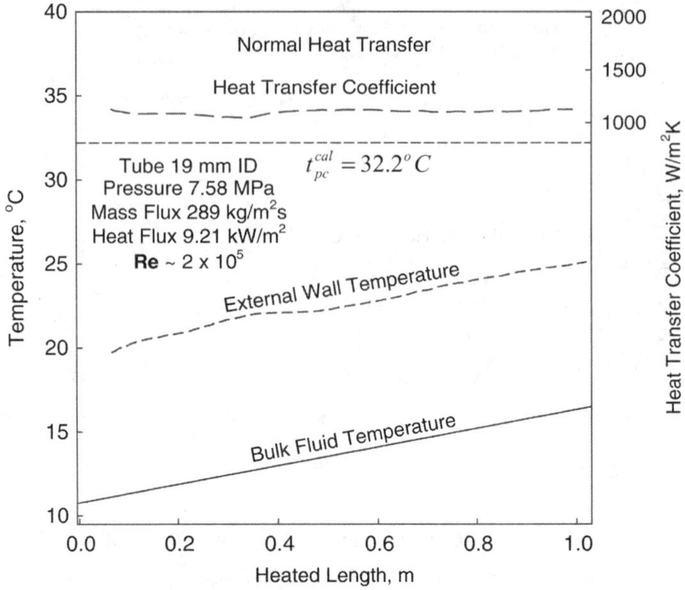

Figure 6.5. Variations in fluid and wall temperatures and in HTC along heated length (Fewster 1976): CO_2, normal heat transfer, circular tube, and upward flow.

Kondrat'ev (1971) and Protopopov and Silin (1973) proposed non-dimensional correlations to estimate the starting point of the deteriorated heat transfer, but these correlations have not been checked independently in supercritical water and carbon dioxide.

Bourke and Pulling (1971a,b) investigated the deterioration of heat transfer in supercritical carbon dioxide. They found that in the upstream part of the tube there was a reduction in the turbulence level, which caused a local deterioration in heat transfer. Further downstream the turbulence increased, which lead to improved heat transfer.

Tanaka et al. (1971) conducted experiments with supercritical carbon dioxide flowing in vertical smooth and rough tubes. In general, they investigated the deterioration of heat transfer near the pseudocritical temperature. They showed that surface roughness has some effect on heat transfer at supercritical pressures, i.e., with increase in tube surface roughness from 0.2 to 14 μm, the heat transfer also increased.

Silin (1973) investigated heat transfer in forced convection of supercritical carbon dioxide in vertical and horizontal tubes. He found that at $t_b \le t_{pc}$ and $t_w \ge t_{pc}$ a region with improved heat transfer existed. During experiments with a 4-mm ID tube, acoustic effects, such as various noises or whistles, were observed in the improved heat-transfer regime.

Fewster and Jackson (Fewster and Jackson 2004; Fewster 1976) conducted heat-transfer experiments in turbulent flow of carbon dioxide inside vertical tubes at supercritical pressures (Figures 6.5 – 6.8). The objective of these experiments was to investigate various regimes of heat transfer at supercritical conditions. They found that, in general, three modes of heat transfer at supercritical pressures exist: (1) normal heat transfer (Figure 6.5), (2) improved heat transfer, characterized by higher-than-expected HTC values than in the normal heat-transfer regime (Figure 6.6) and (3) deteriorated heat transfer, characterized by lower-than-expected HTC values than in the normal heat-transfer regime (Figures 6.6 – 6.8).

Deteriorated heat transfer may appear at high heat fluxes (Figure 6.8) and in any place along the heated length (Figures 6.6 – 6.8).

The experimental results on heat transfer in supercritical carbon dioxide obtained at Chalk River Laboratories (for details on the experimental setup, see Section 10.3) are presented in Figures 6.9 – 6.13.

The data shown in these Figures are actually averaged values over a one-minute data scan, at steady flow conditions. Figure 6.9 shows uncertainties in the measured bulk-fluid temperatures and outside wall temperatures for references purposes. Figure 6.10 shows a comparison between experimental data obtained on February 13th and those obtained on February 26th. In general, the HTC values show good repeatability.

The following supercritical heat-transfer cases were covered:

Within a certain heated length – $T_w^{in} < T_{pc}$, $T_w^{in} = T_{pc}$, $T_w^{in} > T_{pc}$ and $T_b < T_{pc}$;
$T_w^{in} > T_{pc}$ and within a certain heated length $T_b < T_{pc}$ and $T_b = T_{pc}$;
$T_w^{in} > T_{pc}$ and within a certain heated length $T_b < T_{pc}$, $T_b = T_{pc}$ and $T_b > T_{pc}$; and
T_w^{in} and $T_b > T_{pc}$.

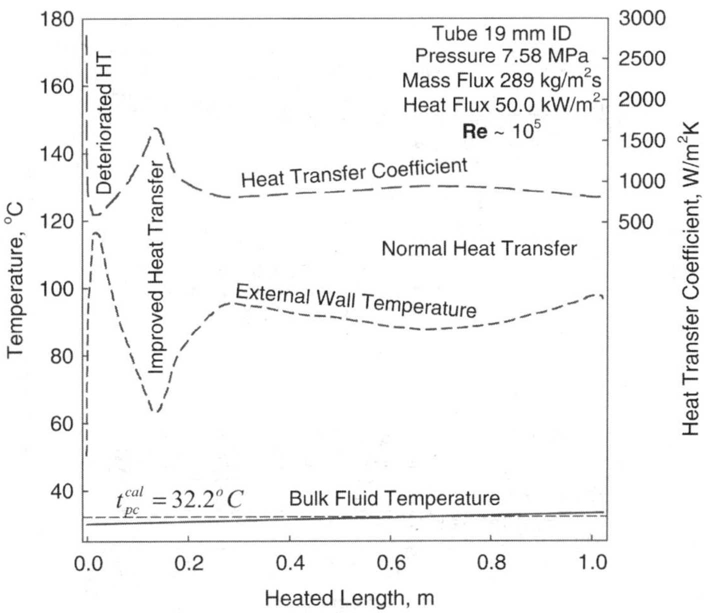

Figure 6.6. Variations in fluid and wall temperatures and in HTC along heated length (Fewster 1976): CO_2, normal, deteriorated and improved heat transfer, circular tube, and upward flow.

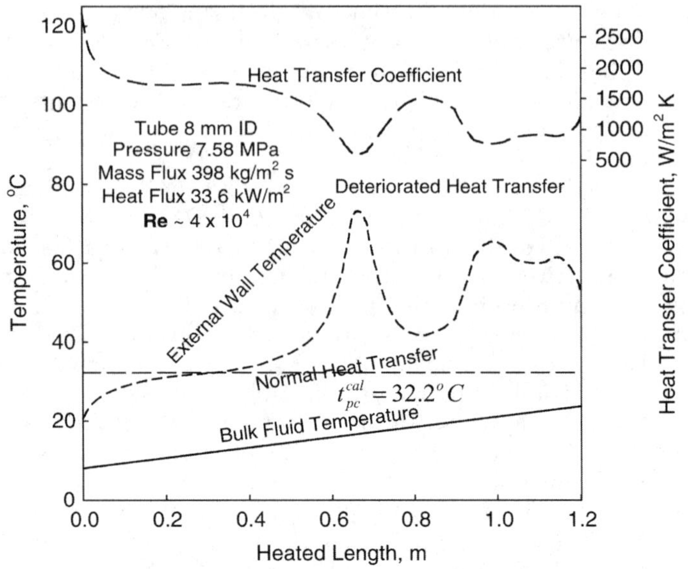

Figure 6.7. Variations in fluid and wall temperatures and in HTC along heated length (Fewster 1976): CO_2, normal and deteriorated heat transfer, circular tube, and upward flow.

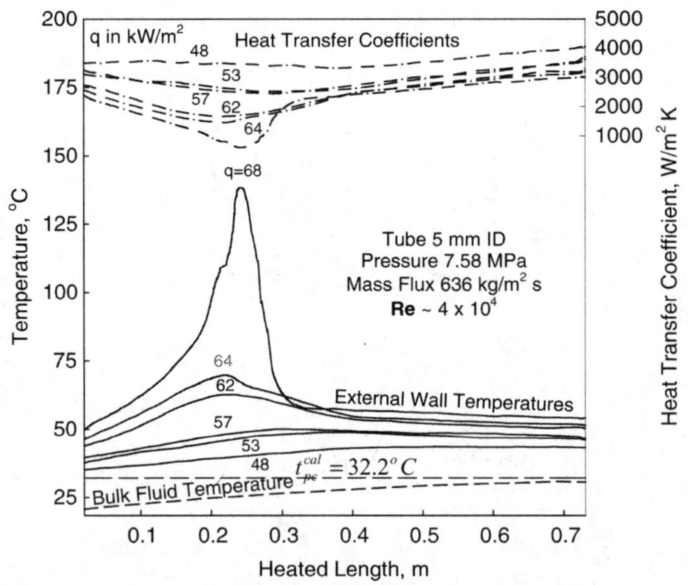

Figure 6.8. Variations in wall temperature and HTC along heated length at various heat fluxes (Fewster 1976): CO_2, circular tube, and upward flow.

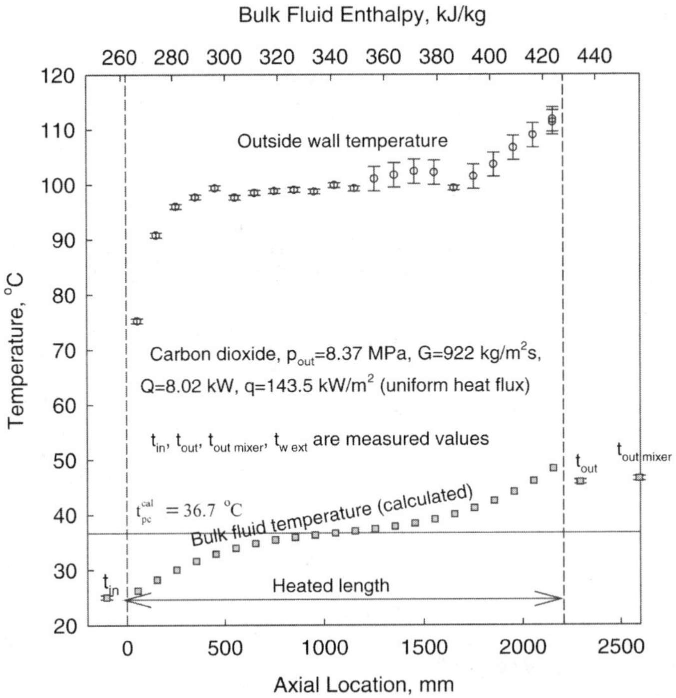

Figure 6.9. Temperature and their uncertainties along vertical circular tube with supercritical CO_2. Flow conditions (nominal values): p_{out}=8.36 MPa, G=900 kg/m²s, t_{in}=25°C, and q=145 kW/m².

Within a short entrance region, the wall temperature rises sharply and HTC drops (Figures 6.9, 6.10, 6.11c, 6.12b,c and 6.13b,c). This is due to the thermal boundary layer development.

Thermocouples TEC07, TEC013, and TEC018 show slightly lower temperature than the neighboring thermocouples just upstream. These thermocouples were located near the pressure drop impulse lines (see Figure 6.11b), which affect the wall temperature measurements just downstream. Structural supports of the test section have no visible impact on wall temperature measurements.

Experimental data of supercritical carbon dioxide obtained at higher mass fluxes (see Figures 6.10, 6.12, and 6.13) show good agreement between the calculated value of the last downstream bulk-fluid temperature, which was calculated from incremental heat balances, and the measured outlet bulk-fluid temperature just downstream of the outlet mixing chamber.

However, at lower mass flux (G = 900 kg/m²s) and with heat flux increase, the difference between the measured and calculated outlet bulk-fluid temperatures increases (see Figures 6.9 and 6.11). This is due to the increased measurement uncertainty at low mass-flow rates (for details, see Appendix D). In general, the difference between the measured and calculated outlet bulk-fluid temperatures was less than 10%.

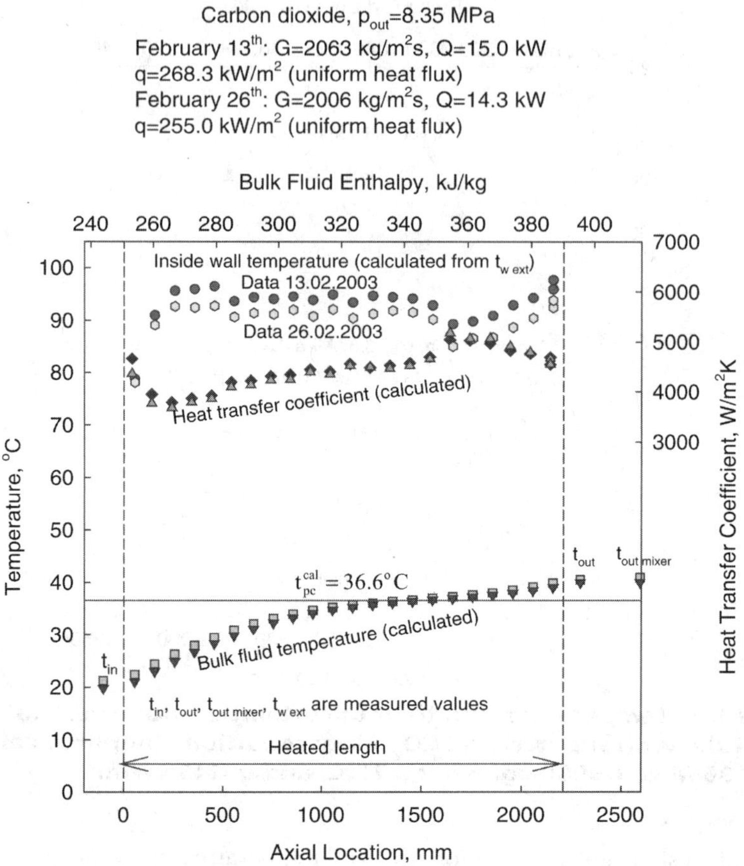

Figure 6.10. Temperatures and HTC profiles along vertical circular tube with supercritical CO_2 (check for data repeatability). Flow conditions (nominal values): p_{out}=8.36 MPa, G=2000 kg/m²s, t_{in}=20°C, and q=265 kW/m².

At high bulk-fluid outlet temperatures, the bulk-fluid temperature measured at the test section outlet was found to be lower than the temperature measured just downstream of the outlet mixing chamber (see Figures 6.9, 6.10, 6.11b-d, 6.12b-d and 6.13a,b,d). This disagreement shows that mixing chambers improve bulk-fluid temperature uniformity in a cross section.

Figures 6.11d, 6.12d, and 6.13d show evidence of deteriorated heat transfer over a certain section of the tube.

6.1.2 Heat Transfer in Horizontal Tubes and Other Flow Geometries

All[11] references with primary experimental data are listed in Table 6.3.

[11] "All" means all sources found by the authors from a total of 650 references dated mainly from 1950 till the beginning of 2006.

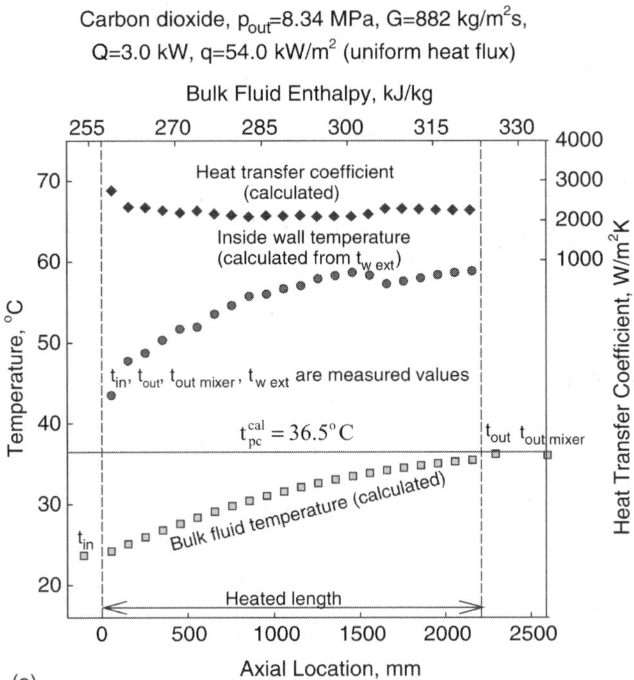

(a)

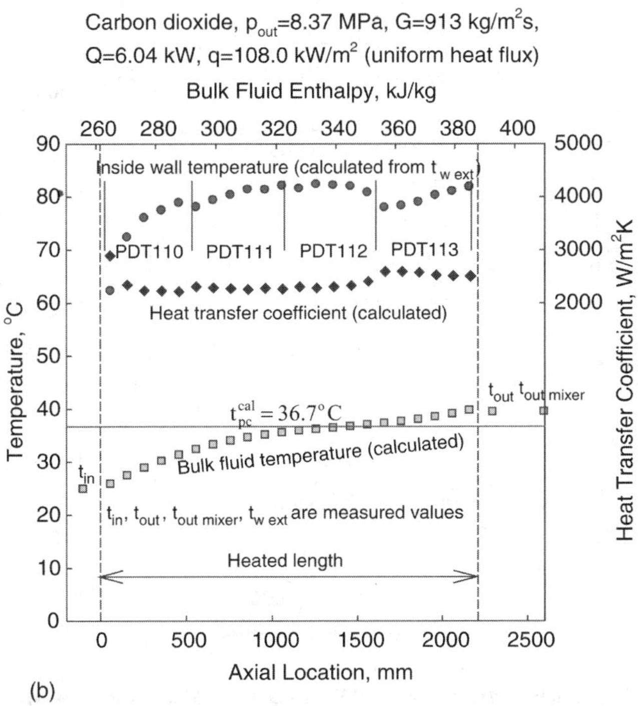

(b)

Figure 6.11. Continued

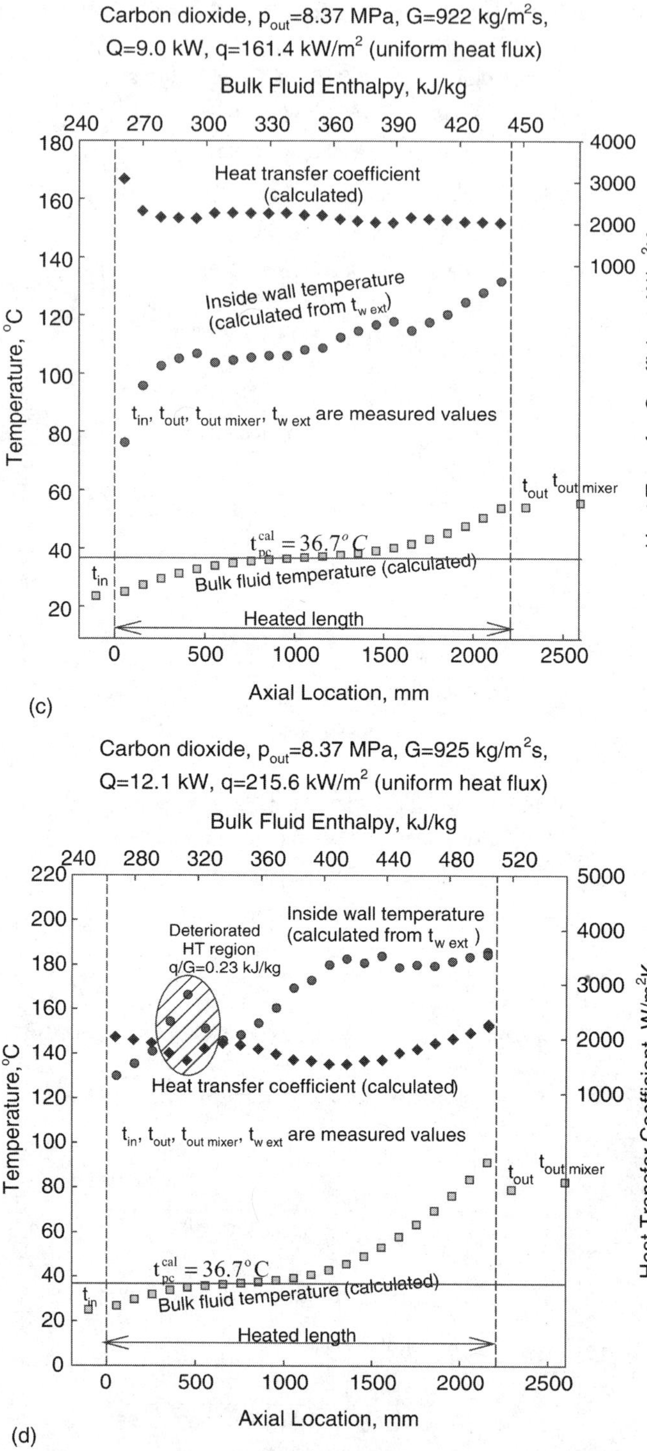

Figure 6.11. Temperatures and HTC profiles along vertical circular tube with supercritical CO_2. Flow conditions (nominal values): p_{out}=8.4 MPa, G=900 kg/m²s, and t_{in}=25°C.

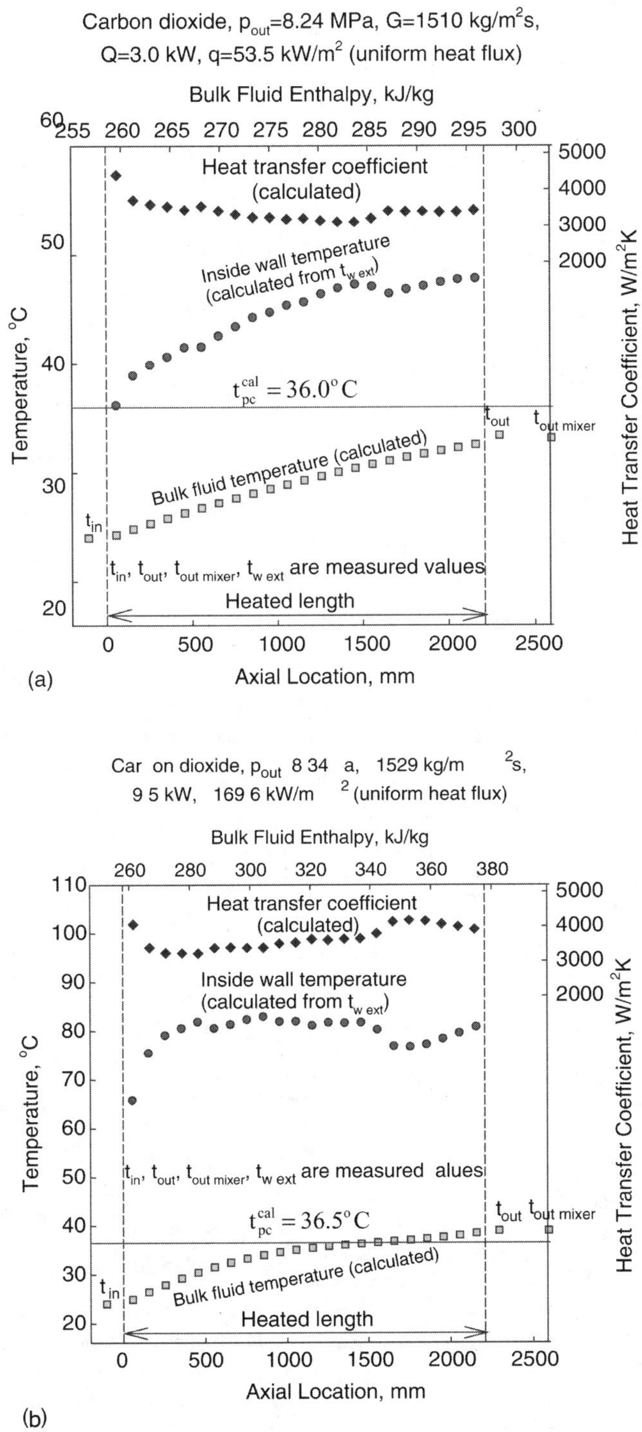

Figure 6.12. Continued

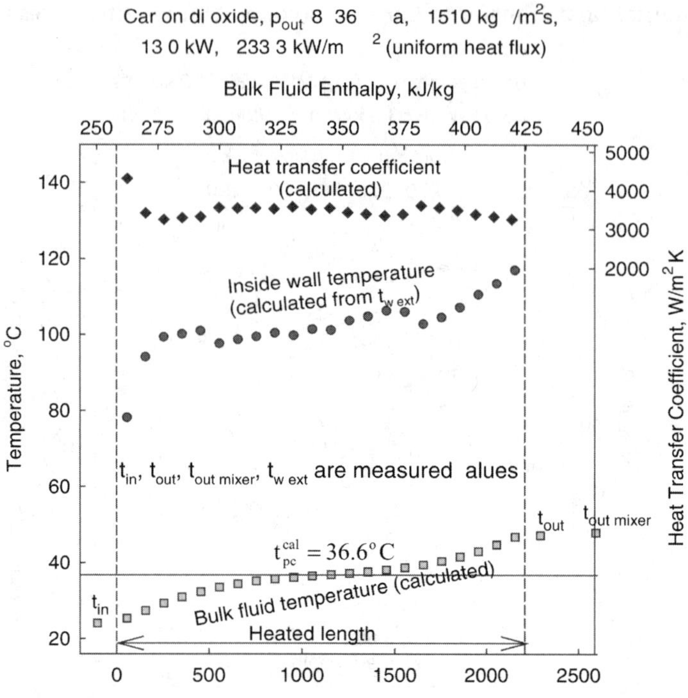

(c)

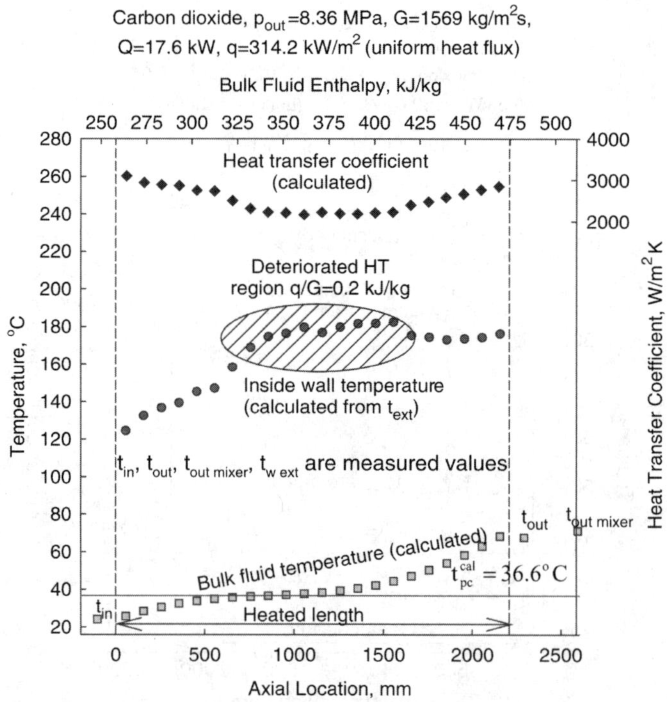

(d)

Figure 6.12. Temperatures and HTC profiles along vertical circular tube with supercritical CO_2. Flow conditions (nominal values): p_{out}=8.4 MPa, G=1500 kg/m²s, and t_{in}=25°C.

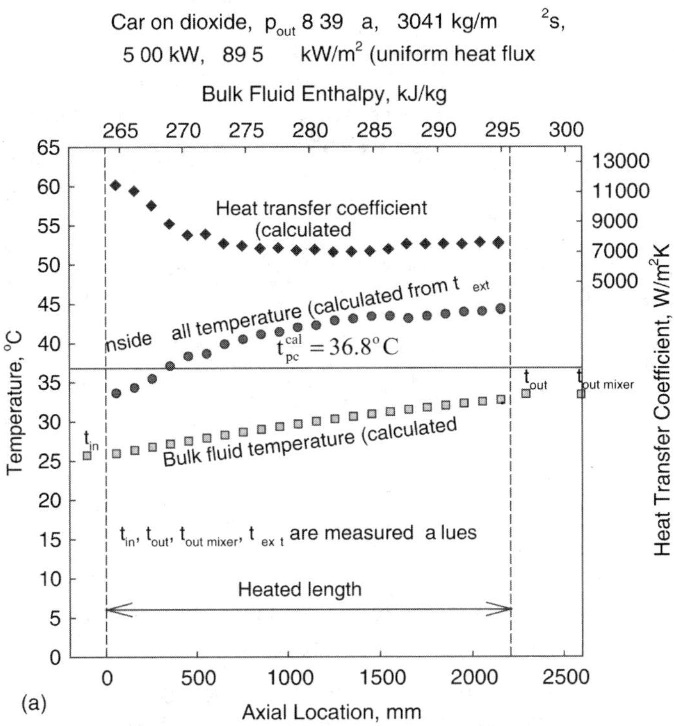

(a)

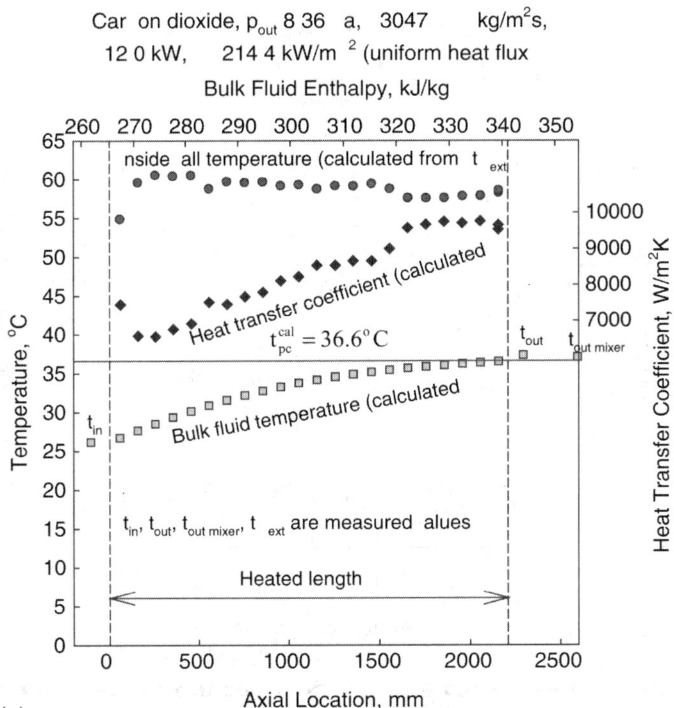

(b)

Figure 6.13. Continued

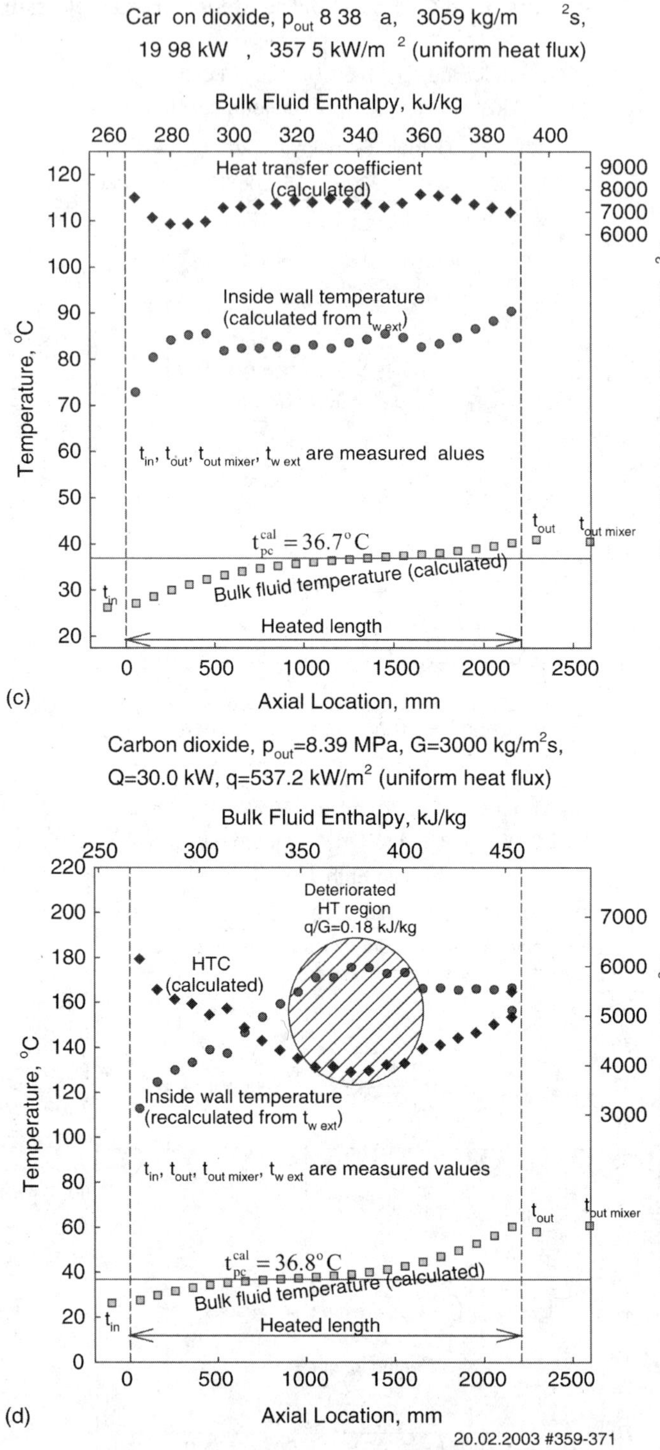

Figure 6.13. Temperatures and HTC profiles along vertical circular tube with supercritical CO_2. Flow conditions (nominal values): p_{out}=8.4 MPa, G=3000 kg/m²s, and t_{in}=25°C.

Table 6.3. Range of investigated parameters for heat-transfer experiments in horizontal tubes and other flow geometries with carbon dioxide at supercritical pressures.

Reference	*p*, MPa	*t*, °C	*q*, kW/m²	*G*, kg/m²s	Re·10⁻³	Flow Geometry
			Horizontal Tubes			
Koppel and Smith 1961	7.38–7.58	t_b=18.3–48.9	62.9–629.1	–	30–300	Horizontal Inconel tube (D=4.93 mm, D_{ext}=6.38 mm, L=457.2 mm)
Melik–Pashaev et al. 1968	9–42	t_b=7–102	up to 8,000	–	150–650	Horizontal tube (D=4.5 mm, L=135 mm)
Krasnoshchekov et al. 1969	8; 10; 12	29–214	120–1110	–	72–94	Horizontal brass tube (D=2.22 mm, L=0.15 m)
Schnurr 1969	7.4–7.7	t_b=21–38	245–887	–	80–680	Horizontal SS tube (D=2.64 mm, L=0.203 m)
Krasnoshchekov et al. 1971	10	t_b=26–45; t_w=160–890	7,500–11,000	22,000	600–1200	Horizontal SS tube (D=2.05 mm, L=95 mm)
Adebiyi and Hall 1976	7.6	10–31	5–40	m=0.035–0.15 kg/s	20–200	Horizontal tube (D=22.1 mm, L=2.4 m) (selected data are shown in Figure 6.14)
Ankudinov and Kurganov 1981	7.7	t_{in}=20	u_o to 1540	2100–3200	230–340	Tube (D=8 mm, L=1.84 m) with helical wire insert, upward, downward, and horizontal flow
Walisch et al. 1997	8–40	50–120	Q=0.4–2 kW	m=0.8–50 g/s	2.3–100	Horizontal, vertical and inclined tubes (D=10 mm, L=1.5 m, Inconel 600)
Pitla et al. (2002, 2001a,b; 2000)	9.4; 10.8; 13.4	100–127	–	m=0.02–0.04 kg/s	–	SS tube (D=6.35 mm)
Liao and Zhao 2002a,b	7.4–12	t_{in}=20–110	$(1–20)\cdot 10^4$	m=0.02–0.2 kg/min	up to 10^5	Horizontal and vertical miniature SS tubes (D=0.70; 1.40; 2.16 mm; L=110 mm)
Yoon et al. 2003	7.5–8.8	t_{in}=50–80	–	225; 337; 450	–	Horizontal copper tube (D=7.73 mm), cooling of CO_2
			Horizontal Rectangular-Shaped Channels			
Sabersky and Hauptmann 1967	7.24; 7.55; 8.27	24; 29; 35; 41	q_{max}=242.2	u_{max}=0.45	–	Horizontal heated plate inside rectangular channel
Protopopov and Igamberdyev 1972	8; 10	t_{in}=18–55	≤2800	1500–4300	150–600	Horizontal copper rectangular channel (cross-section 16 (height)×3.9 mm, D_{hy}=6.26 mm, L=256 mm), one side heated

Green and Hauptmann (1971) conducted experiments with forced-convection heat transfer from a cylinder in carbon dioxide near the critical point.

Adebiyi and Hall (1976) performed heat-transfer experiments in horizontal flow of carbon dioxide at supercritical and subcritical pressures. Axial (Figure 6.14a,b) and circumferential (Figure 6.14c) temperature profiles were obtained. It was found that non-uniform cross-section temperature profile exists in horizontal flow (Figure 6.14c). Comparison with buoyancy free data showed that heat transfer on the bottom of the tube was enhanced by buoyancy forces, but heat transfer on the top was reduced by buoyancy forces (hotter fluid is at the top of a tube).

Figure 6.15 shows a comparison between temperature profiles along horizontal and vertical tubes with upward and downward flow. The data showed that the horizontal flow temperature profiles are more gradual compared to those for vertical upward flows.

Ko et al. (2000) performed flow visualization experiments in a vertical one-side heated rectangular test section cooled with forced flow of supercritical carbon dioxide. They calculated temperature and density profiles of the heated carbon dioxide inside the test section from measured interferometry projections.

A similar investigation was reported by Sakurai et al. (2000).

6.2 FREE-CONVECTION HEAT TRANSFER

All[12] references of primary experimental data for free convection of carbon dioxide are listed in Table 6.4.

Beschastnov et al. (1973) conducted an experimental study of free-convection heat transfer to CO_2 from vertical and inclined surfaces with and without an initially unheated section. The investigation showed that, at turbulent free convection, the effect of the initial section is minor. The heat transfer from the plate decreases as the angle of inclination of the plate was varied from –90° to +90° from the vertical orientation. They also investigated the temperature profile along the plate. The HTC decreased slightly with distance from the leading edge of the heated section, both with and without an initial unheated section of the plate.

Neumann and Hahne (1980) presented experiments on free-convective heat transfer from electrically heated platinum wires and a platinum strip to supercritical carbon dioxide. It was shown that heat transfer could be predicted by a conventional Nusselt-type correlation, if the dimensionless numbers were based on average thermophysical properties, in order to account for large changes in these properties.

[12] "All" means all sources found by the authors from a total of 650 references dated mainly from 1950 till the beginning of 2006.

Table 6.4. Range of investigated parameters for free-convection experiments on heat transfer with carbon dioxide at supercritical pressures.

Reference	*p*, MPa	*t*, °C	Flow Geometry
		Vertical Plates	
Simon and Eckert 1963	7.56–8.97	30.47–33.65	Vertical plate immersed in pool
Beschastnov et al. 1973	7.9–8.8	24–64	Inclined and vertical plates
Sharma and Protopopov 1975	7.5–10	15–54	Vertical surface
		Vertical Tubes	
Protopopov and Sharma 1976	7.5; 8; 9; 10	14–54	Vertical tubes (D_{ext}=8; 18; 19.6 mm, L=160 mm)
Kuraeva et al. 1985	8; 9; 10	4–90	Vertical SS tubes (D_{ext}=19.6 mm, L=190 mm; D_{ext}=13.04 mm, L=226 mm)
Klimov et al. 1985	8; 9; 10	4–90	Vertical tubes (D_{ext}=13.04; 19.6 mm, L=190 mm)
		Horizontal Tubes	
Kato et al. 1968	7.8; 9.8	15–50	Horizontal SS tube (D=2 mm, L=40 mm); vertical plate (height 20 mm, width 100 mm, thickness 50 μm)
Beschastnov and Petrov 1973	7.4–8.8	25–50	Horizontal tubes (D_{ext}=2; 3; 6; 9 mm)
Petrov et al. 1976	7; 8; 9; 10	40–80	Horizontal copper tube (D=6 mm, δ_w=1 mm, L=400 mm)
Tkachev 1981	7.45–8.62	–	Horizontal tube (D_{ext}=3 mm, L=200 mm)
		Wires (horizontal and vertical)	
Knapp and Sabersky 1966	7.58–10.3	9.44–58.3	Horizontal nichrome wire immersed in pool (D_{wire}=0.254 mm)
Goldstein and Aung 1967	6.89–8.96	8.9–57.8; t_{wire}<871	Horizontal platinum wire immersed in pool (D_{wire}=0.38 mm)
Dubrovina and Skripov 1967	6–10	31.5–37	Horizontal and vertical platinum wire immersed in pool
Dubrovina et al. 1969	7.5; 9	–	Platinum wire immersed in pool
Abadzic and Goldstein 1970	5.9–8.1	–	Horizontal platinum wire
Nishikawa et al. 1973	7.58	25	Nichrome and alumel horizontal wires
Hahne et al. 1974	7.4–9.5	25–35	Horizontal wires (D=0.1 mm, L=100 mm)
Beschastnov et al. 1976	7.85	–	Horizontal platinum and aluminium wires (D=0.3 mm)
Neumann and Hahne 1980	7.4–9	10–50	Platinum wires (D=0.05; 0.1; 0.3 mm) and strip (5 mm height), w=67 mm

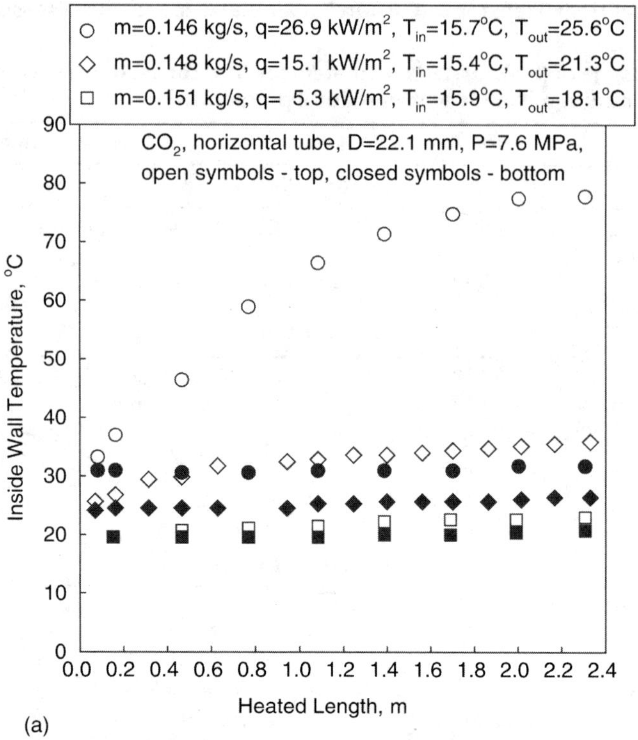

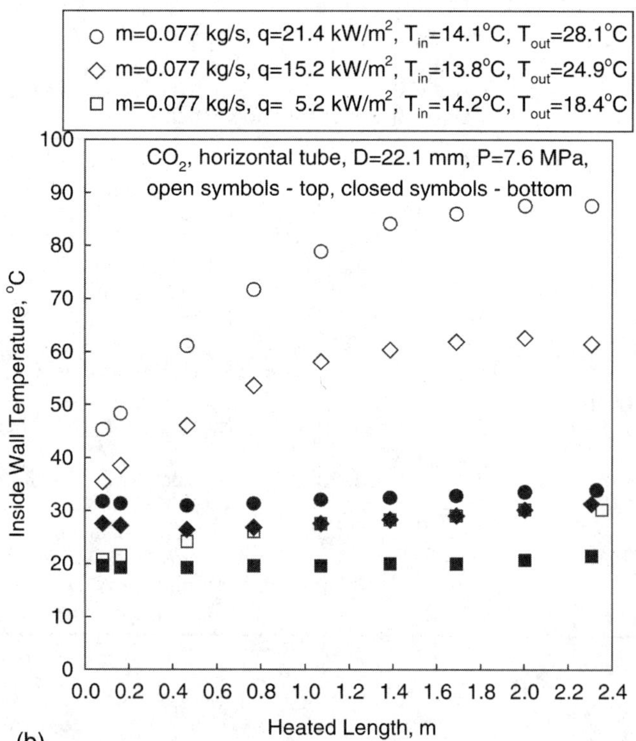

Figure 6.14. Continued

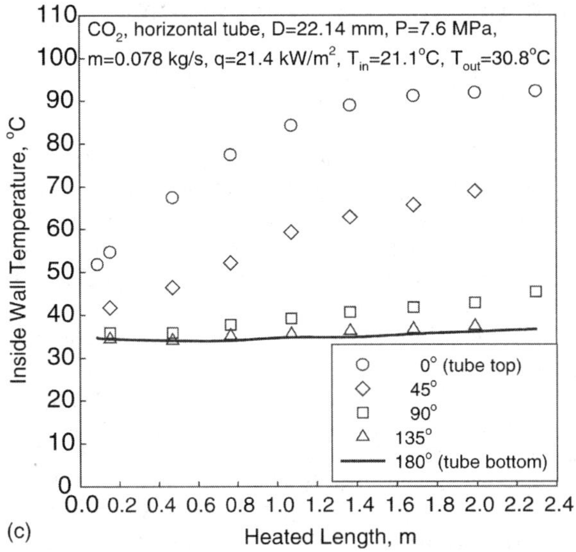

Figure 6.14. Temperature profiles along horizontal circular tube (Adebiyi and Hall 1976): t_{pc}=32.3°C and H_{pc}=337.0 kJ/kg.

Curve	Data	D, mm	T_{in}, °C	m, kg/s	q_w, kW/m^2
Horizontal	Adebiyi, Hall, 1976	22	20	0.121	27
Vertical$_2$	Weinberg, 1972	19	20	0.124	50
Vertical$_1$	Weinberg, 1972	19	20	0.124	30

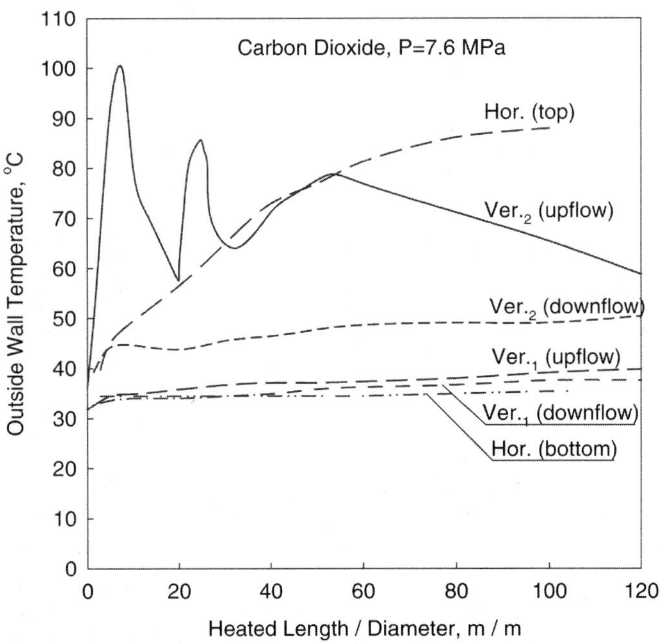

Figure 6.15. Temperature profiles along horizontal and vertical circular tubes (Weinberg 1972): t_{pc}=32.3°C and H_{pc}=337.0 kJ/kg.

6.3 FINAL REMARKS AND CONCLUSIONS

In general, conclusions for supercritical carbon dioxide heat transfer are similar to that of supercritical water.

1. The majority of the experimental studies deal with heat transfer of supercritical carbon dioxide in vertical and horizontal circular tubes. A few studies were devoted to heat transfer in rectangular channels.
2. In general, experiments showed that there are three modes of heat transfer in fluids at supercritical pressures: (1) normal heat transfer; (2) deteriorated heat transfer with lower values of the HTC within some part of a test section; and (3) improved heat transfer with higher values of the HTC.
3. The deteriorated heat transfer usually appears at higher heat fluxes and lower mass fluxes.
4. Heat transfer at supercritical pressures is affected with flow orientation (upward, downward, and horizontal). Horizontal flows show non-uniform cross-section temperature profile with higher temperature and, therefore, lower HTC values, at the top of the channel.

Chapter 7

EXPERIMENTAL HEAT TRANSFER TO HELIUM AT SUPERCRITICAL PRESSURES

As mentioned previously, helium is also used as a coolant in some thermal units. However, a number of papers devoted to heat transfer to supercritical helium is significantly less than that devoted to water and carbon dioxide. Therefore, these papers are listed below just for completeness of the current review and reference purposes.

7.1 FORCED-CONVECTION HEAT TRANSFER

All[13] primary data for forced convective heat transfer of helium are listed in Table 7.1.

[13] "All" means all sources found by the authors from a total of 650 references dated mainly from 1950 till the beginning of 2006.

Table 7.1. Range of investigated parameters for heat-transfer experiments with helium at supercritical pressures.

Reference	*p*, MPa	*T*, K	*q*, kW/m²	*G*, kg/m²s	Re·10⁻³ (if specified)	Flow Geometry
			Vertical Tubes			
Giarratano et al. 1971	0.3–2	4.4–30	1.7–21.5	–	10–380	Vertical SS tube (D=20.8 mm L=1000 mm)
Ogata and Sato 1972	0.2–1.5	4.2–11	1.7–8.75	–	50–90	Vertical SS tube (D=10.9 mm L=85 mm), upward flow
Malyshev and Pron'ko 1972	0.3–1.5	4.5–9.5	1.8–5	–	30–42	Tubes with small diameters
Pron'ko and Malyshev 1972	0.4; 0.6; 0.8	5–9.4	3	$m=(3.6–10)\cdot10^{-5}$ kg/s	$\mathbf{Re}_{in}$=30	Tubes (D=0.7; 1.04 mm)
Petersen and Kaiser 1974	2	up to 1277	–	–	2–70	SS tubes (D/L=12/4; 17/3.1 mm/m) and trefoil–shaped channel (D_{hy}/L=8.9/1.7 mm/m)
Giarratano and Jones 1975	0.25	T_{in}=4.05; 5.04	0.008–7.13	70; 120; 220	–	Vertical SS tube (D=2.13 mm, L=100 mm), downward
Pron'ko et al. 1976	0.3–2	4.5–10	1.8–5.2	42–180	16–57	Nickel tubes (D=1.04 and 0.7 mm)
Brassington and Cairns 1977	0.22–1.4	4.4–15	≤2.5	–	50–1000	Vertical aluminium tube (D=18 mm, L=987; 995 mm), upward and downward flows
Bogachev et al. 1983b	0.23–3	T_{in}=4.2	0.1–1.85	–	36–90	SS tube (D=1.8 mm, L=400 mm)
Bogachev et al. 1983a	0.25–0.4	4.2–1.4	–	–	0.6–3.7; **Ra**=(1.5–510)·10³	Vertical tube, downward flow
Kasatkin et al. 1984	0.253	T_{in}=4.5–5.0	0.5–18	100–180	–	Vertical SS tube (D=0.86 mm, L=86 mm), upward flow
Ito et al. 1986	0.3; 0.5; 0.8	T_{in}=4.7; 10.8	0.5; 1; 2; 4	20; 40; 80	–	Vertical tube (D=1.25 mm, L=200 mm), upward and downward flows
Bogachev et al. 1986	0.25–0.4	T_w=4.2–10	–	$m=(0.7\cdot10^{-2}–0.12)\cdot10^{-3}$ kg/s	3–20	Vertical tube (D=1.8 mm), downward flow

Reference	p, MPa	T, K	q, kW/m²	G, kg/m²s	Re·10⁻³ (if specified)	Flow Geometry
Rectangular-Shaped Channels						
Giarratano and Steward 1983	0.1–1	4–10	1–100	–	0–800	Vertical rectangular channel heated on one side
Bloem 1986	0.3–1	4.2	≤9.8	–	15–200	Rectangular copper channel (6.4 (vert) × 8.4 (hor) mm, D_{hy}=5 mm, L=154.6 mm)
Trefoil–Shaped Channel						
Petersen and Kaiser 1974	2	up to 1277	–	–	2–70	Trefoil-shaped channel (D_{hy}=8.9 mm, L=1.7 m)
Horizontal Tubes						
Dolgoy et al. 1983	0.25; 0.35; 0.50	T_{in}=4.3–8.0	0.1–20	65–500	–	Horizontal SS tube (D=1.6 mm, L=300 mm)
Valyuzhinich et al. 1985a,b,c, 1986	0.24–0.68	T_{in}=4.4–7.1	0.9–6	m=(3.7–20)·10^{-5} kg/s	8–65	Horizontal SS tube (D=1.4 mm, L=545 mm)
Valyuzhin and Kuznetsov 1986	0.24–0.68	T_{in}=4.41–7.1	0.9–6	m=(0.037–0.2)·10^{-3} kg/s	8–65	Horizontal SS tube (D=1.4 mm, L=545 mm)

7.2 FREE-CONVECTION HEAT TRANSFER

The ranges of investigated parameters are listed in Table 7.2.

Table 7.2. Range of investigated parameters for free-convective heat transfer experiments with helium at supercritical pressures.

Reference	*p*, MPa	*T*, K	*q*, kW/m²	Ra (if specified)	Flow Geometry
			Vertical Surfaces		
Deev et al. 1978	0.23–0.46	4.5–10	0.06–5	**Ra**=10^{10}–10^{14}	Vertical copper plate 30 mm height
Mori and Ogata 1991	0.25–0.8	4.2	–	–	Vertical surface with and without channel (channel width 10 mm, length 130 mm, gap 0.4 mm)
			Spheres		
Hilal and Boom 1980	0.23–3.55	5.3–25	–	–	Copper sphere

Chapter 8

EXPERIMENTAL HEAT TRANSFER TO OTHER FLUIDS AT SUPERCRITICAL PRESSURES

Critical parameters and chemical formulas of the following fluids are listed in Table 2.1.

8.1 LIQUIFIED GASES

Air

Budnevich and Uskenbaev (1972) performed heat-transfer experiments in *liquified air* at pressures of p/p_{cr} = 1.175 and 1.47 flowing *inside a stainless steel tube* (D = 10.01 mm) at a heat flux of 3.15 kW/m^2.

Argon

Budnevich and Uskenbaev (1972) investigated heat transfer to *liquified argon* within the range of pressures $1.175 \leq p/p_{cr} \leq 2.04$ and temperatures $0.87 \leq T/T_{cr} \leq 1.19$ flowing *inside a stainless steel tube* (D = 2.8 mm) at heat fluxes from 0.55 to 11.5 kW/m^2.

Hydrogen

Thompson and Geery (1962) conducted experiments with *cryogenic hydrogen* at supercritical pressures *in a SS 347 tube* (D = 4.93 mm, L = 406.4 mm) within the following range: p = 4.69 and 9.27 MPa, T_{in} = 30.6 – 56.7 K, $T_{w\,outer}^{max}$ = 1000 K and **Re** = (260 – 1740)·10^3.

Hendricks et al. (1966) performed heat–transfer experiments in *cryogenic hydrogen* flowing in *electrically heated vertical tubes* (ID 4.78 – 12.9 mm, heated length 406 – 610 mm). The range of investigated parameters was from subcritical to supercritical pressures (0.55 – 5.5 MPa), mass fluxes from 488 to 4880 kg/m^{2}s and heat fluxes of up to 34 kW/m^2. They noticed the similarities in the behavior of the near-critical to two-phase data, including a minimum in the HTC near the saturation and pseudocritical temperatures. Flow oscillations were noted at an inlet temperature below the pseudocritical temperature or saturation temperature. Preliminary results were obtained with a non-uniform axial heat flux. They also discussed the technique for correlating Nusselt numbers and published excessive tables with primary data.

Nitrogen

Uskenbaev and Budnevich (1972) investigated *free-convection* heat transfer to supercritical *nitrogen* (p = 3.9 – 7.7 MPa). The test section was a *vertical stainless steel tube* of 2.8 mm in diameter and 80 mm in height.

Budnevich and Uskenbaev (1972) performed heat-transfer experiments in *liquified nitrogen* within the range of pressures $1.175 \le p/p_{cr} \le 2.92$ and temperatures $0.87 \le T/T_{cr} \le 1.19$ flowing *inside horizontal stainless steel tubes* (D = 2.8 and 10.01 mm) at heat fluxes from 0.55 to 11.5 kW/m^2.

Nitrogen tetroxide

McCarthy et al. (1967) investigated heat transfer in supercritical *nitrogen tetroxide* flowing *in axially curved channels.* They observed a significant increase in HTC on the outside of the curved surface. Also, improved heat transfer occurred due to endothermic equilibrium dissociation of the coolant near the hot wall.

Nesterenko et al. (1974) conducted experiments with *nitrogen tetroxide* (p = 11.8 – 14.7 MPa, T_b = 450 – 570 K, T_w = 460 – 590 K, q = 140 – 500 kW/m^2) flowing *in two horizontal* (D = 6.85 mm, L = 1.44 m; D = 3.8 mm, L = 1.44 m) and one *vertical* (D = 2.05 mm, L = 0.79 m, *upward flow*) *tubes.*

Oxygen

Powel (1957) investigated forced-convective heat transfer to *oxygen* flowing *inside a vertical stainless steel tube* (D = 4.93 mm, L = 0.152 – 1.83 m). He found that the minimum values of HTC were in a vicinity of the critical point.

Sulphur hexa-fluoride

Tanger et al. (1968) investigated heat transfer near the critical point of *sulphur hexa-fluoride in the two closed natural circulation loops* (D = 10.92 mm) within the following range: p = 3.69 – 6.18 MPa, T_b = 49.9 – 89.4°C and q = 5.98 – 28.4 kW/m^2. They found that the highest HTCs were obtained within the range slightly above the critical point.

8.2 ALCOHOLS

Ethanol

Alad'yev et al. (1967, 1963b) investigated heat transfer to *ethanol* at pressures from 30.4 to 81.1 MPa flowing *in a stainless steel tube* (D = 0.6 – 2.1 mm, L/D = 20 – 175). The range of investigated parameters was a bulk-fluid temperature of 15°C

– 350°C, a wall temperature up to 700°C, a fluid velocity of 5–60 m/s and a heat flux up to 40.7 MW/m^2.

Kafengauz (1983) conducted heat-transfer experiments with *ethanol* at supercritical pressures and *pseudo-boiling conditions*.

Methanol

Alad'yev et al. (1963) investigated heat transfer to *methanol* at pressures from 9.8 to 39.2 MPa flowing *in a stainless steel tube* (D = 1.55 – 3.45 mm, L/D = 20 – 40). The range of investigated parameters was a bulk-fluid temperature of 15–165°C, a wall temperature up to 800°C, a fluid velocity of 2 – 60 m/s and a heat flux up to 58 MW/m^2.

8.3 HYDROCARBONS

n-Pentane

Bonilla and Sigel (1961) investigated *turbulent natural-convection* heat transfer *from a horizontal heated plate* to a pool of liquid *n-pentane* near the critical point.

n-Heptane

Kaplan and Tolchinskaya (1974a) carried out an experimental investigation on heat transfer and hydraulic resistance of *n-heptane* flowing *through stainless steel tubes* (D = 2.02; 2.4 mm, L = 40 mm, L_{tot} = 160 mm) within the range of velocities 5, 10 and 30 m/s and at pressures of 2.94, 4.02, 8.43 and 8.82 MPa.

Alad'ev et al. (1976) conducted experiments with *n-heptane* within the range of *pseudo-boiling* phenomenon. They found that the enhanced heat transfer in pseudo-boiling was due to self-excited thermoacoustic oscillations.

Isayev (1983) investigated heat transfer to *n-heptane* flowing *in a stainless steel tube* (D = 2 mm, δ_w = 0.5 mm, and L = 224 mm), which could be inclined at any *angle from vertical to horizontal*. The flow conditions were as the following: p =3 MPa, t_{in} = 15°C, q = 2.05; 1.64; 1.05 and 0.78 MW/m^2 and G = 1690 kg/m^{2}s.

Di-iso-propyl-cyclo-hexane

Kafengauz and Fedorov (1966) investigated heat transfer of *di-iso-propyl-cyclo-hexane* accompanied with *high-frequency pressure oscillations*.

Kafengauz (1967) conducted forced-convective heat transfer experiments with *di-iso-propyl-cyclo-hexane* flowing *in a horizontal tube* (D = 1.6 mm and L = 30 mm) within the following range of parameters: p = 4.56 MPa, t_b = 20 – 60°C, and u = 2; 6 and 15 m/s.

Kafengauz and Fedorov (1968) investigated the onset of *pseudo-boiling* in *di-iso-propyl-cyclo-hexane* (p = 2.9; 3.9; 4.4 and 4.9 MPa, t_{in} = 17 – 47°C, and u = 2 – 50 m/s) flowing *in stainless steel tubes* (D = 0.8; 1.6 and 2.3 mm, and L = 30 mm).

Kafengauz (1969) analyzed data of heat transfer to supercritical *di-iso-propyl-cyclo-hexane* and found that *pseudo-boiling phenomenon* is quite similar to nucleate boiling at subcritical pressures.

Kafengauz (1983) conducted heat-transfer experiments with *di-iso-propyl-cyclo-hexane* and *n-heptane* at supercritical pressures and *pseudo-boiling conditions*.

8.4 AROMATIC HYDROCARBONS

Benzene and toluene

Mamedov et al. (1976) investigated heat transfer to *benzene* and *toluene* flowing *in a vertical stainless steel tube* (D = 2.1 mm, L = 170 mm) at near-critical pressures. The investigated flows were upward and downward covering Reynolds numbers from 8,000 to 12,000.

Kalbaliev (1978) performed heat-transfer experiments in *benzene* and *toluene* in *laminar and turbulent flows* at supercritical conditions. The test section in his experiments was a *stainless steel tube* with 170-mm heated length and ID of 3 mm for *laminar* and 2 mm for *turbulent flows*.

Isaev and Kalbaliev (1979) investigated heat transfer to *benzene* and *toluene* *flowing downward* at supercritical conditions.

Kalbaliev and Babayev (1986) investigated heat transfer to *toluene* flowing *through a vertical tube* within the following range of parameters: $p = 4.36 - 9.29$ MPa, $t_b = 156 - 354$°C and $q = 0.19 - 3.9$ kW/m^2.

Abdullaeva et al. (1991) conducted *free-convection* heat-transfer experiments with *toluene near vertical and horizontal surfaces*. The range of investigated parameters was $p = (1.01 - 3.09)\ p_{cr}$, $T_w = (0.5 - 3.0)\ T_{cr}$, $T_b = (0.4 - 0.7)\ T_{cr}$, $q = 20 - 550$ kW/m^2 and $\mathbf{Ra} = 10^5 - 10^8$ for horizontal surfaces and $\mathbf{Ra} = (9.2 - 920)\cdot 10^{10}$ for vertical surfaces.

Kalbaliev et al. (2002) and Kelbaliev et al. (2003) studied heat transfer to *toluene* flowing *in coils* with ID of 4 and 7.6 mm at supercritical pressures. They found that HTC inside coils is higher than that inside a straight vertical tube.

Verdiev (2002) performed heat-transfer experiments in toluene and benzene at supercritical pressures. Specific of these experiments was application of high frequency thermoacoustic instabilities.

Rzaev et al. (2003) investigated methods for predicting the deteriorated heat-transfer regimes in supercritical *toluene* and *water* upward and downward flows in a vertical tube.

8.5 HYDROCARBON FUELS AND COOLANTS

Kafengauz (1983) conducted heat-transfer experiments with *kerosene* at supercritical pressures and *pseudo-boiling* conditions.

Yanovskii (1995) investigated heat transfer to *hydrocarbon fluids* such as a mixture of standard rocket fuel RT with ethyl acetate at supercritical pressures (p_{in} = 5 MPa, t_{in} = 20; 35; 100°C, and $q = 440 - 940$ kW/m^2) flowing *inside tubes* (D = 1 mm and L = 1 m).

In general, the hydrocarbon fluids have very strong dependence of dynamic viscosity on temperature (in the investigated temperature range μ changed in two-three orders) compared to that of water, CO_2, helium and other fluids (Yanovskii 1995, 1987; Polyakov 1991). This effect results in significant variations in Reynolds number.

Valueva et al. (1995) carried out a numerical modeling of hydrodynamics and heat-transfer processes at *transient and turbulent conditions* for *hydrocarbons* flowing *through heated channels* at supercritical pressures. The experimental data were obtained on heat transfer, pressure drop, and friction resistance, as well as on profiles of the temperature, velocity, tangential stress, and turbulent viscosity coefficient at transient flow conditions. Calculated and experimental wall temperatures in the region of transient and turbulent Reynolds numbers were compared. The working fluids were the standard jet propulsion *fuels RT* and *T-6*. Calculations were made for *fuel RT* at $t_{pc} = 393°C$ and for *fuel T-6* at $t_{pc} = 447°C$ within the following range of parameters: $p_{in} = 3 - 5$ MPa, $t_{in} = 12 - 97°C$, $t_w = 97 - 697°C$, **Re** = 1000 – 20,000, and $q_w = 0.2 - 1.2$ MW/m^2. At these conditions **Gr/Re**$^2 < 10^{-3}$, and the influence of free convection may be neglected. The appearance of a peak in the wall temperature was observed during experiments with hydrocarbon fuels at small Reynolds numbers, which correspond to a transition to turbulent flow. The transition took place due to an increase in the Reynolds number, because of liquid heating. Deteriorated heat transfer was encountered, but it was not very pronounced.

Kalinin et al. (1998) summarized findings for heat transfer in supercritical *hydrocarbons* flowing *inside smooth and ribbed tubes* (for details, see Chapter 9).

8.6 REFRIGERANTS

Freon-12

Holman and Boggs (1960) investigated heat transfer to *Freon-12* within the critical region ($p = 3.45 - 6.55$ MPa and $t_b = 65.6 - 204.4°C$) *in a closed natural-circulation loop* ($D = 10.92$ mm).

Nozdrenko (1968) investigated forced-convective heat transfer to supercritical *Freons*.

Beschastnov et al. (1973) conducted heat-transfer experiments with *thin platinum wires* and *stainless steel tubes submerged in a pool* of *Freon-12*. Their results showed that the heat transfer was at maximum at pseudocritical heater temperatures. Also, it was found that the HTC was higher near the critical point in tubes as compared with wires.

Gorban' et al. (1990) investigated heat transfer to *Freon-12* flowing at subcritical and supercritical pressures *inside a circular tube* ($D = 10$ mm and $L = 1$ m). The range of investigated parameters was as follows: $p = 1.08$ and 4.46 MPa, $G = 500 - 2000$ kg/m^{2}s, $t_{in} = 20 - 140°C$ and $q = 6 - 290$ kW/m^2.

Freon-22

Tkachev (1981) performed *free-convection* experiments *on a horizontal tube* (D_{ext} = 3 mm, $L = 200$ mm) submerged in a pool of *Freon-22* ($p = 4.87 - 6.27$ MPa and $q = 2.02 - 11.7$ kW/m^2).

Yamashita et al. (2003) conducted experiments on heat transfer and pressure drop *in a vertical smooth tube* (Inconel 600, 4.4-mm ID and 6.4-mm OD, and 2-m heated length) with upward flow of *Freon-22*. An operating pressure was 5.5 MPa. For heat-transfer experiments, the following range of parameters was tested: mass flux from 400 to 2000 kg/m^{2}s; heat flux from 10 to 170 kW/m^2; and bulk-fluid enthalpy from 215 to 360 kJ/kg. For pressure drop experiments, the following range of parameters was tested: mass flux—700 kg/m^{2}s; heat flux from 0 to 60 kW/m^2; and bulk-fluid enthalpy from 225 to 395 kJ/kg. They found that the heat transfer in their small diameter tube was similar to that in larger diameters tubes. The effect of inside diameter was found within the normal heat-transfer regime and was absent in the deteriorated regime. The minimal deteriorated heat flux became larger in smaller diameter tube compared to that of larger diameters. The Watts and Chou heat-transfer correlation (1982) showed the best prediction accuracy for the normal heat-transfer data.

Mori et al. (2005) simulated heat transfer and pressure drop in a SCWR fuel bundle with *Freon-22* flowing in a test section consisted of a *single rod*. The heater rod OD was 8.0 mm and the heated length was about 1.8 m. A flow tube was 10-mm ID. Freon-22 flowed upward in a vertical test section. The hydraulic-equivalent diameter was 2.0 mm and thermal-equivalent diameter was 4.5 mm. Pressure was about 5.5 MPa, bulk-fluid enthalpy—218 – 385 kJ/kg, mass flux—400, 700 and 1000 kg/m^{2}s, and heat flux—15 – 100 kW/m^2. They found that deteriorated heat transfer in the single-rod tests was delayed compared to that of a circular tube. The frictional pressure drop was not affected with heat flux.

Kitou et al. (2005) evaluated heat transfer *in a circular tube* cooled with supercritical *Freon-22* and *water*, and *in single-rod* experiments with Freon-22 using unified analysis model. This evaluation is based on the experiments performed at the Kyushu University: (a) a single tube with ID of 4.4 mm and heated length of 2 m, system pressure was 5.5 MPa, mass flux—700 kg/m^{2}s, and q = 20; 60 kW/m^2; (b) a single rod with OD of 8.0 mm and unheated pressure tube with ID of 10.0 mm and heated length of 1 m, system pressure was 5.5 MPa, mass flux—700 kg/m^{2}s, and q = 20–60 kW/m^2.

Freon-114a

Griffith and Sabersky (1960) conducted experiments on *free convection from heated wires* inside *Freon-114a* at supercritical pressures. They mentioned that heat transfer to fluids at the near-critical point is important, because of various applications in industry, including the cooling of rocket engines with hydrocarbon fuels.

Freon-134a

Hong et al. (2004; 2003) performed experiments *in a vertical annulus* cooled with *Freon-134a* within a range of pressure transients from supercritical pressure to subcritical pressure. They found that critical heat flux decreased very fast with pressure approaching the critical pressure, and critical-heat-flux phenomenon disappeared completely at supercritical pressures.

8.7 SPECIAL FLUIDS

Poly-methyl-phenyl-siloxane

Kaplan et al. (1974b) conducted *forced-convection* heat-transfer experiments *inside tubes* ($D = 1.7 - 4.0$ mm, $L = 15 - 120$ mm) with *poly-methyl-phenyl-siloxane* liquid near its critical point. The range of investigated parameters was $p = 0.343 - 2.16$ MPa and $u = 1.3 - 11$ m/s. They found that enhanced heat transfer existed in laminar flow at pressures near the critical pressure.

Chapter 9

HEAT-TRANSFER ENHANCEMENT AT SUPERCRITICAL PRESSURES

Being influenced by the turbulent structure, it is expected that the heat transfer can be enhanced by having ribs, spacers, grids, etc. placed in the flow. Some sources of the primary data on heat-transfer enhancement at supercritical pressures in various fluids are listed in Table 9.1. The main idea of these experiments was to find the most effective ways to enhance supercritical heat transfer and to delay or to avoid the deteriorated heat-transfer regimes in various flow geometries. These studies were important to supercritical "steam" generators where heat fluxes are quite high. Therefore, supercritical water flowing in circular tubes was the primary choice for investigations. Some studies were performed in supercritical carbon dioxide as the modeling fluid.

Also, due to limited space and weight, the supercritical heat-transfer enhancement in various hydrocarbon fluids and fuels was important for aircraft applications.

Shiralkar and Griffith (1968) found that the twisted tape installed inside a bare tube cooled with supercritical carbon dioxide significantly improved the heat transfer (Figure 9.1).

Ackerman (1970) investigated heat transfer to water at supercritical pressures flowing in smooth vertical tubes and tubes with internal ribs over a wide range of pressures, mass fluxes, heat fluxes, and diameters. The experiments with a ribbed tube showed that pseudo-film boiling was suppressed (see Figure 9.2). This suppression permitted operation at higher heat fluxes, compared to operation with smooth tubes as might be expected for a "single-phase" fluid when turbulence is enhanced.

Lee and Haller (1974) conducted experiments with a multi-lead ribbed tube (twisted ribs) in supercritical water and found that the ribbed tube was very efficient in suppressing temperature peaks encountered in smooth tubes. They explained this by suggesting that the ribbed tubes caused the flow to spin. Therefore, centrifugal forces caused the lower temperature and denser fluid to move to the heated wall. These tubes were tested at much higher heat fluxes up to 50% – 100% than smooth tubes without any signs of deteriorated heat transfer.

Results of Kamenetskii (1980) are presented in Figure 9.3. This figure shows that a horizontal tube with twisted tape has a much smaller temperature difference between the tube top and tube bottom compared to a tube without tape. In general, tapes are very efficient heat-transfer enhancing devices in circular tubes, but it may not be so in more complex flow geometry such as a fuel bundle.

Table 9.1. Range of investigated parameters for heat-transfer experiments in various fluids at supercritical pressures flowing in circular flow geometry with turbulizers.

Reference	p, MPa	t, °C	q, kW/m²	G, kg/m²s	Flow Geometry
			Water		
Shitsman 1967	24.3–25.3	t_b= 300–400	90–730	600–690	Horizontal and vertical tubes (D_{ext}=16 mm) with twisted tapes
Ackerman 1970	22.8–41.3	t_b= 77–482	126–1730	136–2170	Vertical smooth (D=9.4; 11.9 and 24.4 mm, L=1.83 m; D=18.5 mm, L=2.74 m) and ribbed tubes (D=18 mm (from rib valley to rib valley), L=1.83 m, helical six ribs, pitch 21.8 mm) (selected data are shown in Figure 9.2)
Kamenetsky and Shitsman 1970	24.5	H_b=100–600	190–1300	540	SS tube (D=22 mm, L=3 m) with and without twisted tape (δ_w=0.8 mm, spiral shape with relative pitch s/D=15), non-uniform circumferential heat flux
Lee and Haller 1974	24.1	260–383	250–1570	542–2441	Vertical SS tubes (D=38.1; 37.7 mm, L =4.57 m), tube with ribs
Kamenetskii 1980	24.5	H_b=100–2200	370–1300	300–1700	SS tube with and without flow spoiler (D=22 mm, L=3 m) (selected data are shown in Figure 9.3)
Chen 2004	24	H_b=1300–2800	300	400	SS tube with internal helical ribs (D_{ave}=15.24 mm, D_{ext}=28 mm, rib height 0.81 mm, rib pitch 20.5 mm, rib width 9 mm), vertical and inclined tubes 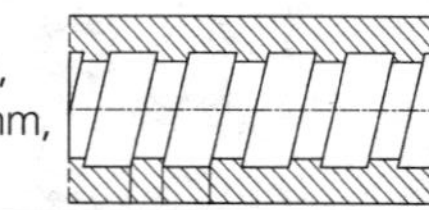
			Carbon Dioxide		
Shiralkar and Griffith 1968, 1969	7.6; 7.9	–	–	–	SS tube with twisted tape (D=6.22 and 3.175 mm, L=1.52 m), upward and downward flow (selected data are shown in Figure 9.1)
Shiralkar and Griffith 1970	7.6; 7.9	t_{in}=–18–31	50–453	–	SS tube with twisted tape (D=3.18; 6.35 mm, L=1.52 m), upward and downward flows
Ankudinov and Kurganov 1981	7.7	t_{in}=20	u_o to 1540	2100–3200	Tube (D=8 mm, L=1.84 m) with helical wire insert, horizontal, upward, and downward flows
			Hydrocarbons		
Fedorov et al. 1986	3–5	t_{in}=35–150	221–1690	m=(2–25)·10^{-3}	SS tubes (D = 4 mm, L = 1 m), smooth and with ribs (d/D=0.93 and t/D=1)
Kalinin et al. 1998	5	t_{in}=100	–	3–800	Tube (D=1–4 mm, L=1 m), coolant kerosene RT (p_{cr}=2.5 MPa, t_{cr}=393°C) (selected data are shown in Figures 9.4–9.9)

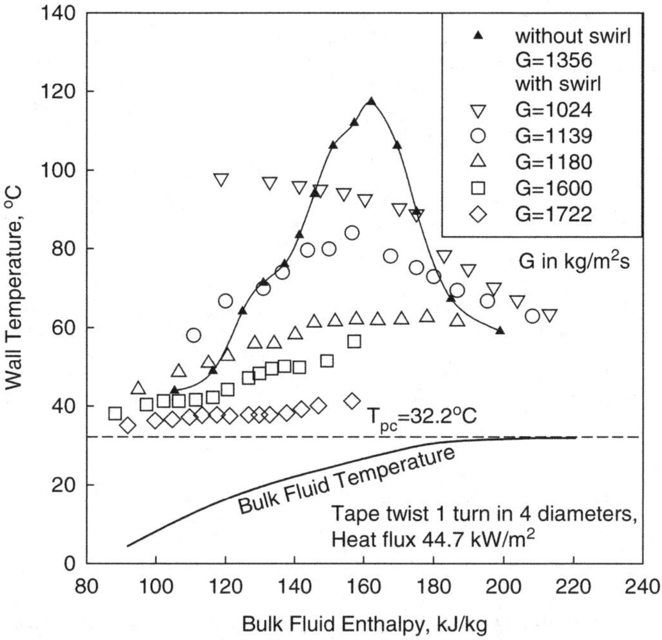

Figure 9.1. Variations in wall temperature with enthalpy at various mass fluxes for bare tube and tube with twisted tape (Shiralkar and Griffith 1968): Carbon dioxide, p_{out}=7.58 MPa, upward flow, D=6.35 mm, and H_{pc}=337.9 kJ/kg.

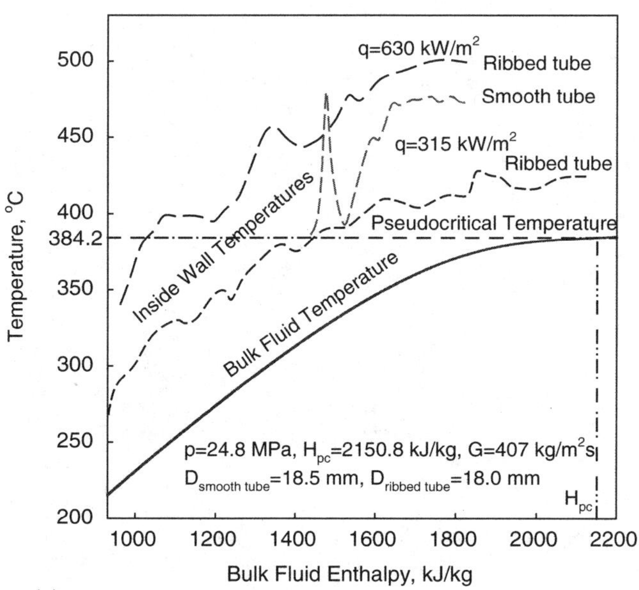

(a)

Figure 9.2. Continued

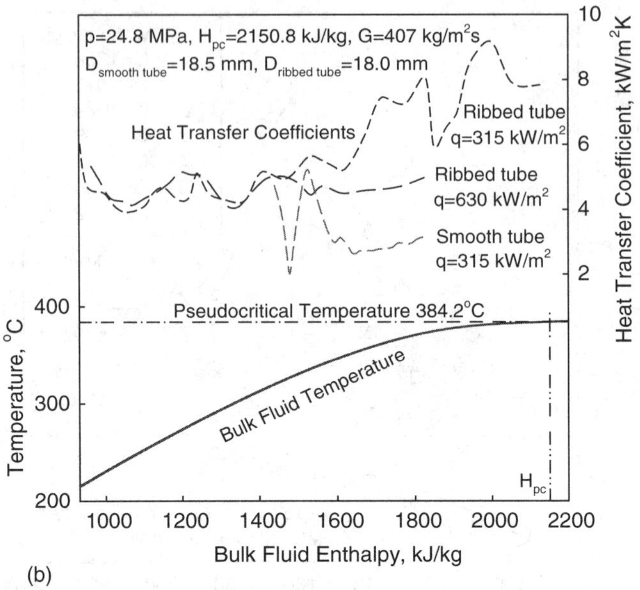

Figure 9.2. Temperature and HTC variations along vertical smooth and ribbed tubes cooled with supercritical water (Ackerman 1970): HTC values were calculated by the authors of the current monograph using the data from figure (a).

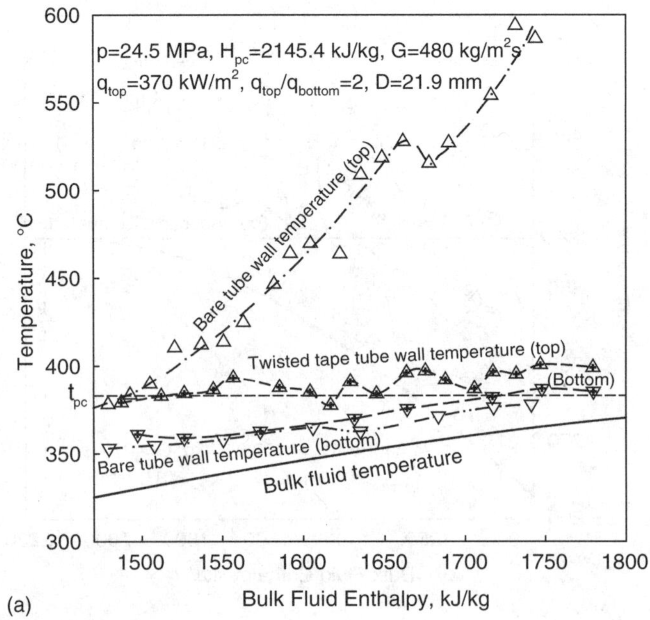

Figure 9.3. Continued

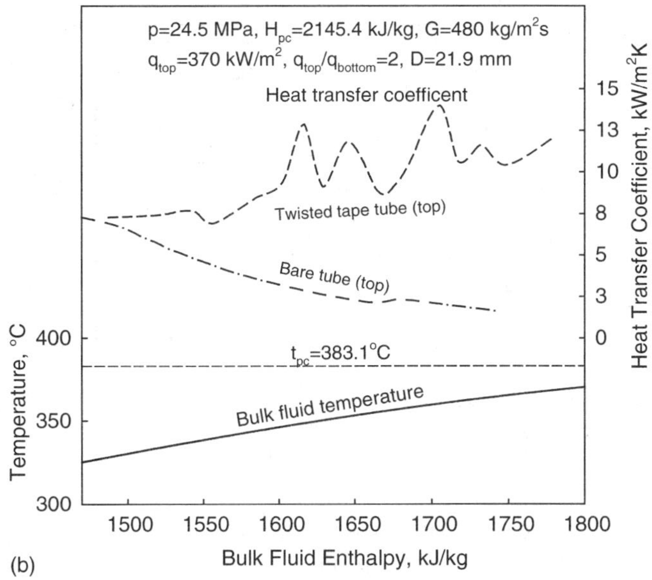

Figure 9.3. Temperature profiles (a) and HTC values (b) along horizontal tubes with and without twisted tape (non-uniform circumferential heat flux) cooled with supercritical water (Kamenetskii 1980): HTC values were calculated by the authors of the current monograph using the data from figure (a).

Additional information on effectiveness of ribs at supercritical conditions can be obtained from experiments in supercritical hydrocarbons.

Fedorov et al. (1986) investigated heat-transfer enhancement in forced convection of supercritical hydrocarbon fluid flowing through stainless steel smooth tubes and tubes with internal circumferential ribs. They found that a tube with concentric ribs (the ribs were manufactured by a roll forming from the outside) had more uniform wall temperature profile along the heated length as well as circumferentially as compared to a smooth tube. An enhancement increases as heat flux increases.

Kalinin and Dreitser with coauthors (Kalinin et al. 1998) conducted experiments with hydrocarbons flowing in tubes at supercritical pressures in smooth and ribbed tubes (for details, see Figures 9.4–9.8 and Table 9.1). The ribbed tubes were manufactured from smooth tubes with roll forming from outside (Figure 9.9). The roll forming or knurling created grooves on the external tube surface with ribs on the internal surface. Main parameters of these tubes were: D—internal smooth tube diameter, d—minimum internal diameter of a ribbed tube, and s—pitch between ribs in the axial direction. They found that the heat-transfer enhancement was the most efficient in suppressing the regimes with the deteriorated heat transfer (Figure 9.7).

Bastron et al. (2005) studied different methods on heat transfer enhancement in fuel assemblies for a HPLWR (Squarer et al. 2003). They found that an introduction of the artificial surface roughness or use of the staircase type grid spacer should increase the HTC by factor of 2, which results in peak cladding temperature by at least 50°C.

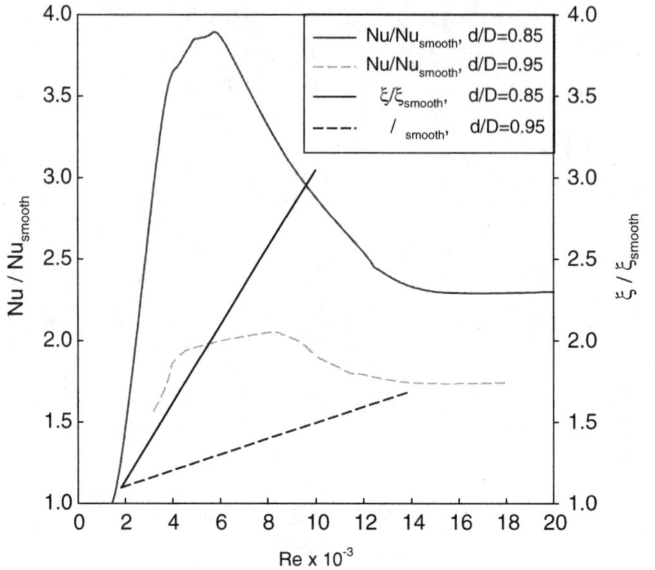

Figure 9.4. Effect of Re on $\mathrm{Nu}/\mathrm{Nu}_{\mathbf{smooth}}$ and ξ/ξ_{smooth} at heating of supercritical kerosene ("normal heat transfer") flowing in tubes (Kalinin et al. 1998): *s*/*D*=1.5.

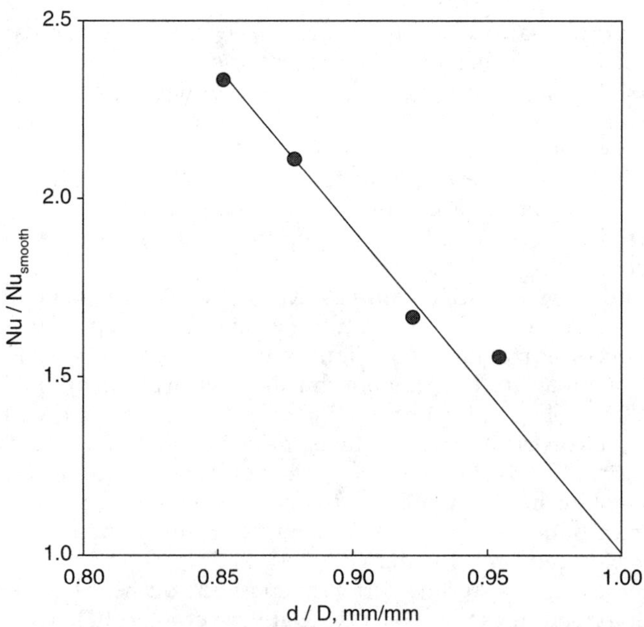

Figure 9.5. Effect of rib depth on $\mathrm{Nu}/\mathrm{Nu}_{\mathbf{smooth}}$ at heating of supercritical kerosene flowing in tubes (Kalinin et al. 1998): Re=10^4 and *s*/*D*=2.

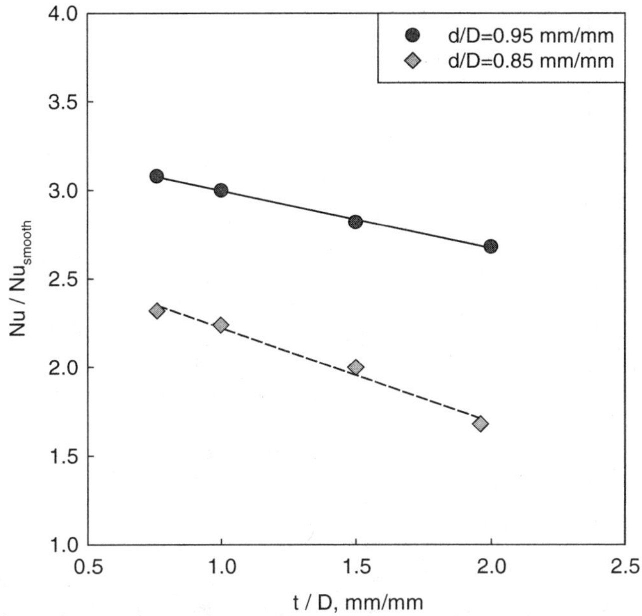

Figure 9.6. Effect of rib pitch on $\mathrm{Nu}/\mathrm{Nu}_{\mathrm{smooth}}$ at heating of supercritical kerosene flowing in tubes (Kalinin et al. 1998): Re=10^4.

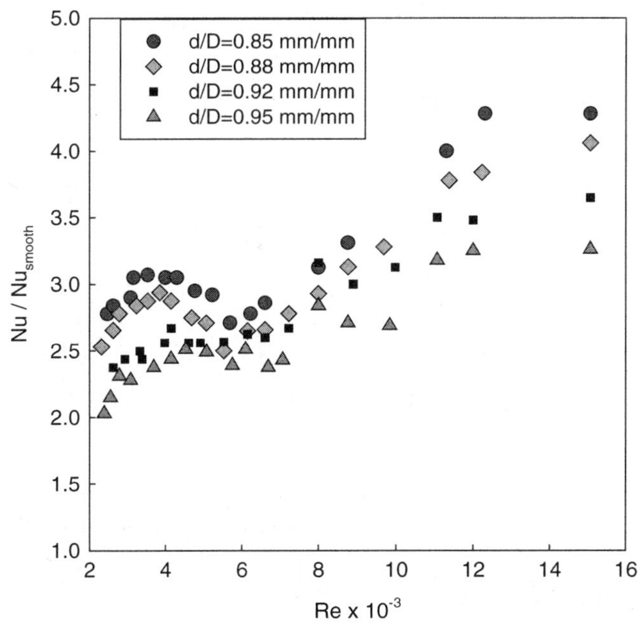

Figure 9.7. Effect of Re on $\mathrm{Nu}/\mathrm{Nu}_{\mathrm{smooth}}$ at heating of supercritical kerosene ("deteriorated heat transfer") flowing in tubes (Kalinin et al. 1998): *s/D*=0.75.

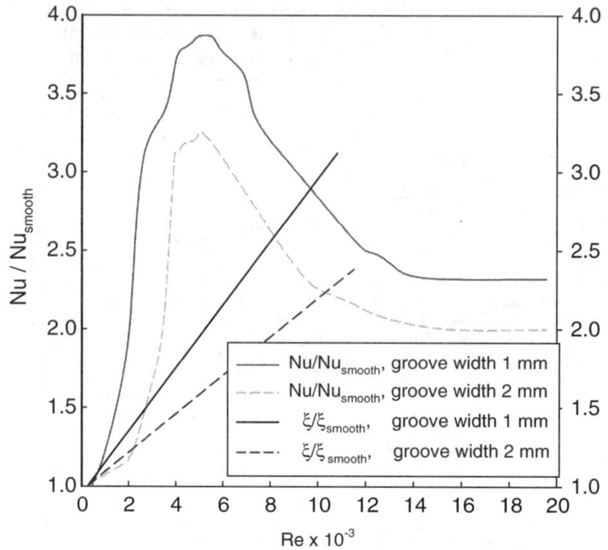

Figure 9.8. Effect of Re on Nu/Nu_{smooth} and ξ/ξ_{smooth} at heating of supercritical kerosene flowing in tubes (Kalinin et al. 1998): *d*/*D*=0.05 and *s*/*D*=1.5.

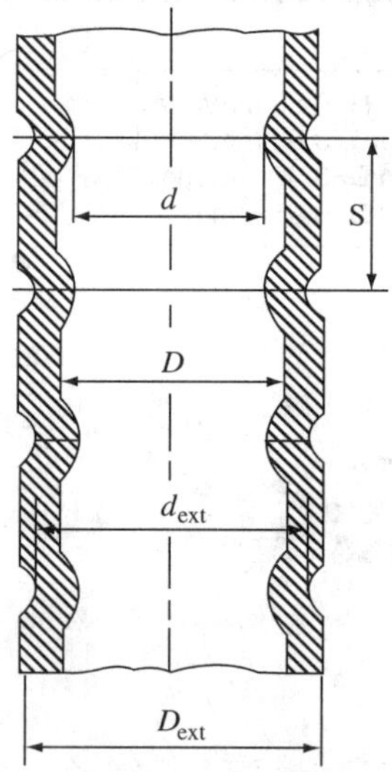

Figure 9.9. Ribbed tube with internal flow (Kalinin et al. 1998).

9.1 CONCLUSIONS

1. The majority of the experimental studies deal with heat-transfer enhancement of supercritical water, carbon dioxide and hydrocarbon fluids in vertical and horizontal circular tubes.

2. In general, various turbulence-enhancing devices such as ribs, twisted tapes etc. increase supercritical HTC up to several times compared to that of bare tubes and can suppress or significantly delay the deteriorated heat-transfer regime in circular flow geometry. However, the heat-transfer enhancing effect in other flow geometries including fuel bundles is mainly unknown.
3. A horizontal tube with twisted tape has a much smaller temperature difference between the tube top and tube bottom compared to a tube without tape. In general, tapes are very efficient heat-transfer enhancing devices in circular tubes, but it may not be so in more complex flow geometry such as a fuel bundle.

Chapter 10

EXPERIMENTAL SETUPS, PROCEDURES AND DATA REDUCTION AT SUPERCRITICAL PRESSURES

In this chapter, we examine and comment on the relevant features of experiments used to obtain heat-transfer and pressure-drop data in water and carbon dioxide based on experience at the IPPE (Kirillov et al. 2005), KPI (Pis'mennyy et al. 2005) and CRL (Pioro and Khartabil 2005; Pioro and Duffey 2003b). Additional information on this topic is given in Appendix E.

10.1 IPPE SUPERCRITICAL WATER TEST FACILITY

10.1.1 Loop and Water Supply

The SKD-1 loop (see Figure 10.1) is a high-temperature and high-pressure pumped loop. The operating pressure range is up to 28 MPa at outlet water temperatures up to 500°C. Distilled and de-ionized water is used in the loop.

Water passes from the pump through a flowmeter, a preheater, a test section, a mixer, main coolers and back to the pump. Pressurization is achieved with a high-pressure gas (N_2).

The test section is installed vertically with upward flow. Power is delivered to the test section by a 600 kW (AC) power supply, and cooling is achieved just downstream of the test section using a mixing cooler. While some of the heat from the test section is removed using this mixing cooler, a large amount is removed using the main loop heat exchangers in the discharge circuit of the pump.

10.1.2 Test-Section Design

The test section is a vertical stainless steel[14] circular tube (10-mm ID, 2-mm wall thickness and tube internal arithmetic average surface roughness R_a = 0.63 – 0.8 μm). The diameter of the test section is close to the hydraulic-equivalent

[14] 12Cr18Ni10Ti stainless steel was used (content similar to SS 304).

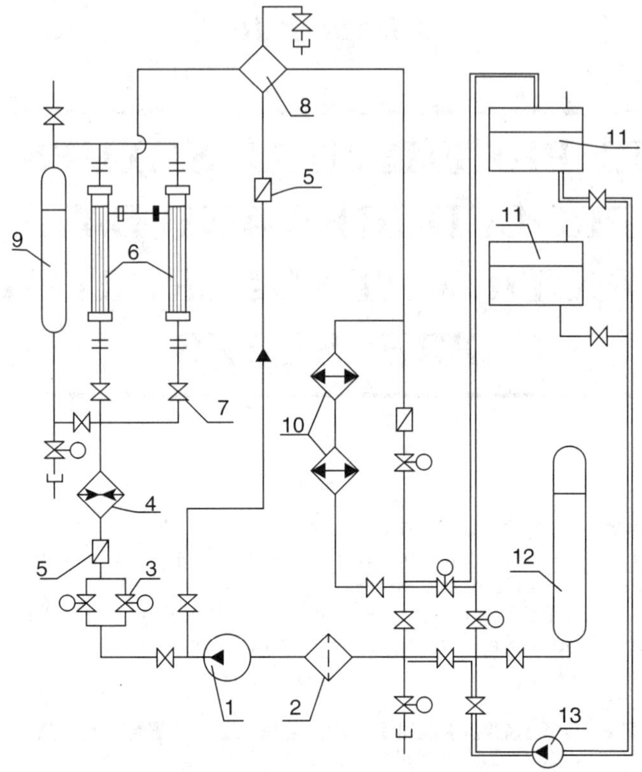

Figure 10.1 SKD-1 loop schematic (Kirillov et al. 2005): 1—Circulating pump, 2—mechanical filter, 3—regulating valves, 4—electrical heater, 5—flowmeter, 6—test section, 7—throttling valve, 8—mixer-cooler, 9—discharge tank, 10—heat exchangers—main coolers, 11—feedwater tank, 12—volume compensator, and 13—feedwater pump.

diameter of a SCWR fuel bundle. Two heated lengths were used: 1-m-long and 4-m-long[15]. Water is heated by means of AC electrical current passing through the tube wall from the inlet to the outlet power terminals (copper clamps). The test section is wrapped with thermal insulation to minimize heat loss.

10.1.3 Instrumentation and Test Matrix

The following test-section parameters were measured or calculated during the experiments:

- Test-section current and voltage were used to calculate the power.
- Pressure at the test-section inlet.

[15] It is important to perform supercritical heat-transfer experiments in one sufficiently long heated test section, which can represent full bundle length (for further explanations, see Section 5.1). However, a CANDU 12-bundle fuel string may be considered as twelve 0.5-m-long independent test sections.

Table 10.1. Uncertainties of primary parameters (Kirillov et al. 2005).

Parameter	Maximum Uncertainty
Test-section power	±1.0%
Inlet pressure	±0.25%
Wall temperatures	±3.0°C
Mass-flow rate	±1.5%
Heat loss	≤3%

- Temperatures at the test-section inlet and outlet. These temperatures were measured using ungrounded sheathed thermocouples inserted into the fluid stream. The thermocouples were installed just downstream of the mixing chambers, which were used to minimize non-uniformity in the cross-sectional temperature distribution. The thermocouples were calibrated *in situ*.
- Wall temperatures at equal intervals (50 mm) along the test section. Twenty-one thermocouples for the 1-m-long tube and 81 thermocouples for the 4-m-long tube were contact welded to the tube outside wall. All these thermocouples were calibrated *in situ*.
- Water mass-flow rate was calculated based on the measured pressure drop over a small orifice plate, which was monitored with a differential-pressure (DP) cell. And
- Ambient temperature.

The instrumentation used to measure the loop parameters was thoroughly checked and calibrated. Uncertainties of primary parameters are summarized in Table 10.1.

Special software was used for the data acquisition system (DAS) in the SKD-1 loop. The experimental data were recorded by the DAS when the desired flow conditions and power level were reached and stabilized. After that, a new power level or/and new set of flow conditions was set. The test matrix covered in the experiments is listed in Table 10.2.

The heat-loss tests, conducted at the beginning of the experimental program, were used to determine the heat-loss characteristics of the test section. The test results showed that heat loss from the test section was minor. However, the power used in the heat-transfer calculations was adjusted for the heat loss.

Data reduction is similar to that listed in Section 10.3.8.

Table 10.2. Test matrix (Kirillov et al. 2005).

p	t_{in}	t_{out}	t_w	q	G
MPa	°C	°C	°C	kW/m^2	kg/m^{2}s
24.5–25	300–380	360–390	<700	90–1050	200; 500; 1000
24	320–350	380–406	<700	160–900	500; 1000; 1500

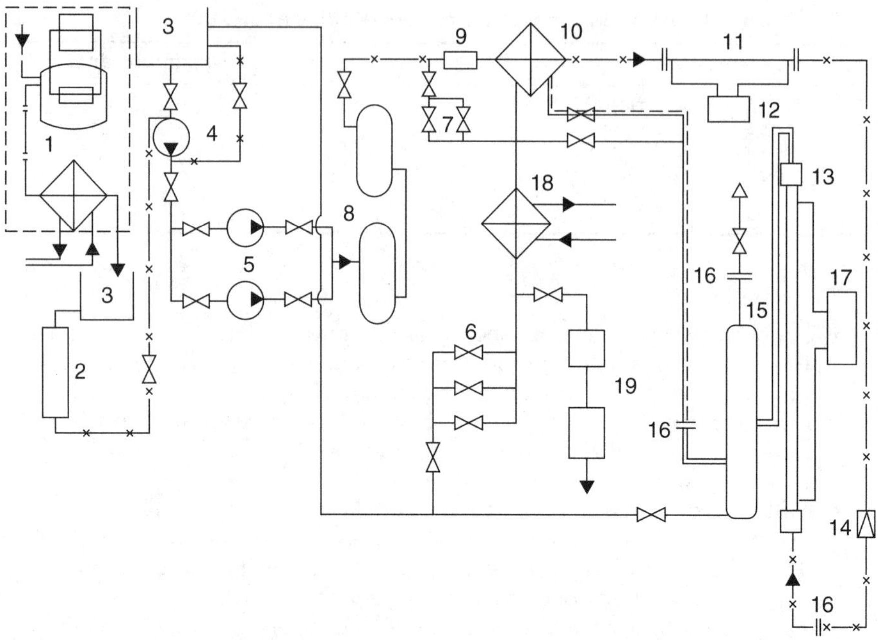

Figure 10.2. General schematic of supercritical water experimental setup (Pis'mennyy et al. 2005): 1 electro-distillator; 2 ion-exchange filter; 3 accumulator reservoirs; 4 boosting pump; 5 plunger pumps; 6 and 7 regulating valves; 8 damping reservoirs; 9 turbine flowmeter; 10 heat exchanger; 11 electrical preheater; 12 electrical generator(s); 13 test section; 14 throttling valve; 15 damping reservoir; 16 electro-isolating flanges; 17 main power supply; 18 cooler; and 19 throttling valves.

10.2 KPI SUPERCRITICAL WATER TEST FACILITY

10.2.1 Loop and Water Supply

Figure 10.2 shows a general schematic of the experimental setup. The supercritical water loop is an "open" stainless-steel loop operating at pressures up to 28 MPa and at temperatures up to 600°C. Chemically desalinated water (pH=7.2 and average hardness of 0.2 μg-eqv./ℓ) was used as a coolant. From a reservoir, fluid passes through pumps (boosting and two main plunger pumps connected in parallel[16]), a set of pressure regulating valves, a turbine-type flowmeter, a tube-in-tube preheater, a 75-kW electrical preheater, a test section, a tube-in-tube cooler, and then it returns to the reservoir through a set of the throttling valves. The reservoir is open to the atmosphere.

The test section is installed vertically for upward or downward flow of water and heated by a 120 kW (60 V × 2000 A, AC) or 90 kW (18/36 V × 5000/2500 A, DC) power supplies.

[16] The boosting pump increases pressure from atmospheric to about 0.2 MPa. Each of the two plunger pumps can increase pressure of up to 70 MPa.

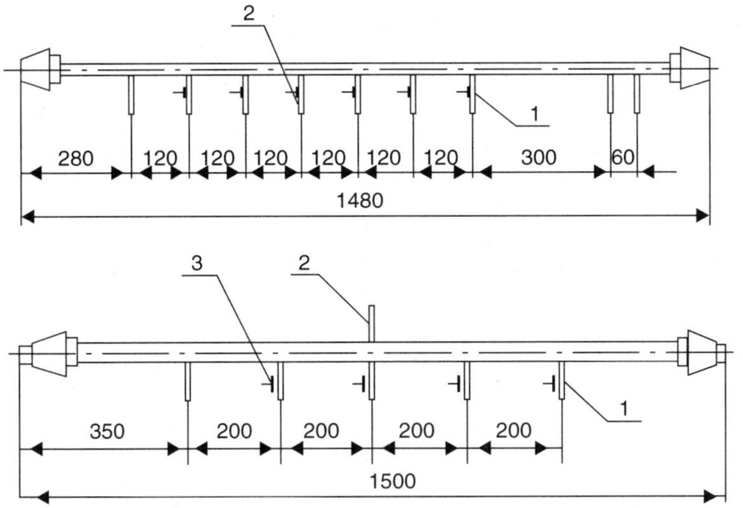

Figure 10.3. Schematic of test sections (Pis'mennyy et al. 2005): (a) *D*=6.28 mm; (b) *D*=9.50 mm; 1—static-pressure taps; 2—pressure-pulsations tap; and 3—voltage-measuring tap.

10.2.2 Test Matrix

The test matrix is listed in Table 10.3. This test matrix was chosen to simulate conditions of a partial loss-of-coolant accident (LOCA), which can be encountered in SCWRs.

10.2.3 Test-Section Design

Figure 10.3 shows a schematic of the test sections (for test-sections details, see Table 10.3); and Figure 10.4 shows flow arrangement in the test sections. Two thin-wall stainless steel tubes were used as the test sections. These tubes simulated bare subchannels of SCWR fuel bundles.

Table 10.3. Test matrix and test-section details (see also Figures 10.3 and 10.4) (Pis'mennyy et al. 2005).

p	t_{in}	H_{in}	q	G
MPa	**°C**	**kJ/kg**	**kW/m²**	**kg/m²s**
23.5	20–380	106–2222	≤515	250; 500
Test Sections				
Vertical stainless-steel (1Cr18Ni9Ti) smooth tubes: D=6.28 mm, δ_w=0.91 mm, L_h=600; 360 mm; and D=9.50 mm, δ_w=1.39 mm, L_h=600; 400 mm; L_h/D=95.5; 57.3 and 63.2; 42.1, respectively; tubes internal surface roughness R_a=0.25÷0.5 μm; upward and downward flows; inlet adiabatic stabilization section L/D=102; 64 and 79; 58, respectively; inlet and outlet mixing chambers with bulk-fluid temperature thermocouples.				

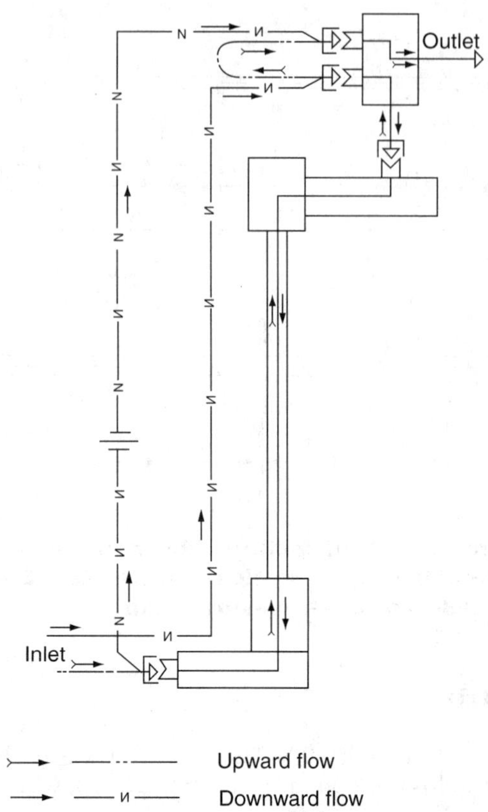

Figure 10.4. Flow arrangement in test sections (Pis'mennyy et al. 2005).

Water is heated by means of an alternative or a direct electrical current passing through the tube wall from the inlet to the outlet power terminals. The test section and mixing chambers are wrapped with thermal insulation to minimize heat loss. An inlet adiabatic stabilization section (for details, see Table 10.3) is installed just upstream of the test section.

10.2.4 Instrumentation

The following test-section parameters were measured or calculated during the experiments:

- Test-section current (based on readings from a measuring transformer in case of AC or on voltage readings from a calibrated shunt for DC) and voltage. These parameters were used to calculate the power.
- Pressure at the test-section outlet[17].
- Temperatures at the test-section inlet and outlet. These temperatures were measured using Chromel-Alumel ungrounded sheathed

[17] Because of the short test sections and lower range of flows, pressure drop was negligible.

Table 10.4. Uncertainties of measured and calculated parameters (Pis'mennyy et al. 2005).

Parameter	Uncertainty
Measured Parameters	
Outlet pressure	±0.22%
Bulk-fluid temperatures	±(0.11 – 0.21)%
Wall temperature	±(0.11 – 0.21)%
Calculated Parameters	
Mass-flow rate	±(0.76 – 2.28)%
Power	±(1.15 – 2.40)%
HTC	±(2.37 – 12.7)%
Heat losses	<2%

thermocouples of 0.2-mm diameter (wire) inserted into the fluid stream inside the mixing chambers. The mixing chambers were used to minimize non-uniformity in the cross-sectional temperature distribution and to decrease pressure pulsations within the test section. The thermocouples were calibrated within the temperature range of 20° to 400°C.

- Thirteen wall thermocouples for $D = 9.50$ mm and 16 thermocouples for $D = 6.28$ mm were welded (if AC current was used for the test-section direct heating) or fitted through mica to the tube outer wall (if DC current was used for the test-section direct heating). The temperature trip for the external wall was set at 600°C. All thermocouples were located on one side of the test section. All wall thermocouples were calibrated within the range of 20° to 600°C. And
- Ambient temperature.

The instrumentation used to measure the loop parameters was thoroughly checked and calibrated. Uncertainties of primary parameters are summarized in Table 10.4.

The experimental data were recorded by a DAS when the desired flow conditions and power level were reached and stabilized. After that, a new power level or/and new set of flow conditions were established. Increase in power was limited with the maximum wall temperature of 540°C.

The heat-loss tests, conducted at the beginning of the experimental program, were used to determine the heat-loss characteristics of the test section. The power used in the heat-transfer calculations was adjusted for the heat loss.

Data reduction is similar to that listed in Section 10.3.8.

10.3 TYPICAL SUPERCRITICAL CARBON DIOXIDE FACILITY AND PRACTICAL RECOMMENDATIONS FOR PERFORMING EXPERIMENTS

As mentioned previously, carbon dioxide has been often used as a supercritical working fluid in various thermodynamic cycles and heat exchangers and as a modelling fluid instead of supercritical water, because of its lower critical

Table 10.5. Basic scaling parameters for fluid-to-fluid modelling at supercritical conditions.

Parameter	Scaling Law
Pressure	$\left(\frac{p}{p_{cr}}\right)_{CO_2} = \left(\frac{p}{p_{cr}}\right)_{Water}$
Bulk-fluid temperature (K)	$\left(\frac{T_b}{T_{cr}}\right)_{CO_2} = \left(\frac{T_b}{T_{cr}}\right)_{Water}$
Mass flux	$\left(\frac{GD}{\mu_b}\right)_{CO_2} = \left(\frac{GD}{\mu_b}\right)_{Water}$

pressure and temperature (see Tables 2.1; 10.6, and Table B.1 and Figure B1 in Appendix B). Therefore, there is a significant interest to this modelling fluid, especially for SCWR's application because of high cost and technical difficulties in performing supercritical water tests in bundle geometry. Also, using a modelling fluid such as CO_2 allows conducting experiments within a wider range of operating parameters than in water.

AECL at Chalk River Laboratories has built a supercritical carbon dioxide loop[18] to perform preliminary heat-transfer and pressure-drop tests instead of supercritical water. Details of the CRL supercritical carbon dioxide experimental setup are listed below.

10.3.1 Scaling Parameters

Scaling parameters used for modelling nominal supercritical water-operating conditions of the SCW CANDU reactor concept into supercritical carbon dioxide equivalent values are listed in Table 10.5. These scaling parameters (Pioro and Duffey 2003b) were deduced from those originally proposed by Jackson and Hall (1979) and Gorban' et al. (1990).

Table 10.6 lists critical parameters of water and carbon dioxide and nominal operating parameters of the SCW CANDU reactor concept in water and carbon dioxide equivalent values.

Recommendation 1

Put safety first: follow local safety codes, rules and regulations in an experimental setup design, construction, and operation.

Recommendation 2

NIST (2002) thermophysical properties software is recommended for property calculations of various fluids. Scaling parameters (see Table 10.5) can only be

[18] The authors of the current monograph express their appreciation to B. Addicott and J. Martin (CRL AECL) for their technical help during preparation and conducting the supercritical heat-transfer experiments.

Table 10.6. Critical parameters of water and CO_2 (NIST 2002) and nominal operating parameters of the SCW CANDU reactor concept and their equivalent values in CO_2.

Parameter	Unit	Water	CO_2
	Critical Parameters		
Critical pressure	MPa	22.1	7.3773
Critical temperature	°C	374.1	31.0
Critical density	kg/m³	315	467.6
	Operating Parameters		
Operating pressure	MPa	25	8.34
Inlet temperature	°C	350	20
Outlet temperature	°C	625	150
Mass flux at inlet	kg/m²s	1185	1630
Mass flux at outlet	kg/m²s	1185	800

used for scaling operating conditions from one fluid to another for reference purposes. Due to scaling parameters simplicity, special behavior of thermophysical properties at supercritical pressures and complexity of the processes involved, some discontinuities may exist. For example, there can be a mass discontinuity, i.e., a value of the mass flux scaled at inlet conditions may not be the same as that scaled at outlet conditions (for details, see Table 10.6).

Heat flux has not been scaled, because it was proposed to change the heat flux from the minimal value to the maximum possible during the experiments.

10.3.2 Loop and CO_2 Supply

As a typical example, the MR-1 loop at the CRL (see Figure 10.5) is a high-temperature and high-pressure pumped loop. The operating pressure range is up to 10.3 MPa at temperatures up to 310°C. The maximum mass-flow rate of supercritical carbon dioxide for the current tests is about 0.153 kg/s (G = 3000 kg/m²s inside an 8-mm ID tube). Carbon dioxide (99.9% purity, content of hydrocarbons 0.8 ppm) is charged into the loop from a tank installed outside the laboratory building. Power is delivered to the test section by a 350 kW (175 V × 2000 A) direct current (DC) power supply.

Recommendation 3

DC power supply is more preferable for direct-heating applications compared to that of alternative current (AC), especially for thin-walled test sections[19]. However, test-section surface temperature measurements with DC power supply

[19] It was found that critical heat flux (CHF) values at boiling obtained in or on a test section with DC direct heating can differ from those obtained in or on a test section with AC direct heating. This difference is larger for thin-walled test sections. At supercritical pressures, there is a similar phenomenon, which is called pseudo-boiling. Therefore, the above-mentioned findings may apply to the supercritical case.

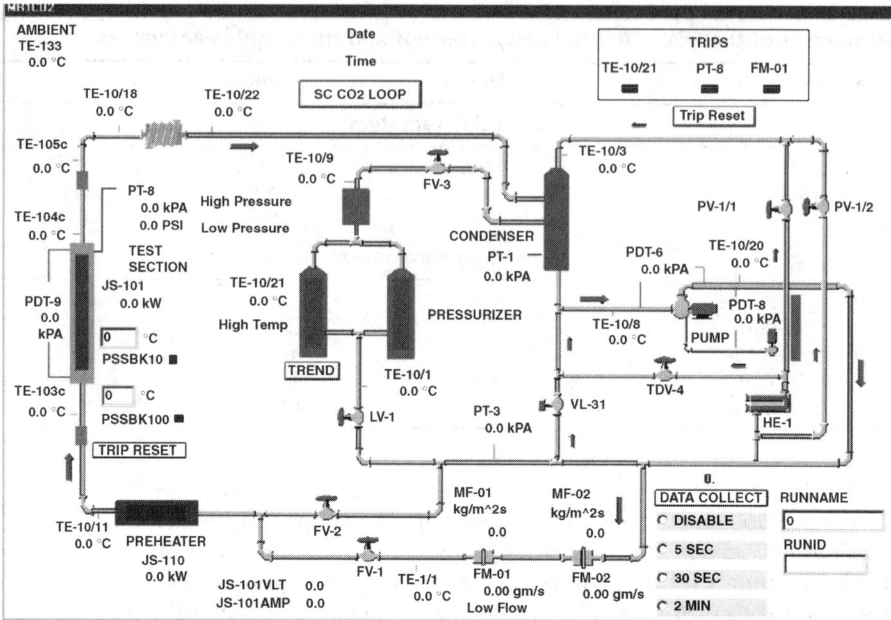

Figure 10.5. $SCCO_2$ loop schematic (as shown on DAS display): FM—flowmeter, FV—flow valve, HE—heat exchanger, JS — power, LV—level valve, MF—mass flux, PDT—pressure differential transducer, PSSBK—pressure set, PT—pressure transducer, PV—pressure valve, $SCCO_2$—supercritical carbon dioxide, TDV—temperature difference valve, TE—temperature, and VL — valve for low flow.

will be more complicated, because they require insulating the thermocouple measuring tip from the electrically heated surface (for details, see below). With an AC power supply, a thermocouple-measuring tip may be in contact with an electrically heated surface as there is no average voltage offset across the probe junction. Nevertheless, special precautions should be taken to avoid spurious electromagnetic induction (emf), for example in the thermocouple wiring, which can be generated by the pulsating electromagnetic field around the test section.

Fluid passes from the pump through an orifice flowmeter, a preheater, a test section, a cooler and back to the pump. The centrifugal pump that circulates CO_2 through the loop is fitted with a special seal and barrier-fluid cooling system. This seal and fluid are needed to compensate for the poor lubricity of the CO_2.

Recommendation 4

A better solution for the circulation pump is a vertical canned-type pump manufactured by commercial suppliers, which does not require any lubrication and is suitable for many fluids within a wide range of pressures and temperatures. For lower mass-flow rates, an electromagnetic driven head pump can be used.

The CO_2 fluid passes through a 25-kW preheater into the test section. The preheater consists of electrically heated elements, installed in a pressure tube, that are controlled from the loop control panel.

Recommendation 5

To minimize thermal inertia and pressure drop of the preheater, the directly heated tube or tubes can be used with an additional power supply.

Cooling is achieved just downstream of the test section using helicoil coolers or heat exchangers (the coolant is near-ambient temperature river water). The temperature of the river water varies between 4°C and 20°C, depending on the season.

Recommendation 6

For CO_2 application, lower inlet temperatures may be required, so a refrigeration unit may be needed in the cooling system.

While some of the heat from the test section can be removed using the small helicoil coolers, large amounts are removed using the main loop heat exchanger in the discharge circuit of the pump. Pressurization is achieved by applying an electrical power to the heating elements in two vessels. Pressure is controlled automatically from the panel by using the outlet test-section pressure (PT-8) as a set point for regulating the power input to the heating elements.

Recommendation 7

Pressurization with electrically heated elements installed inside a pressure vessel is more preferable than the pressurization with an inert gas through a membrane or with direct contact between the gas and the loop working fluid. The volume of the vessel should be at least several times larger than the loop total internal volume to assure stable pressure during operation.

Test matrix is listed in Table 10.7

Table 10.7. Test matrix.

No of Points	p_{out}	G	t_{in}	t_{out}	q
–	MPa	kg/m²s	°C	°C	kW/m²
158	7.36	900, 2000	25, 30	29–82	37.4–447
129	8.36	2000	21	29–94	115–480
664		295–2060	21–30	35–124	16–225
598		900, 2780–3200	25–40	32–139	26.5–616
416		1500, 2020	30–40	33–131	30–460
581		784–2000	20–40	39–120	221–466
182		750	31–40	35–136	18–209
108	8.8	900, 2000	31	37–107	28–225
181	8.83	2002, 3020	30	35–89	30–536

10.3.3 Test-Section Design

The test section was installed vertically with an upward flow of carbon dioxide. The current test-section design (see Figure 10.6) consists of a 2.4-m-long tubular section of 8-mm ID Inconel-600 tubing, with a 2.208-m heated length (for additional information on the test section, see Appendix C). The diameter is close to the hydraulic-equivalent diameter of a typical subchannel in a 43-element fuel bundle that is considered for use in the SCW CANDU reactor.

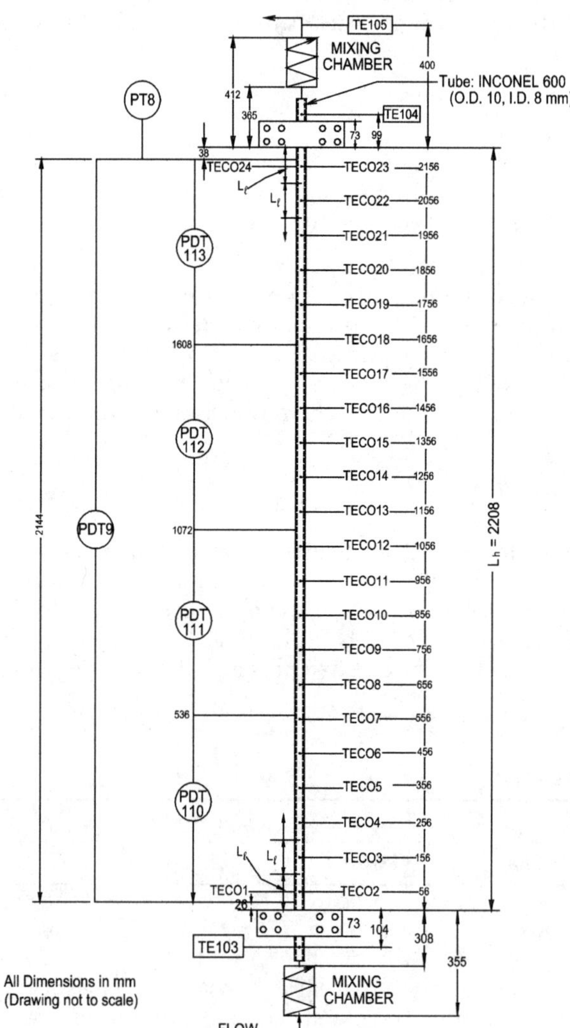

Figure 10.6. SCCO2 test-section thermocouples and pressure-taps layout.

Recommendation 8

It is unlikely that a simple circular geometry can fully model the thermal-hydraulic behavior of any actual subchannel geometry. However, it is a good practice to have a tube ID close to a hydraulic-equivalent diameter of the average subchannel; and a tube heated length equals to that of a fuel bundle.

The fluid is heated by means of a direct electrical current passing through the tube wall from the inlet to the outlet power terminals (copper clamps).

Recommendation 9

In general, an electrical resistance of the heated part of the test section should be equal to or at least to be close to the nominal electrical resistance of the power supply, i.e., for example, for the typical power supply – $R_{el}^{TS} = R_{el}^{nom} = 175\text{V}/2000A = 0.0875$ Ohm. Only in this particular case, the nominal power of the power supply (i.e., 350 kW) can be applied to the test section. Otherwise, the limiting factors on the applied power will be the nominal voltage of the power supply (175 V) if $R_{el}^{TS} > R_{el}^{nom}$ or the nominal current (2000 A) if $R_{el}^{TS} < R_{el}^{nom}$. The power deposited in the test section can be compared to the loop heat balance from the known flow and enthalpy (temperature rise) across the test section.

Recommendation 10

Material for a directly heated test section should be chosen with a relatively high value of the specific electrical resistivity, which in turn should ideally be invariant or nearly independent of wall temperature. In this case, there is a wider range for changing the test-section wall thickness and heated length. Also, the local heat flux will be more uniform along the test section in spite of variable wall temperature if the test-section diameter and wall thickness are constant along the heated length. Therefore, Inconel 600 or Inconel 718 will be the best choice for material of the test section (Figure 10.7). Additional characteristics of the test section, such as the chemical content, precise measurements of the inside/outside tube diameters, the tube burst pressure, tube electrical resistance, and the electrical resistivity and thermal conductivity of the tube material are also essential for the complete performance and safety analysis.

Recommendation 11

For a uniform temperature profile in a cross section of the test section at the inlet and for accurate bulk-fluid temperature measurements, mixing chambers should be used together with sheathed ungrounded thermocouples installed just downstream of them. It is especially important for supercritical pressures applications. Also, between the inlet mixing chamber and the test section, the flow stabilization part should be installed with the same diameter as that of the test section, acting as a settling length. In general, the length of this part should be about (20 – 50) × D

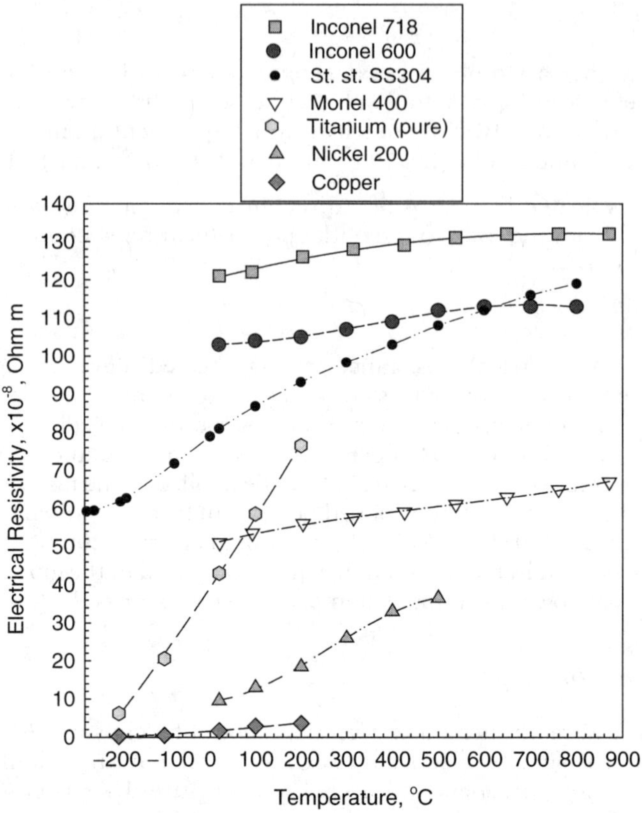

Figure 10.7. Effect of temperature on electrical resistivity of alloys and pure metals.

(Pis'menny et al. 2005; Kirillov et al. 2005). The outlet mixing chamber should be installed just downstream of the test-section outlet to reduce uncertainties in the outlet bulk-fluid temperature measurement, and hence in the loop heat balance.

Recommendation 12

The test section (see Figure 10.6) and mixing chambers should be wrapped with thermal insulation to minimize heat loss. In general, circular insulation should be checked for the optimal insulation thickness value (see Incropera and DeWitt (2002), pp. 107–110) to be effective.

10.3.4 Instrumentation

The following test section parameters were measured during the experiment (see Table 10.5 and Figures 10.5 and 10.6):

- The test-section current (the current was calculated using a measured voltage drop on a calibrated shunt) and voltage (to give power, JS-101).
- The pressure at the test section outlet (PT-8).
- Four pressure drops over equal lengths (536 mm) along the test section (PDT-110 to PDT-113), and an overall test-section pressure drop (PDT-9).
- Temperatures at the test-section inlet and outlet (TE-103, TE-104 and TE-105).
- Wall temperatures at equal intervals[20] (100 mm) along the test section (TEC01 to TECO24).
- The CO_2 mass-flow rate. And
- The ambient temperature.

Measurement uncertainties for all instruments are listed in Table 10.8.

It was decided to provide a sample of the actual uncertainty analysis (for details, see Appendix D) in this monograph because, usually, only simple examples of uncertainty calculations can be found in the open literature.

The total test-section power is a calculated value that is based on the measured values of current and voltage drop across the test section.

The test section has equally spaced pressure taps for pressure loss readings (see Figure 10.6), which allows the determination of the test-section local and total pressure drops.

The sum of the local pressure drops measured with PDT-110 – PDT-113 was used to check the total test-section pressure drop measurement from PDT-9 for an instrument consistency.

The test-section inlet and outlet bulk-fluid temperatures (TE-103, TE-104 and TE-105) are measured using 1/16″ K-type ungrounded sheathed thermocouples inserted into the fluid stream (see Figures 10.5 and 10.6). The thermocouples are used with electronic reference units. Also, the thermocouples were calibrated *in situ*, i.e., by comparing the reading on the DAS with the calibration standard (for measurement uncertainties, see Appendix D). Therefore, the appropriate temperature-dependent correction was applied to each temperature value measured by thermocouples within the calibration temperature range of 0°C – 100°C. Beyond this temperature range, no temperature corrections were applied to the measured temperature values.

Table 10.8. Uncertainties of measured and calculated parameters (for details, see Appendix D).

Parameter	Uncertainty
Test-section power/heat flux	±0.5%
Inlet/outlet pressure	±0.2%
Local pressure drops	±0.8% at $\Delta P = 30$ kPa
	±5.0% at $\Delta P = 5$ kPa
Bulk-fluid and wall temperatures	±0.3°C within 0 – 100°C
	±2.2°C beyond 100°C
Mass-flow rate/mass flux	±1.6% at $m = 155$ g/s ($G = 3084$ kg/m²s)
	±12.5% at $m = 46$ g/s ($G = 915$ kg/m²s)

[20] Smaller pitch is recommended. A pitch of 50 mm is reasonable in this case.

Recommendation 13

Calibration of various measuring instruments *in situ* is the best way to minimize measurement uncertainties compared to other calibration techniques (calibration at instrument shop, calibration check, etc.), because a calibration curve will take care of uncertainties of measuring, signal transmitting, signal converting devices, etc. (for details, see Appendix D).

To measure the enthalpy increase, thermocouples TE-103 and TE-105 (see Figures 10.5 and 10.6) are installed just downstream of the mixing chambers (see Figure 10.7). To assess the impact of the mixing chambers, an additional thermocouple TE-104 was installed just downstream of the test section but upstream of the outlet mixing chamber (see Figures 10.5 and 10.6).

Recommendation 14

For bulk-fluid temperature measurements, Resistance Temperature Detectors (RTDs) are more preferable to use compared to thermocouples because they are usually more accurate. However, it is a good practice to use thermocouples in addition to RTDs as the back-up temperature devices.

For wall temperature estimates, 24 fast-response K-type thermocouples with self-adhesive fibreglass backing were attached to the tube outer wall at equal intervals of 100 mm (see Figure 10.6). The thermocouples were additionally secured by wrapping them with Teflon tape and fibreglass string to achieve proper contact with the wall. The fibreglass backing could withstand temperatures of up to 260°C for extended periods of time, and up to 370°C for a short duration. Therefore, the temperature trip for the external-wall temperature was set at 250°C. Thermocouples TECO2 to TECO23 are located on one side of the test section. Thermocouples TECO1 and TECO24 are located in the same cross-sections as the thermocouples TECO2 and TECO23, but 180°apart.

Recommendation 15

For relatively low wall-temperature measurements, i.e., up to 260°C, fast-response thermocouples are the easiest way to go. However, beyond this range sheathed ungrounded thermocouples should be used welded to the surface and calibrated *in situ*. This recommendation applies to flow geometry cooled from inside (circular tubes, etc.).

10.3.5 Loop Control, Data Acquisition, and Processing

Modern DAS are all computer based. Personal computer (PC) software is used for trip control, together with a programmable logic controller (PLC). The trips are intended to enhance a protection to the personnel and equipment in the event of system malfunctions. The trips are selected for high pressure, high temperature and low flow. The PLC provides power trips to the pressurizer heaters, the preheater, the test-section power supply, the pump and the test-section shutoff valves. An operator can program trip set points into the DAS.

Recommendation 16

Always use multiple trips setup in DAS for safe and reliable test rig operation.

Recommendation 17

Do not rely solely on a DAS and other electronic measuring devices, but also use, in parallel, simple devices for measuring or just for indication of the most important/safety/limiting parameters: for example, high-accuracy mechanical gauges (for pressure), arrow-type voltmeters and ammeters, plus use remote video cameras for monitoring heated parts of a test section, etc.

Figure 10.5 shows the typical DAS display that is available to the experimenter and the operator during loop operation. For steady-state testing, at constant flow and power, data can be collected every five seconds during a one-minute scan. However, data on the screen were updated every second. Data were retrieved and converted into an EXCEL (.xls) file format for post-test processing.

10.3.6 Quality Assurance

To ensure the accuracy of the experimental data:

- Instruments (thermocouples, pressure transmitters and differential pressure (DP) cells) were calibrated by qualified instrument technicians, using traceable standards.
- DAS-to-instrument connections and data storage were verified. And
- Calibration records for instrumentation are stored in a filing cabinet.

10.3.7 Conduct of Tests

To establish a heat balance, a heat-loss test with power applied to the test section (the loop is vacuumed) was conducted (for details, see Section D.10 in Appendix D). Such test is more accurate compared to that with the zero power (i.e., test based on heat-balance technique, when hot fluid is cooled flowing through a test section; power are not applied to the test section), because it eliminates uncertainties that are related to an estimation of the thermophysical properties of CO_2. This test also eliminates flow-measurement uncertainties and uncertainties that are incurred when measuring very small temperature differences (0.5°C – 1°C) between inlet and outlet bulk-fluid temperatures.

Based on these heat-loss tests, it was concluded that heat loss from the test section was small, i.e., about 0.4% at the highest wall-to-ambient temperature difference of 220°C in the $SCCO_2$ loop. Nevertheless, the power used in the heat transfer calculations was adjusted for the heat loss.

The experimental data were recorded by the DAS when the desired flow conditions and power level were reached and stabilized. Operating conditions that were judged to be steady are as follows:

- Outlet-pressure fluctuations were within ±0.2%;
- Mass-flux fluctuations were generally within ±1.0% (in several cases, they fluctuated up to ±1.7%);
- Heat-flux fluctuations were within ±0.2%;
- Inlet-temperature fluctuations were generally within ±1.0% (in several cases, they fluctuated up to 1.5%);
- External wall-temperature fluctuations were within ±0.7%; and
- Pressure-drop fluctuations were generally within ±2.0% (in several cases, they fluctuated up to ±2.8%). The data were recorded[21] over one minute in five-second intervals.

The test procedure was as follows:

1. Increase the loop pressure with electrical heaters to the required value.
2. Start the circulation pump and establish the required mass-flow rate by adjusting the regulating valve installed just upstream of the test section.
3. Use the preheater to establish the required inlet temperature.
4. Adjust all controlled parameters—pressure, mass-flow rate and inlet temperature (for adjustment details, see steps 1 to 3)—to the required nominal values (if needed).
5. Start the power supply and establish the required power level by increasing power in small increments.
6. Re-adjust all controlled parameters (see Step 4) to the required nomi nal values (if needed).
7. Monitor measured parameters via the DAS display. When steady-state conditions are achieved, scan all data in five-second intervals over a one-minute period.
8. Repeat Steps 3 to 7 for other flow conditions.

10.3.8 Data Reduction

The data reduction procedure is based on local parameters, which were measured or calculated at each cross-section (or vertical elevation) corresponding to the external-wall thermocouples. The external-wall temperatures, inlet and outlet bulk-fluid temperatures and electrical current were used as the basis for local-parameter calculations. These local parameters include thermal conductivity and electrical resistivity of the wall material, electrical resistance, power, heat flux, volumetric heat flux, internal wall temperature, heat loss, bulk-fluid temperature, and pressure.

It should be noted that when the measured inlet bulk-fluid temperature is close to pseudocritical or critical temperatures, the value of inlet temperature *cannot be reliably used* as the starting point for downstream bulk-fluid temperature calculations. There are significant variations in thermophysical properties near the pseudocritical and critical points, which lead to increased uncertainties in calculations (for additional information, see Section D.11 in Appendix D). Therefore, a value of the measured test-section outlet bulk-fluid temperature should be used for upstream bulk-fluid temperature calculations

[21] Input signals from all 64 channels were read over a period of 10 ms.

instead of the inlet bulk-fluid temperature. A similar conclusion was drawn by Yamagata et al. (1972).

The same explanation applies to the temperature on which fluid density for the flow-rate calculation is based, i.e., thermocouple TE-1/1 located near the orifice-plate flowmeter. In this case, this temperature should be below or beyond the critical or pseudocritical temperature regions.

The general and local parameters are defined as follows:
General parameters:

- Flow area: $A_{fl} = \pi D^2/4$, where D is the inside tube diameter.
- Mass flux: $G = m/A_{fl}$, where m is the mass-flow rate.
- Total heated area: $A_h = \pi D L$, where L is the total heated length.
- Measured power: $POW = U I$, where U is the test-section voltage drop, and I is the electrical current.
- Average heat flux: $q = (POW - HL)/A_h$.
- Inlet pressure: $p_{in} = p_{out} + \Delta p_{TS}$, where Δp_{TS} is the total pressure drop across the test section.
- Thermophysical properties of carbon dioxide: fluid density, specific heat, enthalpy, thermal conductivity, and dynamic viscosity. These properties were calculated using NIST software (2002). The thermophysical properties at a particular cross-section were calculated based on the local pressure and local bulk-fluid temperature. The pseudocritical temperature was evaluated at the outlet pressure. However, there is a range of pseudocritical temperatures for each set of flow conditions along the test section, because t_{pc} depends on pressure and pressure changes from the inlet to the outlet. Nevertheless, the changes in pseudocritical temperature are small (less than 0.6°C) within the heated length of 2.2 m. For example, within the range of pressures p_{in} = 8.3 MPa and p_{out} = 8.4 MPa, i.e., Δp_{TS} = 100 kPa, the difference in pseudocritical temperature is only Δt_{pc} = 0.55°C.

Local parameters:

- Heated area: $A_{h\ell} = \pi D L_\ell$, where L_ℓ is the local heated length (see Figure 10.6). For thermocouples TECO3 to TECO22 the local heated length is 100 mm, for TECO1 and TECO2 it is 106 mm and for TECO23 and TECO24 it is 102 mm. It was assumed that within the local heated length the external wall temperature is constant and equals the value measured by wall thermocouples.
- Power[22]: $POW_\ell = I^2 \cdot R_{el\ell}$, where $R_{el\ell}$ is the local electrical resistance[23] within the local heated length calculated using a local value of electrical resistivity.

[22] Sum of local powers were checked against the total power applied to the test section for each run, to ensure that these values do not differ significantly. In general, the difference was within 2.5%.

[23] Sum of $R_{el\ell}$ was checked against the value of the total electrical resistance of the test section, calculated as U_{TS}/I, for each run, to ensure that these values do not differ significantly. In general, this difference was within 2.5%.

- Heatflux: $q_\ell = (POW_\ell - HL_\ell)/A_{h\ell}$, where HL_ℓ is the local heat loss based on the corresponding external wall-temperature measurements. There is a minor change in axial heat flux due to direct heating and the effect of wall temperature on electrical resistivity of material. Inconel 600, which has one of the weakest effects of temperature on electrical resistivity, was used as the test section material to minimize this effect.
- Tube wall thermal conductivity ($k_{w\ell}$) was calculated using the average wall temperature $t_w^{ave} = (t_w^{ext} + t_w^{int})/2$.
- Volumetric heat flux: $q_{v\ell} = POW_\ell / (\pi/4(D_{ext}^2 - D^2))$.
- Internal wall temperature (Incropera and DeWitt 2002): $t_w^{int} = t_w^{ext} + q_{vl}/(4k_{wl})\left[(D_{ext}/2)^2 - (D/2)^2\right] - q_{vl}/(2k_{wl})(D_{ext}/2)^2 \ln(D_{ext}/D)$. This equation accounts for the uniformly distributed heat-generating sources inside the tube wall, i.e., heating with direct current passing through the tube wall.
- Local pressure: $p_\ell = p_{in} - \Delta p_{PDT110}/536 \cdot (z_{TECOn} - 26)$, where z is the external wall thermocouple location in mm from the beginning of the heating zone, $n = 1 - 7$ and is the thermocouple number. For thermocouples TECO8 –TECO12:$p_\ell = p_{in} - \Delta p_{PDT110} - \Delta p_{PDT111}/536 \cdot Z_{TECOn}$, where $n = 8 - 12$. The same approach is applied to other thermocouples.
- Local inlet bulk-fluid enthalpy is calculated through the NIST software (2002) using the measured inlet temperature and the calculated inlet pressure.
- Local bulk-fluid enthalpies in the cross-section, where the external wall thermocouples were located, are calculated using the heat-balance method: $H_{b\ell}\big|_{i+1} = H_{b\ell}\big|_i + (POW_\ell - HL_\ell)/m$.
- Local bulk-fluid temperature (t_b) was calculated through the NIST software (2002) using the local pressure and local enthalpy in the cross-section, where the external wall thermocouples were located.
- Local HTC: $HTC_l = q_\ell / (t_w^{int} - t_b)$.

Chapter 11

PRACTICAL PREDICTION METHODS FOR HEAT TRANSFER AT SUPERCRITICAL PRESSURES

11.1 WATER

11.1.1 Forced Convection

Circular tubes

Unfortunately, satisfactory analytical methods have not yet been developed due to the difficulty in dealing with the steep property variations, especially in turbulent flows and at high heat fluxes. Therefore, empirical generalized correlations based on experimental data are used for HTC calculations at supercritical pressures.

McAdams (1942) proposed to use the Dittus and Boelter (1930) equation in the following form for forced convective heat transfer in turbulent flows and subcritical pressures (this statement is based on the recent study by Winterton (1998)):

$$\mathbf{Nu}_b = 0.0243\,\mathbf{Re}_b^{0.8}\mathbf{Pr}_b^{0.4}. \tag{11.1}$$

Later, Equation (11.1) was also used for supercritical heat transfer. According to Schnurr et al. (1976) Equation (11.1) shows good agreement with the experimental data of supercritical water flowing inside circular tubes at pressure of 31 MPa and low heat fluxes. However, Equation (11.1) may give unrealistic results in some flow conditions, especially, near the critical and pseudocritical points, because it is sensitive to properties variations. In general, this classical equation was used intensively as the base for modified supercritical heat-transfer correlations.

Bringer and Smith (1957) developed the following correlation for supercritical water up to p = 34.5 MPa and for carbon dioxide:

$$\mathbf{Nu}_x = C\,\mathbf{Re}_x^{0.77}\,\mathbf{Pr}_w^{0.55} \tag{11.2}$$

where $C = 0.0266$ for water, $C = 0.0375$ for carbon dioxide, and $\mathbf{Nu}_x$ and $\mathbf{Re}_x$ are evaluated at t_x. Temperature t_x is defined as t_b if $(t_{pc} - t_b)/(t_w - t_b) < 0$, as t_{pc} if $0 \le (t_{pc} - t_b)/(t_w - t_b) \le 1.0$ and as t_w if $(t_{pc} - t_b)/(t_w - t_b) > 1.0$. However, they did not account for the peak in thermal conductivity near the pseudocritical temperature.

Shitsman (1959, 1974), analysing the heat-transfer experimental data of supercritical water (Miropol'skiy and Shitsman 1957) flowing inside tubes, generalized these data with the Dittus-Boelter type correlation (the following correlation was proposed first by Miropol'skiy and Shitsman (1957, 1958a,b)):

$$\mathbf{Nu_b} = 0.023\,\mathbf{Re_b^{0.8}Pr_{min}^{0.8}} \tag{11.3}$$

where "min" means minimum **Pr** value, i.e., either the **Pr** value evaluated at the bulk-fluid temperature or the **Pr** value evaluated at the wall temperature, whichever is less. However, Shitsman, based on the knowledge at that time, assumed that thermal conductivity was a smoothly decreasing function of temperature near the critical and pseudocritical points.

Krasnoshchekov and Protopopov (1959, 1960) proposed (later, together with Petukhov (Petukhov et al. 1961)) the following correlation for forced convective heat transfer in water and carbon dioxide at supercritical pressures:

$$\mathbf{Nu} = \mathbf{Nu_0}\left(\frac{\mu_b}{\mu_w}\right)^{0.11}\left(\frac{k_b}{k_w}\right)^{-0.33}\left(\frac{\bar{c}_p}{c_{pb}}\right)^{0.35}, \tag{11.4}$$

where according to Petukhov and Kirillov (1958):

$$\mathbf{Nu_0} = \frac{\frac{\xi}{8}\mathbf{Re_b}\,\overline{\mathbf{Pr}}}{12.7\sqrt{\frac{\xi}{8}}\left(\overline{\mathbf{Pr}}^{\frac{2}{3}} - 1\right) + 1.07} \tag{11.5}$$

$$\text{and}\quad \xi = \frac{1}{(1.82\log_{10}\mathbf{Re_b} - 1.64)^2} \tag{11.6}$$

In effect, the **Pr** and c_p were averaged over the ranges to account for the thermophysical properties variations. The majority of their data (85%) and data of others (water data within $p = 22.3 - 32$ MPa obtained by Miropol'skiy and Shitsman (1957), data of Dickinson and Welch (1958), Petukhov and Kirillov (1958); and carbon dioxide data at $p = 8.3$ MPa obtained by Bringer and Smith (1957)) were generalized using Equation (11.4) and showed discrepancies within ±15%. Equation (11.4) is valid within the following ranges:

$2{\cdot}10^4 < \mathbf{Re_b} < 8.6{\cdot}10^5$, $0.85 < \overline{\mathbf{Pr}}_\mathbf{b} < 65$, $0.90 < \mu_b/\mu_w < 3.60$, $1.00 < k_b/k_w < 6.00$, and $0.07 < \bar{c}_p/c_{pb} < 4.50$.

Domin (1963) performed experiments with supercritical water flowing inside horizontal tubes ($D = 2$ mm and $L = 1.075$ m, and $D = 4$ mm and $L = 1.233$ m) and proposed the following correlations:

$$\mathbf{Nu_b} = 0.1\,\mathbf{Re_b^{0.66}Pr_b^{1.2}} \tag{11.7}$$

$$\mathbf{Nu_b} = 0.036\,\mathbf{Re_b^{0.8}Pr_b^{0.4}}\left(\frac{\mu_w}{\mu_b}\right). \tag{11.8}$$

Equation (11.7) is valid for $t_w \geq 350°C$ and Equation (11.8) – for $t_w = 250 - 350°C$. Both equations were obtained within the following range: $p = 233 - 263$ MPa and $q = 0.58 - 4.65$ MW/m^2.

Bishop et al. (1964) conducted experiments with supercritical water flowing upward inside tubes and annuli within the following range of flow and operating parameters: pressure 22.8 – 27.6 MPa, bulk-fluid temperature 282°C – 527°C, mass flux 651 – 3662 kg/m^{2}s and heat flux 0.31 – 3.46 MW/m^2. Their data for heat transfer in tubes were generalized using the following correlation, with a fit of ±15%:

$$\mathbf{Nu_x} = 0.0069\,\mathbf{Re_x^{0.9}}\,\overline{\mathbf{Pr}}_\mathbf{x}^{\,0.66}\left(\frac{\rho_w}{\rho_b}\right)_x^{0.43}\left(1 + 2.4\frac{D}{x}\right) \tag{11.9}$$

where x is the axial location along the heated length.

Swenson et al. (1965) investigated local forced-convection HTCs in supercritical water flowing inside smooth tubes. They found that, due to rapid changes in thermophysical properties of supercritical water near the pseudocritical point, conventional correlations did not work well. They recommended the following correlation:

$$\mathbf{Nu_w} = 0.00459\,\mathbf{Re_w^{0.923}}\,\overline{\mathbf{Pr}}_\mathbf{w}^{\,0.613}\left(\frac{\rho_w}{\rho_b}\right)^{0.231} \tag{11.10a}$$

$$\frac{h\,D}{k_w} = 0.00459\left(\frac{D\,G}{\mu_w}\right)^{0.923}\left(\frac{H_w - H_b}{T_w - T_b}\frac{\mu_w}{k_w}\right)^{0.613}\left(\frac{\rho_w}{\rho_b}\right)^{0.23}. \tag{11.10b}$$

Equation (11.10) was obtained within the following range: $p = 22.8 - 41.4$ MPa, $G = 542 - 2150$ kg/m^{2}s, $t_w = 93 - 649°C$, and $t_b = 75 - 576°C$; and re-produced the data to within ±15%. Also, this correlation predicted the data of carbon dioxide with good accuracy.

However, Swenson et al. assumed that thermal conductivity was a smoothly decreasing function of temperature near the critical and the pseudocritical points. According to their experimental data, the HTC in the pseudocritical region is strongly affected by heat flux. At low heat fluxes, the HTC had a sharp maximum near the pseudocritical temperature. At high heat fluxes, the HTC was much lower and did not have a sharp peak.

Krasnoshchekov et al. (1967) modified their original correlation for forced-convective heat transfer in water and carbon dioxide at supercritical pressures (see Equation (11.4)) to the following form:

$$\mathbf{Nu} = \mathbf{Nu_0} \left(\frac{\rho_w}{\rho_b} \right)^{0.3} \left(\frac{\overline{c}_p}{c_{pb}} \right)^n, \tag{11.11}$$

where $\mathbf{Nu}_0$ is defined in Equation (11.5). Exponent n is 0.4 at $T_w / T_{pc} \le 1$ or $T_b / T_{pc} \ge 1.2$; $n = n_1 = 0.22 + 0.18 T_w / T_{pc}$ at $1 \le T_w / T_{pc} \le 2.5$; and $n = n_1 + (5 \cdot n_1 - 2) \times (1 - T_b / T_{pc})$ at $1 \le T_b / T_{pc} \le 1.2$. Equation (11.11) is accurate within ±20% and is valid within the following range:

$8 \cdot 10^4 < \mathbf{Re}_b < 5 \cdot 10^5$, $0.85 < \overline{\mathbf{Pr}}_b < 65$, $0.90 < \rho_w / \rho_b < 1.0$, $0.02 < \overline{c}_p / c_{pb} < 4.0$, $0.9 < T_w / T_{pc} < 2.5$, $4.6 \cdot 10^4 < q < 2.6 \cdot 10^6$ (q is in W/m²) and $x / D \ge 15$. Later, Krasnoshchekov et al. (1971) added to Equation (11.11) a correction factor for the tube entrance region in the form of;

$$f\left(\frac{x}{D} \right) = 0.95 + 0.95 \left(\frac{x}{D} \right)^{0.8}. \tag{11.12}$$

Also, this correction factor can be used for a heated tube with abrupt inlet within $2 \le x / D \le 15$.

Kondrat'ev (1969) analyzed experimental data for heat transfer inside vertical (D = 12.02 mm, p = 22.8 – 30.4 MPa, and t_b = 260 – 560°C) and horizontal tubes (D = 7.62 mm, p = 25.2, 32.0 MPa, and t_b = 105 – 537°C) and inside vertical annular channels (D = 9.73/6.35 mm, p = 24.3 MPa, and t_b = 220 – 545°C) and proposed the following correlation:

$$\mathbf{Nu_b} = 0.020\, \mathbf{Re_b^{0.8}} \tag{11.13}$$

Equation (11.13) is valid within the range of $10^4 < \mathbf{Re} < 4 \cdot 10^5$ and t_b = 130 – 600°C. The majority of the experimental points agree with the correlation within ±10% range. However, Equation (11.13) is not valid within the pseudo-critical region.

Ornatsky et al. (1970) correlated experimental data for forced convection inside five parallel tubes at supercritical pressures with the following correlation:

$$\mathbf{Nu_b} = 0.023\, \mathbf{Re_b^{0.8}}\, \mathbf{Pr_{min}^{0.8}} \left(\frac{\rho_w}{\rho_b} \right)^{0.3} \tag{11.14}$$

where $\mathbf{Pr}_{min}$ is the minimum value of $\mathbf{Pr}_w$ or $\mathbf{Pr}_b$.

Khabenskii et al. (1971) compared several correlations with the available experimental data and proposed a new correlation based on thermodynamic analysis.

Yamagata et al. (1972) investigated forced-convective heat transfer to supercritical water flowing in tubes. They recommended the following correlation:

$$\mathbf{Nu_b} = 0.0135\, \mathbf{Re_b^{0.85}}\, \mathbf{Pr_b^{0.8}}\, F_c, \tag{11.15}$$

where $F_c = 1.0$ for $E > 1$, $F_c = 0.67\,\mathbf{Pr}_{\mathbf{pc}}^{-0.05}(\overline{c}_p / c_{pb})^{n_1}$ for $0 \le E \le 1$, $F_c = (\overline{c}_p / c_{pb})^{n_2}$ for $E < 0$, $E = (T_{pc} - T_b)/(T_w - T_b)$, $n_1 = -0.77\,(1 + 1/\mathbf{Pr}_{\mathbf{pc}}) + 1.49$ and $n_2 = -1.44\,(1 + 1/\mathbf{Pr}_{\mathbf{pc}}) - 0.53$.

Jackson and Fewster (1975) modified the correlation of Krasnoshchekov et al. (Equation (11.11)) to employ a Dittus-Boelter type form for $\mathbf{Nu}_0$ (Equation (11.1)). Finally, they obtained a correlation similar to that of Bishop et al. (1964) (see Equation (11.9)) without the effect of geometric parameters and with different values of constant and exponents:

$$\mathbf{Nu} = 0.0183\,\mathbf{Re}_{\mathbf{b}}^{0.82}\,\overline{\mathbf{Pr}}^{0.5}\left(\frac{\rho_w}{\rho_b}\right)^{0.3}. \qquad (11.16)$$

Hence, it can be expected that Jackson and Fewster correlation will follow closely a trend predicted by Bishop et al. correlation (Equation (11.9)).

Kakaç et al. (1987) presented the following correlations obtained in 1975 by J.D. Jackson, W.B. Hall, J. Fewster, A. Watson and M.J. Watts (University of Manchester, Manchester, UK), which define ranges for negligible buoyancy effects:

$$\frac{\overline{\mathbf{Gr}}_{\mathbf{b}}}{\mathbf{Re}_{\mathbf{b}}^{2.7}} < 10^{-5} \text{ for vertical tubes} \qquad (11.17a)$$

and

$$\frac{\overline{\mathbf{Gr}}_{\mathbf{b}}}{\mathbf{Re}_{\mathbf{b}}^{2}} < 10^{-3} \text{ for horizontal tubes.} \qquad (11.17b)$$

Alferov et al. (1975) proposed to calculate heat transfer with mixed convection based on the HTC ratio of forced and free convection.

The analysis performed by Petukhov et al. (1983) involved experimental data from a number of authors on heat transfer to water, carbon dioxide, and helium for the "normal heat- transfer regime" and yielded the following equation:

$$\mathbf{St} = \frac{\dfrac{\xi}{8}}{1 + \dfrac{900}{\mathbf{Re}} + 12.7\left(\dfrac{\xi}{8}\right)^{\frac{1}{2}}\left(\overline{\mathbf{Pr}}^{\frac{2}{3}} - 1\right)} \qquad (11.18)$$

where ξ is calculated according to $\dfrac{\xi}{\xi_0} = \left(\dfrac{\rho_w}{\rho_b}\right)^{0.4}\left(\dfrac{\mu_w}{\mu_b}\right)^{\frac{1}{5}}$ (Popov et al. (1978)).

Kirillov et al. (1990) showed that the role of free convection in heat transfer at the near-critical point can be taken into account through $\mathbf{Gr} / \mathbf{Re}^2$ or

$$k^* = \left(1 - \frac{\rho_w}{\rho_b}\right)\frac{\mathbf{Gr}}{\mathbf{Re}^2}. \qquad (11.20)$$

For $k^* < 0.4$ or $\mathbf{Gr} / \mathbf{Re}^2 < 0.6$, deteriorated heat transfer exists. At larger values of these terms, improved heat transfer occurs.

For the heating of a supercritical fluid flowing inside a circular tube at $q = const$ Kirillov et al. (1990) proposed to use the following equations:

Table 11.1. Values of $\varphi(k^*)$.

k^*	0.01	0.02	0.04	0.06	0.08	0.1	0.2	0.4
$\varphi(k^*)$	1	0.88	0.72	0.67	0.65	0.65	0.74	1

for $k^* < 0.01$,

$$\frac{\mathbf{Nu}}{\mathbf{Nu}_0} = \left(\frac{\bar{c}_p}{c_{pb}}\right)^n \left(\frac{\rho_w}{\rho_b}\right)^m \tag{11(.21)}$$

and for $k^* > 0.01$,

$$\frac{\mathbf{Nu}}{\mathbf{Nu}_0} = \left(\frac{\bar{c}_p}{c_{pb}}\right)^n \left(\frac{\rho_w}{\rho_b}\right)^m \varphi(k^*)\,, \tag{11.22}$$

where values of $\varphi((k^*)$ are listed in Table 11.1 or evaluated from Equation (11.25).

The local value of $\mathbf{Nu}_0$ for smooth circular tubes under turbulent flow can be calculated as follows:

$$\mathbf{Nu}_0 = \frac{\frac{\xi}{8}\,\mathbf{Re}\,\overline{\mathbf{Pr}}}{b + 4.5\,\xi^{0.5}\,(\overline{\mathbf{Pr}}^{\frac{2}{3}} - 1)}\,, \tag{11.23}$$

where T_b is the characteristic temperat$\underline{\text{ure}}$, $b = 1 + 900/\mathbf{Re}$ and $\xi = (1.82 \cdot \log_{10} \mathbf{Re} - 1.64)^{-2}$. Equation (11.23) is valid for $\mathbf{Pr} = 0.1 - 2000$ and $\mathbf{Re} = (4 - 5000)\cdot 10^3$, and has an error of about ±5%; for $\mathbf{Pr} = 0.1 - 2000$ and $\mathbf{Re} = (4 - 5000)\cdot 10^3$, the error is about ±10%.

Equation (11.23) is also valid for calculations of an average value of $\mathbf{Nu}_0$ for tubes with $x/D > 50$. For a more narrow range ($\mathbf{Pr} = 0.7 - 2$ and $\mathbf{Re} = (100 - 1000)\cdot 10^3$), the value of $\mathbf{Nu}_0$ can be found using:

$$\mathbf{Nu}_0 = 0.023\,\mathbf{Re}^{0.8}\,\overline{\mathbf{Pr}}^{0.4}\,C_t\,, \tag{11.24}$$

where T_b is the characteristic temperature and C_t is the correction factor for non-isothermity of flow: for $\mu_w/\mu_b = 0.08 - 40$; $\mathbf{Re} = 10^4 - 1.25 \cdot 10^5$ and $\mathbf{Pr} = 2 - 140$ — $C_t = (\mu_b/\mu_w)^n$, where $n = 0.11$ for heating fluid and $n = 0.25$ for cooling fluid.

Parameter $\varphi(k^*)$ can be calculated through the following equation (developed by the authors):

$$\varphi(k^*) = 0.79782686 - 1.6459037 \cdot \ln k^* - 2.7547316 \cdot (\ln k^*)^2 - 1.7422714 \cdot (\ln k^*)^3 - 0.54805506 \cdot (\ln k^*)^4 - 0.086914323 \cdot (\ln k^*)^5 - 0.0055187343 \cdot (\ln k^*)^6 \tag{11.25}$$

For $k^* > 0.4$ $\quad \varphi(k^*) = 1.4 \cdot (k^*)^{0.37}$.

At $k^* < 0.01$, Equation (11.21) can be used to calculate the deteriorated heat transfer for any value of k^*. A peak in wall temperature usually appears in the tube cross section, where the fluid temperature is lower than the pseudocritical temperature by several degrees. Possibly the deteriorated heat transfer at $k^* < 0.01$ is associated with the effects of acceleration and variability of physical properties over the flow cross-section in the process of turbulent transport. At $k^* = 0.01 - 0.4$, additional deterioration of heat transfer occurs due to the effect of natural convection. Maxima in wall temperature appear in the tube cross-section, where the average flow temperature is lower than the pseudocritical temperature by 15°C – 20°C or more. At $k^* = 0.4$, the heat transfer decreases when the effect of natural convection disappears, and the regime with improved heat transfer starts.

In Equations (11.21) and (11.22), $\mathbf{Nu}$ and $\mathbf{Nu_0}$ are calculated based on the average bulk temperature, and $\bar{c}_p = (H_w - H_b)/(T_w - T_b)$ is the integral average specific heat in the range of $(T_w - T_b)$. The exponent m is 0.4 for upward flow in vertical tubes; $m = 0.3$ for horizontal tubes.

Table 11.2. Values of exponent *n*,

Region	n
$\frac{T_w}{T_{pc}} < 1$ and $\frac{\bar{T}_b}{T_{pc}} > 1.2$	0.4
$\frac{T_w}{T_{pc}} > 1$ and $\frac{\bar{T}_b}{T_{pc}} < 1$	$0.22 + 0.18\frac{T_w}{T_{pc}}$
$\frac{T_w}{T_{pc}} > 1$ and $1 < \frac{\bar{T}_b}{T_{pc}} < 1.2$	$0.9\frac{\bar{T}_b}{T_{pc}}\left(1 - \frac{T_w}{T_{pc}}\right) + 1.08\frac{T_w}{T_{pc}} - 0.68$

For horizontal tubes, the exponent n is calculated using the ratios T_b/T_{pc} and T_w/T_{pc}, where all temperatures are in K. For downward flow in vertical tubes, the exponents m and n are calculated in the same way as those for horizontal tubes.

For upward flow in vertical tubes at $\bar{c}_p/c_{pb} \geq 1$, $n = 0.7$; for $\bar{c}_p/c_{pb} < 1$, the value of n is determined according to Table 11.2, the same as for horizontal tubes.

Equations (11.21) and (11.22) are valid for $\mathbf{Re} = (20 - 800)\cdot 10^3$, $\overline{\mathbf{Pr}} = 0.85 - 55$, $\rho_w/\rho_b = 0.09 - 1$, $\bar{c}_p/c_{pb} = 0.02 - 4$, $q = 0.023 - 2.6$ MW/m², $\pi = 1.01 - 1.33$, $T_b/T_{pc} = 1 - 1.2$, and $T_w/T_{pc} = 0.6 - 2.6$ (all temperatures are in K).

For the cooling of supercritical fluid flowing inside a circular tube at $q = const$ Kirillov et al. (1990) proposed the use of the following correlation:

$$\frac{\mathbf{Nu}}{\mathbf{Nu_0}} = \left(\frac{\bar{c}_p}{c_{pw}}\right)^n \left(\frac{\rho_w}{\rho_b}\right)^m \tag{11.26}$$

where $n = B\left(\frac{\bar{c}_p}{c_{pw}}\right)^s$, and coefficients m, B, s are listed in Table 11.3.

Table 11.3. Values of the coefficients *m*, *B* and *s*.

$\pi = \frac{p}{p_{cr}}$	1.06	1.08	1.15	1.22	1.35	1.63
m	0.3	0.38	0.54	0.61	0.68	0.8
B	0.68	0.75	0.85	0.91	0.97	1
s	0.21	0.18	0.104	0.066	0.04	0

Equation (11.26) is valid for **Re** = (90 – 450)·10³, q = 0.014 – 1.1 MW/m², $\pi = 1.06 - 1.63$, $T_b / T_{pc} = 1 - 1.2$, $T_b / T_{pc} = 0.95 - 1.5$, and $T_w / T_{pc} = 0.9 - 1.2$ (all temperatures are in K).

Gorban' et al. (1990) proposed to calculate the heat transfer to water and R-12 flowing inside circular tubes at temperatures higher than the critical temperature with the following correlations:

Water:

$$\mathbf{Nu_b} = 0.0059\,\mathbf{Re_b^{0.90}}\mathbf{Pr_b^{-0.12}} \tag{11.27}$$

R-12:

$$\mathbf{Nu_b} = 0.0094\,\mathbf{Re_b^{0.86}}\mathbf{Pr_b^{-0.15}} \tag{11.28}$$

Griem (1996) presented correlation for forced convection heat transfer at critical and supercritical pressures in tubes in the following form:

$$\mathbf{Nu_b} = 0.0169\,\mathbf{Re_b^{0.8356}}\mathbf{Pr_b^{0.432}}. \tag{11.29}$$

Equation (11.29) covers the entire enthalpy range, due to a new method for determining a representative specific heat capacity. Heat capacities were computed with semi-empirical equations at five reference temperatures. The two highest values closest to the critical or pseudocritical points were then sorted out. The average of the remaining three values represents a reasonable characteristic heat capacity.

Kitoh et al. (1999) proposed the HTC for forced convection in supercritical water, within the range of bulk temperature from 20°C to 550°C (bulk-fluid enthalpy from 100 to 3300 kJ/kg), mass flux from 100 to 1750 kg/m²s, and heat flux from 0 to 1.8 MW/m², to be:

$$\mathbf{Nu} = 0.015\,\mathbf{Re^{0.85}}\,\mathbf{Pr^{m}} \tag{11.30}$$

where $m = 0.69 - 81{,}000 / q_{dht} + f_c \cdot q$. The heat flux ($q_{dht}$) is that at which deteriorated heat transfer occurs (W/m²). This heat flux is calculated according to

$$q_{dht} = 200\,G^{1.2} \tag{11.31}$$

The coefficient f_c is calculated according to

$$f_c = \begin{cases} 29 \cdot 10^{-8} + \dfrac{0.11}{q_{dht}} & \text{for} \quad 0 \le H_b \le 1500 \text{ kJ/kg} \\ -8.7 \cdot 10^{-8} - \dfrac{0.65}{q_{dht}} & \text{for} \quad 1500 \le H_b \le 3300 \text{ kJ/kg} \\ -9.7 \cdot 10^{-7} + \dfrac{1.30}{q_{dht}} & \text{for} \quad 3300 \le H_b \le 4000 \text{ kJ/kg} \end{cases}$$

Jackson (2002) modified original correlation of Krasnoshchekov et al. (Equation (11.11)) for forced-convective heat transfer in water and carbon dioxide at supercritical pressures to employ the Dittus-Boelter type form for $\mathbf{Nu_0}$ (see Equation (11.1)). Finally, they obtained the following correlation:

$$\mathbf{Nu} = 0.0183\, \mathbf{Re_b^{0.82}}\, \mathbf{Pr_b^{0.5}} \left(\frac{\rho_w}{\rho_b}\right)^{0.3} \left(\frac{\bar{c}_p}{c_{pb}}\right)^{n}. \tag{11.32}$$

Exponent n is:

$n = 0.4$ for $T_b < T_w < T_{pc}$ and for $1.2 \cdot T_{pc} < T_b < T_w$;

$n = 0.4 + 0.2\left(\dfrac{T_w}{T_{pc}} - 1\right)$ for $T_b < T_{pc} < T_w$; and

$n = 0.4 + 0.2\left(\dfrac{T_w}{T_{pc}} - 1\right)\left[1 - 5\left(\dfrac{T_b}{T_{pc}} - 1\right)\right]$ for $T_{pc} < T_b < 1{\cdot}2 \cdot T_{pc}$; and $T_b < T_w$,

where T_b, T_{pc} and T_w are in K. Hence, it can be expected that the Jackson correlation will follow closely a trend predicted by the Krasnoshchekov et al. correlation (Equation (11.11)).

Grabezhnaya and Kirillov (2004) summarized their experience with using several well-known correlations for calculating supercritical water heat transfer in tubes and bundles. They found that the correlations of Bishop (Equation (11.9)) and Swenson (Equation (11.10)) can be used for preliminary heat transfer calculations in tubes. However, the correlation of Krasnoshchekov-Protopopov (Equation (11.4)) gave less accurate prediction.

Chen (2004) performed supercritical heat-transfer experiments in vertical and inclined smooth tubes with uniform and non-uniform radial heating and ribbed tubes and proposed several correlations to calculate HTC at these conditions.

For a smooth tube (D = 26 mm and L = 2 m) inclined on 20°C to horizontal plane:
(a) HTC on the tube upper part

$$\mathbf{Nu_b} = 0.00271\, \mathbf{Re_b^{0.93}}\, \overline{\mathbf{Pr}}^{\mathbf{0.88}} \left(\frac{\rho_w}{\rho_b}\right)^{0.52} \left(\frac{k_w}{k_b}\right)^{0.21}, \tag{11.33}$$

(b) HTC on the tube lower part

$$\mathbf{Nu_b} = 0.0784\, \mathbf{Re_b^{0.72}}\, \overline{\mathbf{Pr}}^{\mathbf{0.79}} \left(\frac{\rho_w}{\rho_b}\right)^{0.75} \left(\frac{k_w}{k_b}\right)^{0.1}. \tag{11.34}$$

where $\overline{\mathbf{Pr}} = \overline{\mathbf{Pr}}_b = \left(\frac{H_w - H_b}{T_w - T_b}\frac{\mu_b}{k_b}\right)$ at $t_b \leq t_{pc}$; and $\overline{\mathbf{Pr}} = \overline{\mathbf{Pr}}_w = \left(\frac{H_w - H_b}{T_w - T_b}\frac{\mu_w}{k_w}\right)$ at $t_b > t_{pc}$. Equations (11.33) and (11.34) were obtained within the following range: pressure of 23 – 30 MPa, mass flux of 600 – 1200 kg/m²s, and heat flux 200 – 600 kW/m². Experimental points have deviation from the correlations within ±7%.

For a vertical tube (D = 15.4 mm and L = 1 m) with helical internal ribs (for details, see Table 9.1):

(a) at $H_b < H_{pc}$

$$\mathbf{Nu} = 1.277\,\mathbf{Re}^{0.416}\left(\frac{H_w - H_b}{T_w - T_b}\frac{\mu_b}{k_w}\right)^{1.114}\left(\frac{\upsilon_b}{\upsilon_w}\right)^{0.458}. \tag{11.35}$$

(b) at $H_b > H_{pc}$

$$\mathbf{Nu} = 66.68\,\mathbf{Re}^{0.178}\left(\frac{H_w - H_b}{T_w - T_b}\frac{\mu_b}{k_w}\right)^{1.191}\left(\frac{\upsilon_b}{\upsilon_w}\right)^{1.706}. \tag{11.36}$$

Equations (11.35) and (11.36) are valid within the following range: pressure of 23 – 27 MPa, mass flux of 400 – 1800 kg/m²s, and heat flux 200 – 800 kW/m². Experimental points have deviation from the correlations within ±15% for Equation (11.35) and ±21% for Equation (11.36).

A new approach in calculating HTC was proposed by Löwenberg et al. (2005). They are developing a look-up table for heat transfer of supercritical water in vertical, smooth tubes. However, only first steps have been done in that direction.

Annuli

McAdams et al. (1950) conducted experiments in an annulus with internal heating. All their data were generalized with the following correlation:

$$\mathbf{Nu_f} = 0.0214\,\mathbf{Re_f^{0.8}}\,\mathbf{Pr_f^{0.33}}\left(1 + \frac{2.3}{\frac{L}{D_{hy}}}\right) \tag{11.37}$$

where D_{hy} is the hydraulic-equivalent diameter. All properties were evaluated at a film temperature of $t_f = (t_b + t_w)/2$. Equation (11.37) was obtained in the following range: D_{hy} = 3.32 mm, L/D_{hy} = 14.7 – 80.0, p = 0.8 – 24 MPa, G = 75 – 224 kg/m²s, t_w = 319 – 698°C, t_b = 221 – 544°C and h = 0.52 – 2 kW/m²K. The equation has a maximum error of 17%.

Bishop et al. (1964) conducted experiments with supercritical water flowing upward inside annuli and tubes and proposed a correlation to calculate heat transfer (for details, see Equation (11.9)).

Kondrat'ev (1969) generalized experimental data for heat transfer inside vertical annular channels and obtained a correlation for forced convective heat transfer (for details, see Equation (11.13)).

Ornatsky et al. (1972) performed experiments in an externally heated annulus (D_{rod} = 10.6 mm, gap 0.7 mm, and L = 0.28 m) cooled with water at p = 23.5 MPa and G = 2000 – 5000 kg/m²s. They generalized the experimental data with the Dittus-Boelter correlation (Equation (11.1)). It was found that Equation (11.1) is valid within the investigated range for $t_w^{\text{int}} < t_{pc}$.

Bundles

Dyadyakin and Popov (1977) performed experiments with a tight 7-rod bundle with helical fins cooled with supercritical water and they correlated their data for the local HTCs as:

$$\mathbf{Nu_x} = 0.021\,\mathbf{Re_x^{0.8}}\,\overline{\mathbf{Pr}}_x^{\,0.7}\left(\frac{\rho_w}{\rho_b}\right)_x^{0.45}\left(\frac{\mu_b}{\mu_{in}}\right)_x^{0.2}\left(\frac{\rho_b}{\rho_{in}}\right)_x^{0.1}\left(1+2.5\frac{D_{hy}}{x}\right) \tag{11.38}$$

where x is the axial location along the heated length in meters, and D_{hy} is the hydraulic-equivalent diameter (equals four times the flow area divided by wetted perimeter) in meters. This correlation fits the data (504 points) to within ±20%. The maximum deviation of the experimental data from the correlating curve corresponds to points with small temperature differences between the wall temperature and bulk temperature. Sixteen experimental points had deviations from the correlation within ±30%.

11.1.2 Comparison of Correlations

Jackson (2002) and previously, Jackson and Hall (1979) assessed the accuracy of the following correlations: Equation (11.32) of Jackson (2002); Equation (11.16) of Jackson and Fewster (1975); Equation (11.15) of Yamagata et al. (1972); Equation (11.14) of Ornatsky et al. (1970); Equation of Miropol'skii et al. (1970); Equation (11.11) of Krasnoshchekov et al. (1967); Equation (11.10) of Swenson et al. (1965); Equation (11.9) of Bishop et al. (1964) and Equation (11.3) of Shitsman (1959, 1974); based on the experimental data (2000 points) of water (75% of 2000 points) and carbon dioxide (25%). They found Equation (11.11) of Krasnoshchekov et al. (1967) and its modified version—Equation (11.32) of Jackson (2002)—to be the most accurate ones. Ninety seven percent of the experimental data were correlated with the accuracy of ±25%.

Figures 11.1 and 11.2 show a comparison between the selected correlations given in Section 11.1.1 and the experimental data of Shitsman (1963), respectively. The experimental data of Shitsman (1963) were mainly obtained within the range of q/G from 0.72 to 0.86, i.e., within a region of deteriorated heat transfer (usually this regime appears at $q/G \geq 0.4$ and only over some part of a heated length). Therefore, the experimental data were used as a reference in these figures.

The majority of the correlations seem to follow the general trend of the experimental data outside the regions of deteriorated or improved heat transfer (see Figures 11.1 and 11.2). However, simpler correlations of Dittus-Boelter (Equation (11.1)), Shitsman (Equation (11.3)), and Ornatsky et al. (Equation 11.14) are sensitive to the significant variations in thermophysical properties near the critical and pseudocritical points, which are beyond conditions in Figures 11.1 and 11.2.

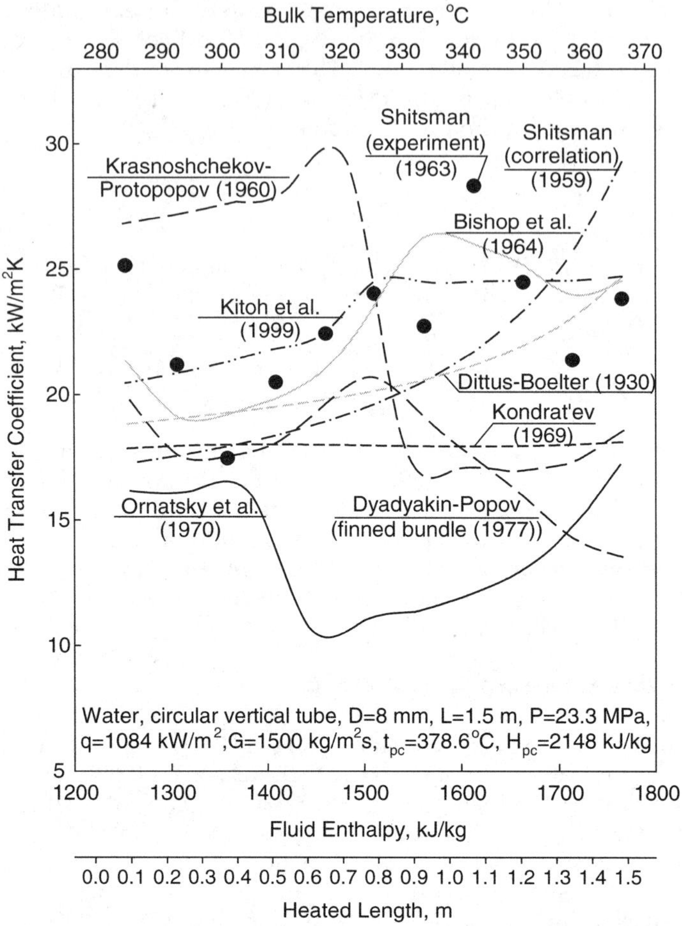

Figure 11.1. Comparison of various correlations with experimental data: Water, p=23.3 MPa, G=1500 kg/m²s, and q=1084 kW/m².

The correlations of Bringer and Smith (Equation (11.2)) (Figure 11.2) and Ornatsky et al. (Equation (11.14)) (Figure 11.1) show a significant deviation from the rest of correlations and experimental data in some flow conditions.

Deteriorated heat transfer may appear in the outlet or in the inlet section (see Figures 11.1 and 11.2). Also, several sections of the tube show improved heat transfer (see Figures 11.1 and 11.2).

Figures 11.3 and 11.4 show a comparison between the correlations and carbon dioxide experimental data recently obtained at Chalk River Laboratories, AECL. These data were obtained at $q / G = 0.096$, i.e., far below the deteriorated heat transfer regime. Only three correlations, Gorban' (Equation (11.27)), Dyadyakin and Popov (Equation (11.38)), and Bringer and Smith (Equation (11.2)), show results that are close to the experimental data.

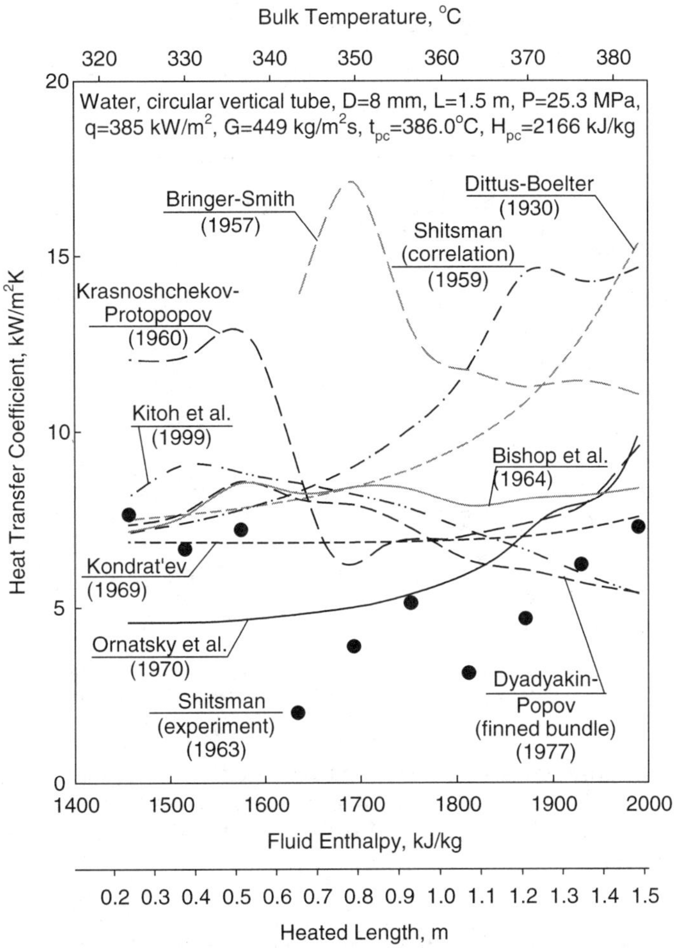

Figure 11.2. Comparison of various correlations with experimental data: Water, p=25.3 MPa, G=449 kg/m²s, and q=385 kW/m².

The general form of correlations (Equation (11.39)) for calculating heat transfer at supercritical pressures in water and other fluids is summarized in Table 11.4. In general, the correlations based on $\overline{c}_p$ instead of c_p have better agreement with the experimental data (see Figures 11.1 – 11.4), as would be expected.

$$\mathbf{Nu}_{\mathbf{t,x}} = C_1\, \mathbf{Re}_{\mathbf{t,x}}^{\mathbf{m_1}}\, \mathbf{Pr}_{\mathbf{t,x}}^{\mathbf{m_2}} \left(\frac{\rho_t}{\rho_t}\right)_x^{m_3} \left(\frac{\mu_t}{\mu_t}\right)_x^{m_4} \left(\frac{k_t}{k_t}\right)_x^{m_5} \left(\frac{\overline{c}_p}{c_{p,t}}\right)_x^{m_6} \left(1 + C_2\frac{D_{hy}}{L_h}\right)^{m_7} \qquad (11.39)$$

However, there is no consensus on the general trends in supercritical heat transfer, therefore, a more consistent experimental approach should be taken.

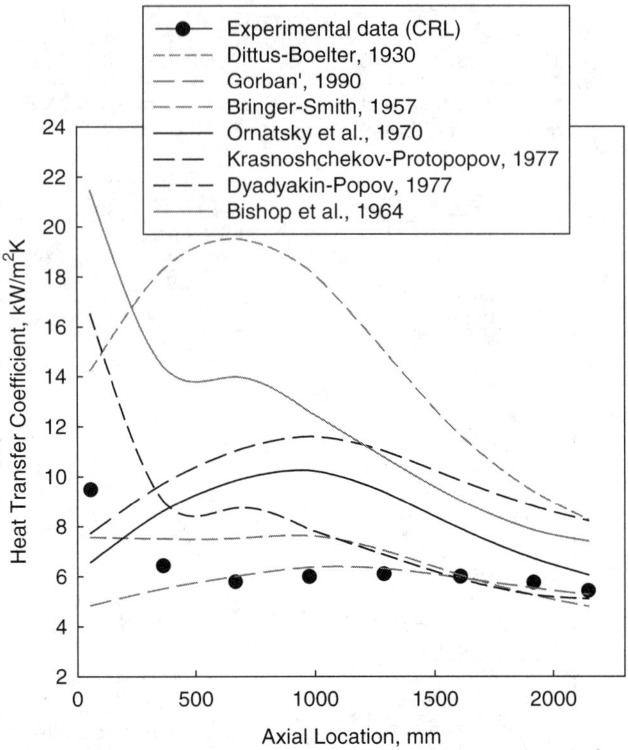

Figure 11.3. Comparison of various correlations with experimental data: Carbon dioxide, vertical tube (upward flow), D=8 mm, L=2.208 m, p=8.2 MPa (t_{pc}=35.8°C), t_{in}=33.4°C, t_{out}=41.5°C, G=1978 kg/m²s, and q_{ave}=189.2 kW/m².

11.1.3 Correlations for Determining Starting Point of Deteriorated Heat Transfer

The literature search turned up several correlations related to determining the starting point of deteriorated heat transfer.

Thus, Kondrat'ev (1971) proposed the following correlation to calculate the maximum heat flux at which deteriorated heat transfer occurs ($q_{HT}^{\max}$):

$$q_{HT}^{\max} = 5.815 \cdot 10^{-17}\,\mathbf{Re}_b^{1.7}\left(\frac{p}{0.101325}\right)^{4.5}, \qquad (11.40)$$

where $q_{HT}^{\max}$ is in kW/m² and p is in MPa. Equation (11.40) was stated as valid within the following range: p = 23.3 – 30.4 MPa, $\mathbf{Re}_b$ = (30 – 100) · 10³ and q = 116.3 – 1163 kW/m².

Protopopov et al. (1973) discussed the problems of heat transfer (mainly deterioration of heat transfer) in the supercritical region. As a result of their analysis of the experimental data for the sections with the deteriorated heat transfer, they proposed a non-dimensional number;

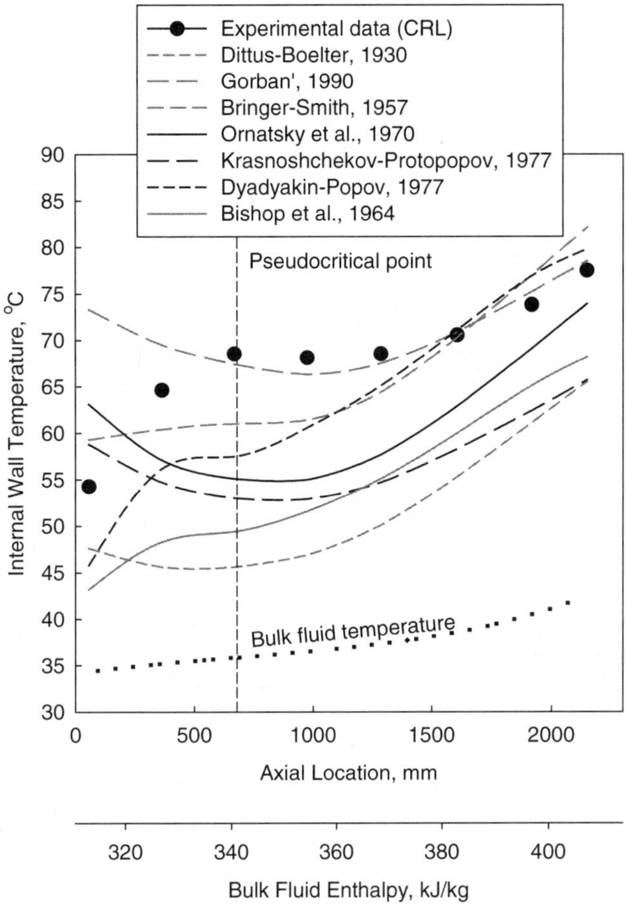

Figure 11.4. Comparison of various correlations with experimental data (temperature profiles along heated length): Carbon dioxide, vertical tube (upward flow), D=8 mm, L=2.208 m, p=8.2 MPa (t_{pc}=35.8°C), t_{in}=33.4°C, t_{out}=41.5°C, G=1978 kg/m²s, and q_{ave}=189.2 kW/m².

$$\mathbf{K} = \frac{C\,(\rho_b - \rho_w)^4 \rho_b\, \mathbf{Pr_b}}{\mu_b\, G \sqrt{\xi} \left(\dfrac{\mathbf{Gr_*}}{\mathbf{Re}^3} \right)^{0.23}}, \tag{11.41}$$

which approaches the value of $K_{cr} \approx 1.35{\cdot}10^4$ within the sections with the deteriorated heat transfer. In Equation (11.41), constant $C = 8{\cdot}10^{-14}$ for water and $C = 1{\cdot}10^{-14}$ for carbon dioxide, ρ_b and ρ_w are the fluid density at bulk and wall temperatures, respectively.

Protopopov and Silin (1973) proposed a correlation to calculate the starting point of deteriorated heat transfer in a tube with supercritical flow at $t_b < t_{pc} < t_w$. This method was based on the following correlation:

Table 11.4. Trends in correlating heat transfer data at supercritical pressures.

Reference	Flow Geometry	Characteristic parameters in Nu, Re and Pr: t, °C	Characteristic parameters in Nu, Re and Pr: Length	m_1	m_2	m_3	m_4	m_5	m_6	m_7
McAdams et al. 1950	Annulus	$t_f = (t_b + t_w) / 2$	D_{hy}	0.8	0.33	0	0	0	0	1
Bringer, Smith 1957	Tube	t_b, t_{pc} or t_w	D	00.77	0.55 t_w	0	0	0	0	0
Shitsman 1959, 1974	Tube	t_b	D	0.8	0.8 t_b or t_w	0	0	0	0	0
Krasnoshchekov, Protopopov 1959	Tube	t_b	D	~0.8	~0.33	0	0.11 t_b / t_w	−0.33 t_b / t_w	0.35	0
Swenson et al. 1965	Tube	t_w	D	0.923	0.613 based on $\bar{c}_p$	−0.231 t_b / t_w	0.231 t_b / t_w	0	0	0
Kondrat'ev 1969	Tube, annulus	t_b	D_{hy}	0.8	0	0	0	0	0	0
Ornatsky et al. 1970	Tube	t_b	D	0.8	0.8 t_b or t_w	−0.3 t_b / t_w	0		0	0
Ornatsky et al. 1972	Annulus	t_b	D_{hy}	0.8	0.4	0	0	0	0	0
Yamagata et al. 1972	Tube	t_b	D	0.85	0.8 and $\mathbf{Pr}^{\mathbf{n}}_{\mathbf{pc}}$	0	0	0	0 or n_1	0
Dyadyakin, Popov 1977	Bundle	t_b	D_{hy}	0.8	0.7 based on $\bar{c}_p$	−0.45 t_b / t_w and $(\rho_b / \rho_{in})^{0.1}_x$	0.2 t_b / t_{in}		0	1
Kirillov et al. 1990	Tube	t_b	D	~0.8	~0.33 or 0.4	$-n_1$	0	0	n_2	0
Gorban' et al. 1990	Tube	t_b	D	0.9	−0.12	0	0	0	0	0

$$\mathbf{Nu}_{\mathbf{t,x}} = C_1\, \mathbf{Re}^{\mathbf{m_1}}_{\mathbf{t,x}}\, \mathbf{Pr}^{\mathbf{m_2}}_{\mathbf{t,x}}\, (\rho_t/\rho_t)^{m_3}_x (\mu_t/\mu_t)^{m_4}_x (k_t/k_t)^{m_5}_x (\bar{c}_p/c_{p,t})^{m_6}_x (1 + C_2\, D_{hy}/L_h)^{m_7} \tag{11.39}$$

$$\left(\frac{q}{G}\right)_{dht} = \frac{1.3}{(t_{pc} - t_b)\, c_{pb} \left(\frac{\xi}{8}\right)\left(\frac{\upsilon_w}{\upsilon_{pc}}\right)^{1.3}}. \tag{11.42}$$

Petukhov and Polyakov (1974) reviewed several correlations for deteriorated heat transfer and proposed new correlations.

Kirillov et al. (1990) showed that the role of free convection in heat transfer at the near-critical point can be taken into account through $\mathbf{Gr} / \mathbf{Re}^2$ or:

$$k^* = \left(1 - \frac{\rho_w}{\rho_b}\right)\frac{\mathbf{Gr}}{\mathbf{Re}^2}, \tag{11.43}$$

where $\mathbf{Gr} = \frac{g\left(1 - \frac{\rho_w}{\rho_b}\right) D^3}{\upsilon_b^2}$ and ρ_b and ρ_w are the fluid densities at bulk and wall temperatures, respectively. For $k^* < 0.4$ or $\mathbf{Gr} / \mathbf{Re}^2 < 0.6$, deteriorated heat transfer exists. At larger values of these terms, improved heat transfer occurs.

Koshizuka and Oka (2000) found that the heat flux at which deteriorated heat transfer occurs is:

$$q_{dht} = 200\, G^{1.2} \tag{11.44}$$

This correlation is based on numerical studies only and can be used within the following range: G = 1000 – 1750 kg/m²s, H_b = 1000 – 3300 kJ/kg (t_b = 20 – 550°C) and q = 0 – 1.8 MW/m².

11.1.4 Preliminary Calculations of Heat Transfer at SCW CANDU Operating Conditions

To estimate the possible HTCs and sheath temperatures along a bundle string in the SCW CANDU reactor (Bushby et al. 2000), nine heat transfer correlations (eight correlations for circular tubes and one for bundles) were compared on the basis of HTC vs. bulk-fluid temperature (see Figure 11.5). Analysis of Figure 11.5 reveals that the following correlations for circular tubes and bundles by Gorban' et al. (Equation (11.27)), Kondrat'ev (Equation (11.13)), Krasnoshchekov and Protopopov (Equation (11.4)), Dyadyakin and Popov (Equation (11.38)), Bishop et al. (Equation (11.9)), Kitoh et al. (Equation (11.30)) show more or less similar results in terms of HTC values.

The correlation by Dyadyakin and Popov (Equation (11.38)) was obtained in a short finned bundle (heated length 0.5 m) and can be used only for that heated length. Therefore, this correlation was applied each 0.5 m along the heated length.

The six correlations mentioned above were used to calculate the wall (sheath) temperature (see Figure 11.6). The calculations showed that at the downstream end of a smooth bundle string (i.e., bundle string without appendages), the outside wall temperature would be less than 790°C.

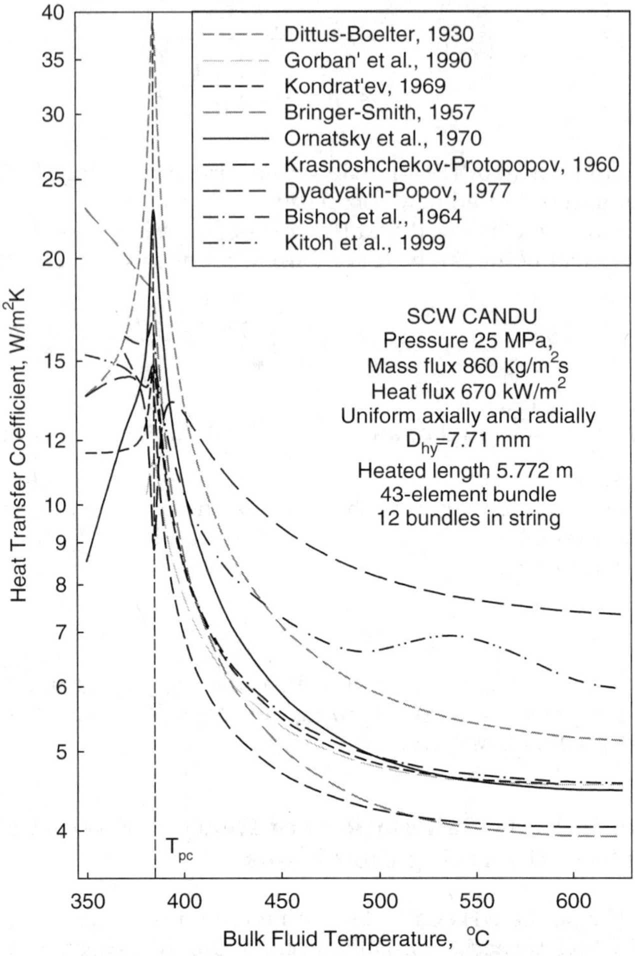

Figure 11.5. Calculated HTC along SCW CANDU reactor bundle string — water (for comparison, HTC in AGR cooled with subcritical carbon dioxide is about 1 kW/m²K (Hewitt and Collier 2000)).

This range of temperatures is not unique in nuclear reactors. For example, reactors cooled with liquid metals (particularly with sodium) have sheath temperatures of about 700°C – 750°C (Thermal and Nuclear Power Plants 1988), and AGRs cooled with subcritical carbon dioxide at a pressure of 4 MPa and an outlet temperature of 650°C have the maximum wall temperature (sheath made of stainless steel) also of about 750°C (Hewitt and Collier 2000).

Moreover, any bundle design usually contains some type of appendages (end plates, spacers, ribs, fins, bearing pads, etc.), which are heat-transfer enhancing devices; therefore, the maximum sheath temperature should be significantly less.

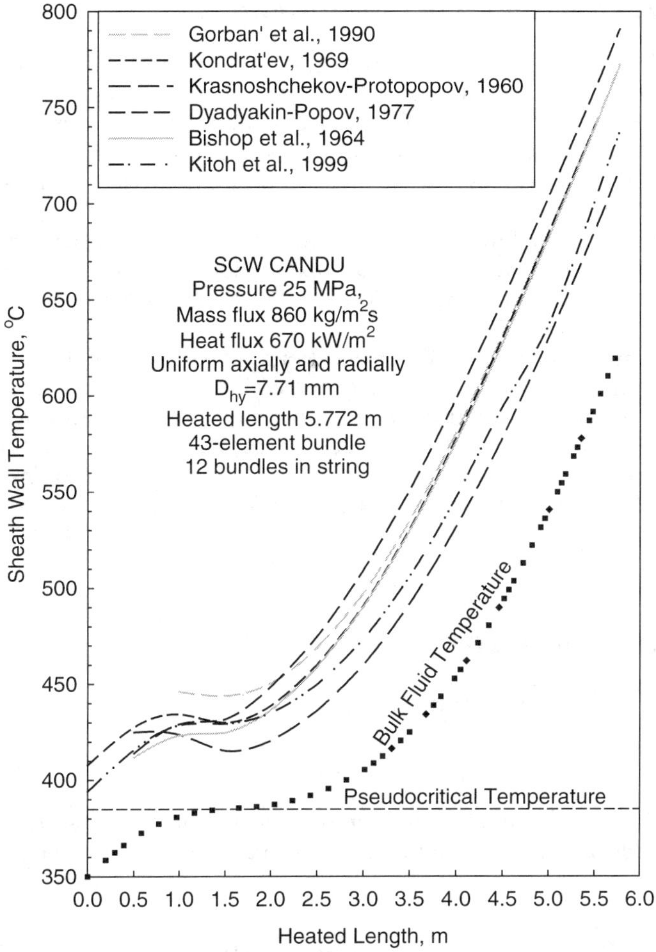

Figure 11.6. Calculated sheath temperatures along SCW CANDU reactor bundle string—water.

In general, there are several ways to enhance heat transfer and lower the wall temperature (and/or centerline fuel temperature) if required:

- By increasing the flow rate and decreasing the outlet fluid temperature.
- By decreasing the diameter of the fuel rod; using hollow fuel pellets (AGR design (Hewitt and Collier 2000)) or using concentric fuel rods (i.e., rods with hollow center cooled from outside and inside) to decrease centerline temperature.
- By introducing turbulence in the coolant flow to enhance heat transfer and to decrease wall temperature.

11.1.5 Final Remarks and Conclusions

1. The literature survey showed that, there are hundreds of publications devoted to the forced convective heat transfer of water at supercritical pressures. The majority of them are related to heat transfer in circular tubes. However, there is no consensus on the general trends in supercritical heat transfer, so a more consistent experimental approach should be taken.
2. There are very few publications that are devoted to heat transfer in bundles cooled with water at supercritical pressures. More work is needed to provide information for designing nuclear reactors cooled with supercritical water.
3. A comparison of various correlations for supercritical heat transfer showed that several correlations could be used for preliminary estimations of heat transfer in tubes and bundles. However, no one correlation is able to describe deteriorated heat transfer in tubes.
4. Preliminary calculations of HTC and temperature profiles in SCW CANDU reactor showed that the proposed concept is feasible for future development.

11.2 CARBON DIOXIDE

11.2.1 Forced Convection

Vertical tubes

Bringer and Smith (1957) conducted experiments with supercritical CO_2 flowing inside a tube (for details, see Table 6.1) and correlated their data as follows:

$$\mathbf{Nu_x} = C\,\mathbf{Re_x^{0.77}}\mathbf{Pr_w^{0.55}} \tag{11.45}$$

where $C = 0.0375$ for carbon dioxide and "x" means that thermophysical properties were evaluated at T_x (°C). Temperature T_x is T_b, if $(T_{pc} - T_b)/(T_w - T_b) < 0$, T_x is T_w if $(T_{pc} - T_b)/(T_w - T_b) > 0$, and beyond this range $T_x = (T_{pc} - T_b)/(T_w - T_b)$. However, thermal conductivity was assumed a smoothly decreasing function with temperature near the critical and pseudocritical points (for details, see Chapter 2).

Shitsman (1959, 1974) analyzed the heat-transfer experimental data of supercritical carbon dioxide (Bringer and Smith (1957), also see Table 6.1) and other fluids flowing inside tubes and proposed a correlation (for details, see Equation (11.3)).

Krasnoshchekov and Protopopov (1959, 1960) and, later on, together with Petukhov (Petukhov et al. 1961) proposed a general correlation for forced-convective heat transfer in carbon dioxide and water at supercritical pressures (for details see Equation (11.4)).

Ikryannikov et al. (1973) proposed the generalized correlation for calculating heat transfer in the near-critical region in a case of the combined effect of forced and free convection in a vertical tube with upward and downward flow of carbon dioxide and other fluids.

Petukhov et al. (1983) obtained a correlation to calculate heat transfer to carbon dioxide and other fluids for the "normal regime" (for details, see Equation (11.18)).

Kurganov and Zeigarnik (2005) analyzed experimental data on the hydraulic resistance and the structure of the supercritical pressure fluid flow at regimes of normal and deteriorated heat transfer and proposed explanations on possible mechanisms of heat transfer deterioration. This work is a continuation of the previous studies by Kurganov (1991; 1998a,b).

Yang and Khartabil (2005) developed correlations for normal and deteriorated heat transfer based on the Petukhov and Kirillov (1958) correlation (11.5). The following correlation for normal heat transfer was derived using 1416 CO_2 data points (for details, see Table 6.1 and Figures 6.11 – 6.13).

$$\mathbf{Nu_b} = 3.4445\left(\frac{p}{p_{cr}}\right)^{-0.4493}\left(\frac{T_b}{T_{pc}}\right)^{1.9737}\left(10000\frac{q}{G\,H_b}\right)^{0.2533}\left(\mathbf{Nu_0}\right)^{1.0649} \times\left(\frac{\mu_b}{\mu_w}\right)^{-1.4362}\left(\frac{k_b}{k_w}\right)^{0.3879}\left(\frac{\overline{c}_p}{c_{pb}}\right)^{0.9771} \tag{11.46}$$

where $\mathbf{Nu_0}$ is the reference Nusselt number calculated through Equation (11.5). The data were correlated with an average error of –0.2% and an RMS error of 11.7%. The ranges of dimensionless parameters in the correlation are: $p\,/\,p_{cr} = 1.03 - 1.2$, $T_b\,/\,T_{pc} = 0.95 - 1.22$, $10{,}000\,q/\,G,\ H_b = 0.98 - 5.78$, $\mathbf{Nu_0} = 79.28 - 1281.55$, $\mu_b/\mu_w = 0.92 - 3.47$, $k_b\,/\,k_w = 0.92 - 4.25$, and $\overline{c}_p\,/\,c_{pb} = 0.05 - 2.22$.

For deteriorated heat transfer, only data with values below the threshold value were used. (The threshold value at p = 8.37 MPa and G = 1180 – 2000 kg/m²s is HTC ≈ 2.5 kW/m²K (for details, see Figure 11.7). However, the authors have not provided threshold values at other conditions.) The resulting correlation for the deteriorated heat transfer, which was based on 1172 points, is

$$\mathbf{Nu_b} = 9.3886\left(\frac{p}{p_c}\right)^{-0.5196}\left(\frac{T_b}{T_{pc}}\right)^{2.5939}\left(10{,}000\frac{q}{G\,H_b}\right)^{-0.2858}\left(\mathrm{Nu_0}\right)^{0.9467} \times\left(\frac{\mu_b}{\mu_w}\right)^{0.2244}\left(\frac{k_b}{k_w}\right)^{-0.4083}\left(\frac{\overline{c}_p}{c_{pb}}\right)^{0.764} \tag{11.47}$$

The correlation was fitted with an average error of –0.5 % and an RMS error of 6.7%. The ranges of dimensionless parameters in the correlation are $p\,/\,p_{cr} = 1.03 - 1.21$, $T_b\,/\,T_{pc} = 0.95 - 1.25$, $10{,}000\,q\,/\,G\,H = 3.49 - 9.67$, $\mathbf{Nu_0} = 74.73 - 1382.93$, $\mu_b\,/\,\mu_w = 0.82 - 3.36$, $k_b\,/\,k_w = 0.8 - 4.25$, and $\overline{c}_p\,/\,c_{pb} = 0.01 - 0.92$.

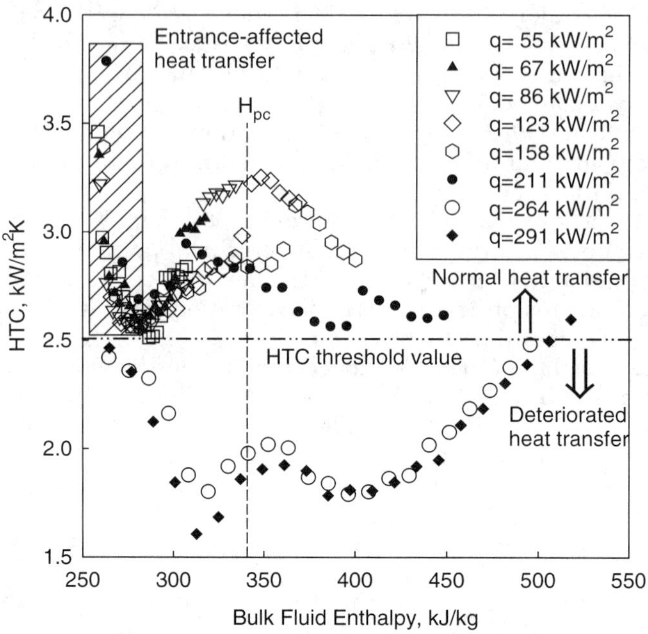

Figure 11.7. Variations in HTC of supercritical carbon dioxide with bulk-fluid enthalpy in 8-mm ID vertical tube with upflow (Yang and Khartabil 2005): *p* = 8.37 MPa and *G* = 1180 – 2000 kg/m²s.

Horizontal tubes

Schnurr (1969) generalized his data (see Table 6.3) obtained at supercritical carbon dioxide flow in a horizontal tube by the following correlation:

$$\mathbf{Nu}_z = 0.0266\,\mathbf{Re}_z^{0.77}\,\mathbf{Pr}_w^{0.55}. \tag{11.48}$$

However, it is not clear how to calculate the reference temperature T_z at which $\mathbf{Nu}_z$ and $\mathbf{Re}_z$ were evaluated.

Pitla et al. (2002; 2001a,b; 2000; 1998) proposed a new correlation on heat transfer for in-tube cooling of turbulent supercritical carbon dioxide based on their experiments and numerical calculations. This correlation is defined as the following:

$$\mathbf{Nu} = \left(\frac{\mathbf{Nu}_w + \mathbf{Nu}_b}{2}\right)\frac{k_w}{k_b}, \tag{11.49}$$

where Nusselt numbers and thermal conductivities are evaluated at wall and bulk-fluid temperatures, respectively. Nusselt numbers are calculated as follows (Gnielinski 1976):

$$\mathbf{Nu} = \frac{\frac{\xi}{8}(\mathbf{Re} - 1000)\,\mathbf{Pr}}{12.7\sqrt{\frac{\xi}{8}}\left(\mathbf{Pr}^{\frac{2}{3}} - 1\right) + 1.07}, \tag{11.50}$$

where the friction factor (Krasnoshchekov et al. 1970) is

$$\xi = \left[0.79 \ln(\mathbf{Re}) - 1.64\right]^{-2}. \tag{11.51}$$

11.2.2 Free Convection

Ghorbani-Tari and Ghajar (1985) proposed the following correlation to calculate the free-convective heat transfer in the near-critical region:

$$\mathbf{Nu} = a\left(\mathbf{Gr}\,\overline{\mathbf{Pr}}\right)^{b} \cdot \left(\frac{\rho_w}{\rho_b}\right)^{c} \left(\frac{\overline{c}_p}{c_{pb}}\right)^{d} \left(\frac{k_w}{k_b}\right)^{e} \left(\frac{\mu_w}{\mu_b}\right)^{f} \tag{11.52}$$

where a, b, c, d, e and f are the curve-fitted constants and

$$\mathrm{Gr} = \left(\frac{g\,D^3\,\rho_b^2}{\mu_b^2}\right)\left(\frac{\rho_b - \rho_w}{\rho_w}\right)$$

For CO_2 data of Dubrovina and Skripov (1967) (see also Table 6.4) within the range of $\mathbf{Ra}_b = 0.2 - 292$, the curve-fitted constants are $a = 1.03$, $b = 0.333$, $c = 10.07$, $d = 0.438$, $e = 0.561$ and $f = -5.6$, and the average deviation between the correlation and experimental data is 10.7%.

For the CO_2 data of Neumann and Hahne (1980) (see also Table 6.4) within the range of $\mathbf{Ra}_b = 13.1 - 1260$, the curve-fitted constants are $a = 0.717$, $b = 0.231$, $c = 0.404$, $d = 0.320$, $e = 0.245$ and $f = 0.007$, and the average deviation between the correlation and experimental data is 6.7%.

For CO_2 data of Hahne et al. (1974) (see also Table 6.4) within the range of $\mathbf{Ra}_b = 88.2 - 10200$, the curve-fitted constants are $a = 1.153$, $b = 0.187$, $c = 0.045$, $d = 0.132$, $e = 0.722$ and $f = -0.110$, and the average deviation between the correlation and experimental data is 7.7%.

For the CO_2 data of Protopopov and Sharma (1976) (see also Table 6.4) within the range of $\mathbf{Ra}_b = 4.66{\cdot}10^{11} - 9.02{\cdot}10^{12}$, the curve-fitted constants are $a = 0.024$, $b = 0.393$, $c = 1.213$, $d = 0.394$, $e = -0.316$ and $f = -0.314$, and the average deviation of the experimental data is 9.8%.

For CO_2 data of Beschastnov et al. (1973) (see also Table 6.4) within the range of $\mathbf{Ra}_b = 7.97{\cdot}10^{11} - 4.04{\cdot}10^{13}$, the curve-fitted constants are: $a = 0.103$, $b = 0.333$, $c = -2.0$, $d = 0.726$, $e = 0.52$ and $f = 1.23$, and the average deviation of the experimental data is 13.5%.

For water data of Fritsch and Grosh (1963) within the range of $\mathbf{Ra}_b = 8.88{\cdot}10^{6} - 4.45 \cdot 10^{8}$, the curve-fitted constants are $a = 0.15$, $b = 0.333$, $c = -0.533$, $d = 0.268$, $e = 0.455$ and $f = 2.24$, and the average deviation of the experimental data is 15.6%.

11.3 HELIUM

11.3.1 Forced Convection

Shitsman (1959, 1974) analyzed the heat-transfer experimental data of supercritical helium (Shitsman, 1974) and other fluids flowing inside tubes and proposed a general correlation (for details, see Equation (11.3)).

Ogata and Sato (1972) correlated their data for forced convection of helium in a tube (also see Table 7.1) with Equation (11.1). All thermophysical properties were evaluated at the bulk-fluid temperature. They noted that the use of Equation (11.1) was possible due to small variations in thermophysical properties of helium at 0.15 MPa. At that pressure, helium is closer to ordinary single-phase fluid rather than to supercritical fluid.

Yaskin et al. (1977) found that available data on heat transfer to supercritical helium in a purely forced-convection flow regime can be correlated on the basis of an analogy with the heat-transfer process accompanying gas injection through a heated wall. They proposed a correlation in the following form:

$$\frac{\mathbf{Nu}}{\mathbf{Nu_0}} = \left[1 - 0.2\,\frac{\mathbf{Nu}}{\mathbf{Nu_0}}\,\beta\,(T_w - T_b)\right]^2 \tag{11.53}$$

where $\mathbf{Nu_0}$ is calculated according to the Dittus-Boelter correlation (see Equation (11.1)).

Alad'ev et al. (1980) proposed to use nomograms for predicting HTC in the heated turbulent flow of supercritical helium I in narrow channels (tubes with an inside diameter of up to 2 mm). To predict the HTC in tubes at pressures of 0.25, 0.3, 0.4, 0.8 and 2 MPa, the authors derived the nomograms, which clearly reflect the behavior of the heat-transfer characteristics within a wide range of parameters. The nomograms correlated quantities independently determined in tests or given in calculations. Use of the nomograms therefore eliminates iterative calculations.

Yeroshenko and Yaskin (1981) analyzed the applicability of the following correlations for the Nusselt number: correlations by Miropol'skiy and Shitsman (1957), Krasnoshchekov and Protopopov (1966), Malyshev and Pron'ko (Pron'ko et al. 1976), Petukhov et al. (1976), and proposed the following correlation:

$$\mathbf{Nu} = 0.023\,\mathbf{Re_b^{0.8}}\,\mathbf{Pr_b^{0.4}}\left[\frac{2}{(0.8\,\psi + 0.2)^{0.5} + 1}\right]^2 F \tag{11.54}$$

The correction factor F accounts for possible heat-transfer enhancement; $(\mathbf{Nu}/\mathbf{Nu_0} > 1)$; $F = (\bar{c}_p / c_{pb})^{0.28}$ at $\bar{c}_p > c_{pb}$, and $F = 1$ at $\bar{c}_p \le c_{pb}$. The Nusselt number ($\mathbf{Nu_0}$) is calculated according to Equation (11.1).

The non-iteration prediction of the supercritical helium heat transfer from a given wall temperature can be performed for values of ψ (where $\psi = 1 + \beta_b\,(T_w - T_b)$ is the non-isothermality parameter) up to 32 ($\mathbf{Nu}/\mathbf{Nu_0} \sim 0.1$) using the authors' correlation, which predicts about 95% of the experimental values of **Nu** within ±20%. At $\psi > 2$, e.g., within the region of deteriorated heat transfer, the following equation must be used:

$$\mathbf{Nu} = \mathbf{Nu_0}\left[\frac{2}{(0.8\,\psi + 0.2)^{0.5} + 1}\right]^2 \tag{11.55}$$

Bogachev et al. (1983b) gave special attention to the conditions of heat-transfer increase in turbulent flow of helium where free-convection effect can be neglected.

The experiments were carried out in a vertical tube ($D = 1.8$ mm, $L_{tot} = 0.51$ m, and $L = 0.4$ m) with $q_w = const$. The range of investigated parameters was $p = 0.23 - 0.3$ MPa, $T_{in} = 4.21 - 4.24$ K $< T_{pc}$, $m = (0.19–0.26)\cdot 10^{-3}$ kg/s, and $q = 0.1 - 1.85$ kW/m^2. Local values of Reynolds number were $(3.6 - 9)\cdot 10^4$ and the parameter $\mathbf{Gr} / \mathbf{Re}^2 < 10^{-2}$, which allowed the consideration of these flow regimes as regimes without the effect of natural convection (**Gr** is the Grashof number $g\,\beta\,(1 - \rho_w / \rho_b)\, D^3 / \upsilon_b^2$. The values for ($\mathbf{Nu} / \mathbf{Nu}_0 > 1$) were described with an accuracy of about $\pm 20\%$ by theProtopopov correlation:

$$\frac{\mathbf{Nu}}{\mathbf{Nu}_0} = \left(\frac{\bar{c}_p}{c_{p\,b}}\right)^{0.35}. \tag{11.56}$$

Equation (11.56) is similar to the well-known power expressions for water and carbon dioxide (Jackson and Hall 1979a,b). The character of the heat-transfer change in terms of quality was similar to the behaviour of water and carbon dioxide (Polyakov 1991).

Petukhov et al. (1983) obtained a correlation to calculate heat transfer to helium and other fluids at the "normal regime" (for details, see Equation (11.18)).

Bogachev et al. (1985) verified several correlations for heat transfer to supercritical helium and came up with some suggestions on how to calculate heat transfer in upward and downward flows of supercritical helium in vertical tubes.

Bogachev and Eroshenko (1986), based on the experimental data for supercritical-pressure helium, verified the validity of a number of equations for mixed convective heat transfer in vertical tubes. These equations can be used for calculations of heat transfer for water and carbon dioxide.

11.3.2 Free Convection

Popov and Yankov (1985) used the procedure of numerical simulation for obtaining the results of the calculation of turbulent free convectionof helium within a wide range of parameters. The proposed equation is as follows:

$$\mathbf{Nu}_{\mathbf{x}} = 0.12\,\mathbf{Ra}_{\mathbf{x}}^{\frac{1}{3}} \left(\frac{\bar{c}_p}{c_{p\,b}}\right)^{n} \left(\frac{\rho_w}{\rho_b}\right)^{0.15} \tag{11.57}$$

where $n = 1$ for $T_b > T_{pc}$, and $n = 0.5$ for $T_b \le T_{pc}$.

11.4 OTHER FLUIDS

11.4.1 Forced convection

Shitsman (1959, 1974), analyzing the heat transfer experimental data of supercritical oxygen (Powel 1957) and other fluids, proposed a general equation (for details, see Equation (11.3)).

Hendrics et al. (1962) conducted experiments with low temperature hydrogen (83.3 – 138.9 K) above the critical pressure. For near-critical pressure heat transfer they used the correlation approach similar to that of hydrogen film boiling.

Szetela (1962) analyzed heat-transfer data of hydrogen near its critical point using equations of Deissler and Hsu. He found that in some cases, predicted results were up to 33% off compared to experimental data.

Melik-Pashaev (1966) presented a calculation of convective heat transfer at supercritical pressure in a stabilized turbulent flow of chemically homogeneous liquid in circular tubes.

Gorban' et al. (1990) proposed to calculate the heat transfer of Freon-12 flowing inside a circular tube at temperatures higher than the critical temperature with the following correlation:

$$\mathbf{Nu_b} = 0.0094\,\mathbf{Re_b^{0.86}Pr_b^{-0.15}} \tag{11.58}$$

Kalbaliev et al. (2002) studied heat transfer of toluene flowing in coils with ID of 4 and 7.6 mm at supercritical pressures. They developed several correlations for predicting the HTC inside coils.

Verdiev (2002) and Verdiev et al. (2001) performed heat-transfer experiments in toluene and benzene at supercritical pressures. Specific of these experiments was application of high frequency thermoacoustic instabilities. They proposed several correlations to calculate the HTC at these conditions.

Komita et al. (2003) investigated heat transfer to supercritical R-22 in a loop. They used 4.4- and 13-mm ID Inconel tubes with a heated length of 2 m. They found that: (a) the HTC for normal heat transfer increased with the diameter decrease; (b) the deteriorated HTC for a small diameter tube is the same as for the larger diameter tube; and (c) for the normal heat transfer the Watt's correlation showed the best accuracy.

Rzaev et al. (2003) and Kelbaliev et al. (2002) investigated methods for predicting the deteriorated heat-transfer regimes of supercritical toluene and water upward and downward flows in a vertical tube.

11.4.2 Free convection

Popov and Yankov (1982) calculated heat transfer in a laminar natural convection near a vertical plate for nitrogen and other fluids in the supercritical region for boundary conditions $T_w = const$ and $q_w = const$.

11.5 FLUID-TO-FLUID MODELING AT SUPERCRITICAL CONDITIONS

In some cases, when modeling fluid (usually carbon dioxide, Freons, etc.) is used instead of a primary coolant (usually water) it is important to scale properly the equivalent conditions of the primary coolant to the equivalent conditions of the modeling fluid. Therefore, fluid-to-fluid modeling techniques or scaling laws should be used.

Jackson and Hall (1979a) proposed about 12 non-dimensional groups to satisfy the complete requirements for similarity between two systems, A and B, at supercritical pressures. However, they stated that it is unlikely all these similarities can be satisfied. Nevertheless, the basic similarities listed in Table 11.5 can be used for fluid-to-fluid modelling at supercritical conditions.

Gorban' et al. (1990) developed a fluid-to-fluid modelling technique for supercritical pressures to scale water-equivalent conditions into Freon-12-equivalent conditions and vice versa.

Another approach can be taken for scaling bulk fluid temperature based on:

$$\left(\frac{\rho_b}{\rho_{cr}}\right)_{at\ T_b\ A} = \left(\frac{\rho_b}{\rho_{cr}}\right)_{at\ T_b\ B}. \tag{11.66}$$

Table 11.5. Basic similarities for fluid-to-fluid modeling at supercritical conditions based on inlet conditions approach.

Similarity Criteria	Equation	
Geometric Similarity	$\left(\frac{x}{D}\right)_A = \left(\frac{x}{D}\right)_B$	(11.59)
Pressure	$\left(\frac{p}{p_{cr}}\right)_A = \left(\frac{p}{p_{cr}}\right)_B$	(11.60)
Bulk Fluid Temperature (all temperatures in K)	$\left(\frac{T_b}{T_{cr}}\right)_A = \left(\frac{T_b}{T_{cr}}\right)_B$	(11.61)
Heat Flux or Wall Superheat	$\left(\frac{q\,D}{k_b\,T_b}\right)_A = \left(\frac{q\,D}{k_b\,T_b}\right)_B$ or	(11.62)
	$\left(\frac{T_w - T_b}{T_{cr}}\right)_A = \left(\frac{T_w - T_b}{T_{cr}}\right)_B$	(11.63)
Mass Flux	$\left(\frac{G\,D}{\mu_b}\right)_A = \left(\frac{G\,D}{\mu_b}\right)_B$	(11.64)
Heat Transfer Coefficient	$\mathbf{Nu}_A = \mathbf{Nu}_B$	(11.65)

11.6 CONCLUSION

Scaling parameters should be used with caution. In general, they can be used for scaling operating conditions from one fluid to another just for reference purposes. Due to scaling parameters simplicity, special behavior of thermophysical properties at supercritical pressures and complexity of the processes involved, some discontinuities may exist (for details, see Section 10.3.1).

Chapter 12

HYDRAULIC RESISTANCE

12.1 GENERAL CORRELATION FOR TOTAL PRESSURE DROP

In general, the total pressure drop for forced convection flow inside a test section installed in the closed-loop system can be calculated according to the following expression:

$$\Delta p = \sum \Delta p_{fr} + \sum \Delta p_{\ell} + \sum \Delta p_{ac} + \sum \Delta p_{g}, \tag{12.1}$$

where Δp is the total pressure drop, Pa.

Δp_{fr} is the pressure drop due to frictional resistance (Pa), which defined as

$$\Delta p_{fr} = \left(\xi_{fr} \frac{L}{D} \frac{\rho u^2}{2} \right) = \left(\xi_{fr} \frac{L}{D} \frac{G^2}{2\rho} \right), \tag{12.2}$$

where ξ_{fr} is the frictional coefficient, which can be obtained from appropriate correlations for different flow geometries. For smooth circular tubes, ξ_{fr} is as follows (Filonenko 1954):

$$\xi_{fr} = \left(\frac{1}{\left(1.82 \log_{10} \mathbf{Re}_{\mathbf{b}} - 1.64\right)^2} \right). \tag{12.3}$$

Equation (12.3) is valid within a range of $\mathbf{Re} = 4{\cdot}10^3 - 10^{12}$.

Usually, thermophysical properties and the Reynolds number in Equations (12.2) and (12.3) are based on arithmetic average of inlet and outlet values.

Δp_{ℓ} is the pressure drop due to local flow obstruction (Pa), which is defined as;

$$\Delta p_{\ell} = \left(\xi_{\ell} \frac{\rho u^2}{2} \right) = \left(\xi_{\ell} \frac{G^2}{2\rho} \right), \tag{12.4}$$

where ξ_{ℓ} is the local resistance coefficient, which can be obtained from appropriate correlations for different flow obstructions.

Δp_{ac} is the pressure drop due to acceleration of flow (Pa) defined as:

$$\Delta p_{ac} = \left(\rho_{out} u_{out}^2 - \rho_{in} u_{in}^2 \right) = G^2 \left(\frac{1}{\rho_{out}} - \frac{1}{\rho_{in}} \right) \tag{12.5}$$

Δp_g is the pressure drop due to gravity (Pa) defined as:

$$\Delta p_g = \pm g \left(\frac{H_{out} \rho_{out} + H_{in} \rho_{in}}{H_{out} + H_{in}} \right) L \sin \theta. \tag{12.6}$$

where θ is the test-section inclination angle to the horizontal plane, sign "+" is for the upward flow and sign "–" is for the downward flow. The arithmetic average value of densities can be used only for short sections in the case of strongly non-linear dependency of the density versus temperature. Therefore, in long test sections at high heat fluxes and within the critical and pseudocritical regions the integral value of densities should be used (see Equation (12.7)).

In general, form of Equation (12.6) depends on a pressure-drop measuring scheme (Figure 12.1). Usually, pressure drop is measured with a

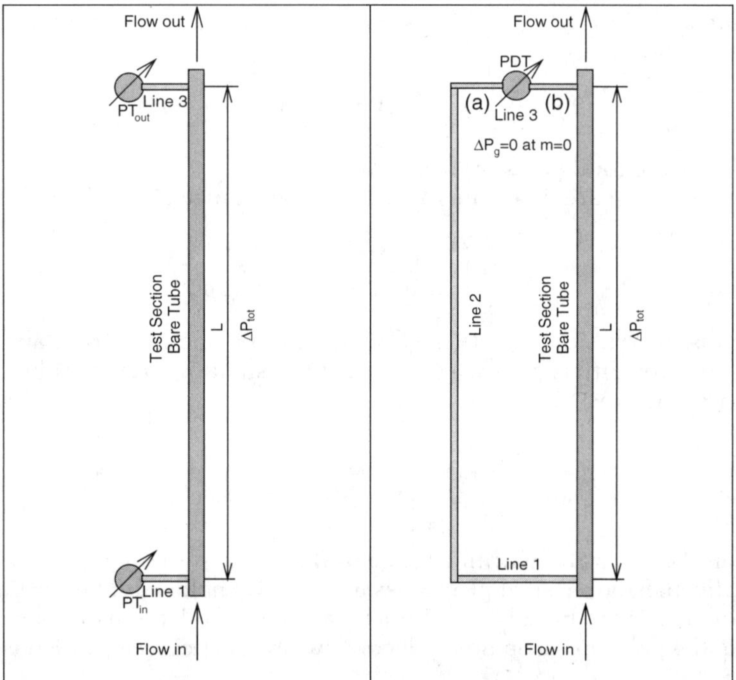

(a)	(b)
At $m = 0$ or at some flow and test section heating: $\Delta p_g = p_{in} - p_{out}$ and $\Delta p_g = g\, \rho_{TS}^{ave}\, L \sin \theta$.	At some flow and test section heating: $\Delta p_g = -g\, (\rho_2^{ave} - \rho_{TS}^{ave})\, L \sin \theta$.

Figure 12.1. Simplified schemes for pressure-drop measurements and corresponding equations for gravitational pressure-drop correction.

differential-pressure gauges or differential-pressure transmitters (Figure 12.1b). Therefore, Equation (Figure 12.1b) is more likely to be used in pressure-drop experiments.

Ornatskiy et al. (1980) and Razumovskiy (2003) proposed to calculate Δp_g at supercritical pressures as the following:

$$\Delta p_g = \pm g \left(\frac{H_{out}\,\rho_{out} + H_{in}\,\rho_{in}}{H_{out} + H_{in}} \right) L \sin\theta \qquad (12.7)$$

In general, Equation (12.1) is applicable for subcritical and supercritical pressures. However, adjustment of this expression to the conditions of supercritical pressures with single-phase dense gas and significant variations in thermophysical properties near the critical and pseudocritical points was the major task for the researchers and scientists.

In general, two major approaches to solve this problem were taken: analytical approach (including numerical approach) and experimental (empirical) approach.

Unfortunately, satisfactory analytical and numerical methods have not yet been developed due to the difficulty in dealing with the steep property variations, especially in turbulent flows and at high heat fluxes. Also, due to limited space it is impossible to present these usually lengthy methods in this monograph. However, for completeness, it was decided to list these references. The following references contain the latest findings in analytical and numerical approaches: Tanaka et al. (1973); Popov (1977); Popov et al. (1978); Petukhov and Medvetskaya (1978, 1979); Sinitsyn (1980); Popov (1983) and Popov and Petrov (1985).

12.2 EXPERIMENTS ON HYDRAULIC RESISTANCE OF WATER AT SUPERCRITICAL PRESSURES

All sources[24] of experimental data on hydraulic resistance of water flowing in tubes are listed in Table 12.1 and selected experimental results are shown in Figures 12.2 – 12.5.

In general, during experiments the total pressure drop was measured across a test section and after that, the components of the total pressure drop according to Equation (12.1) were determined. The vast majority of the test sections were smooth channels and mainly tubes. Therefore, only three components of the total pressure drop, i.e., pressure drop due to frictional resistance, pressure drop due to flow acceleration and pressure drop due to gravity can be considered.

However, in many studies, pressure drops due to acceleration of flow and due to gravity were neglected. In general, pressure drop due to flow acceleration is the most significant when critical or pseudocritical regions exist along the heated length of test section, i.e., when flow undergoes a significant variation in fluid density. Pressure drop due to gravity is the most significant at the

[24] "All" means all sources found by the authors from a total of 650 references dated mainly from 1950 till the beginning of 2006.

Table 12.1. Range of investigated parameters for hydraulic-resistance experiments with water flowing in tubes at supercritical pressures.

Reference	*p*, MPa	*t*, °C (*H* in kJ/kg)	*q*, MW/m²	*G*, kg/m²s	Flow Geometry
		Tubes (vertical)			
Tarasova and Leont'ev 1968	22.6–26.5	t_b or H_b were not provided	0.58–1.32	2000; 5000	SS tube (*D* = 3.34 mm, *L* = 0.134 m; *D* = 8.03 mm, *L* = 0.602 m) (selected data are shown in Figure 12.2)
Krasyakova et al. 1973	23; 25	H_b = 450–2400	0.2–1	500–3000	Vertical (*D* = 20 mm, *L* = 2.2; 7.73 m) and horizontal (*D* = 20 mm, *L* = 2.2; 4.2 m) SS tubes, upward and horizontal flows
Chakrygin et al. 1974	26.5	t_{in} = 220–300	*q* was not provided	445–1270	SS tube (*D* = 10 mm, *L* = 0.6 m), upward and downward flows
Ishigai et al. 1976	24.5; 29.5; 39.2	H_b = 220–800	0.14–1.4	500; 1000; 1500	Vertical (*D* = 3.92 mm, *L* = 0.63 m) and horizontal (*D* = 4.44 mm, *L* = 0.87 m) SS polished tubes
Razumovskiy 1984; Razumovskiy et al.1984, 1985	23.5	H_{in} = 1400; 1600; 1800	0.657–3.385	2190	Smooth tube (*D* = 6.28 mm, *L* = 1.44 m), upward flow (selected data are shown in Figure 12.5)
		Horizontal Tubes			
Kondrat'ev 1969	22.6; 24.5; 29.4	t_b = 105–540	0.12–1.2	**Re** = 10^5	Polished SS tube (*D* = 10.5 mm, *L* = 0.52 m) (selected data are shown in Figure 12.3)
Krasyakova et al. 1973	23; 25	t_b or H_b were not provided	0.2–1	500– 3000	SS tube (*D* = 20 mm, *L* = 2.2; 4.2 m)
Ishigai et al. 1976	24.5; 29.5; 39.2	H_b = 220–800	0.14–1.4	500; 1000; 1500	Horizontal (*D* = 4.44 mm, *L* = 0.87 m) and vertical (*D* = 3.92 mm, *L* = 0.63 m) SS polished tubes (selected data are shown in Figure 12.4)

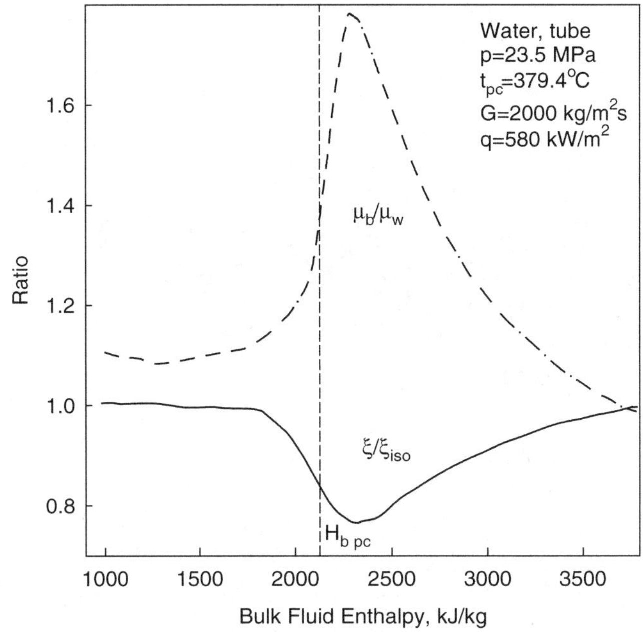

Figure 12.2. Effect of bulk-fluid enthalpy on ratios μ_b/μ_w and ξ/ξ_{iso} (Tarasova and Leont'ev 1968).

vertical orientation of the test section and equals to 0 at the horizontal layout. Neglecting these two pressure-drop components may lead to exaggeration of the frictional pressure drop or friction factor.

For frictional pressure drop (Equation (12.2)), the main objective was to calculate the frictional resistance coefficient and to compare its value with that of isothermal flow.

Results of Tarasova and Leont'ev (1968) (for details, see Table 12.1) are shown in Figure 12.2. They found that a decrease in the value of ξ_{exp}/ξ_{iso} coincides with an increase in the value of μ_b/μ_w. The most significant changes in values of these ratios correspond to the pseudocritical region (see Figure 12.2). Their findings are similar to those recorded at subcritical pressures. However, it seems that they did not account for pressure drop due to acceleration of the flow.

Kondrat'ev (1969) performed pressure-drop experiments in a horizontal tube within a wide range of flow conditions. His findings are shown in Figure 12.3. In his calculations of the frictional resistance coefficient, he accounted for pressure drop due to acceleration. The most significant variations in a value of ξ were within the pseudocritical region (see Figure 12.3).

The pressure-drop experimental data of Ishigai et al. (1976) are shown in Figure 12.4. They found that their experimental pressure-drop values

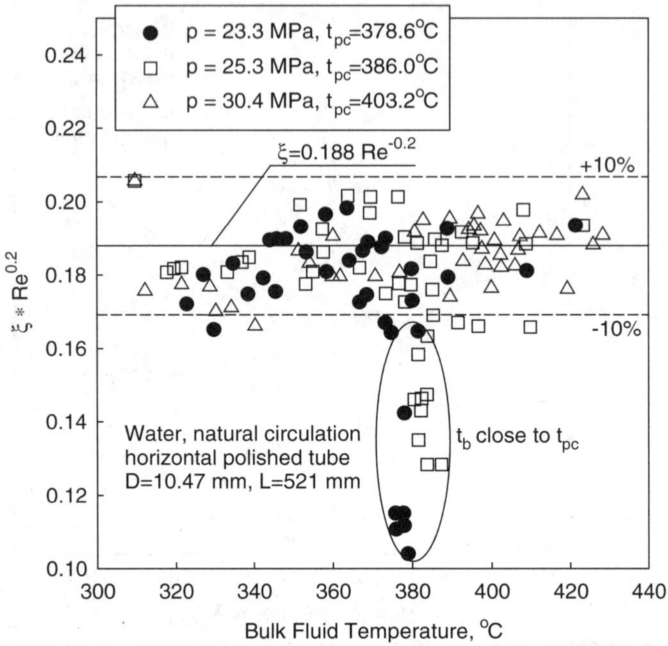

Figure 12.3. Effect of bulk-fluid temperature on frictional resistance coefficient (Kondrat'ev 1969).

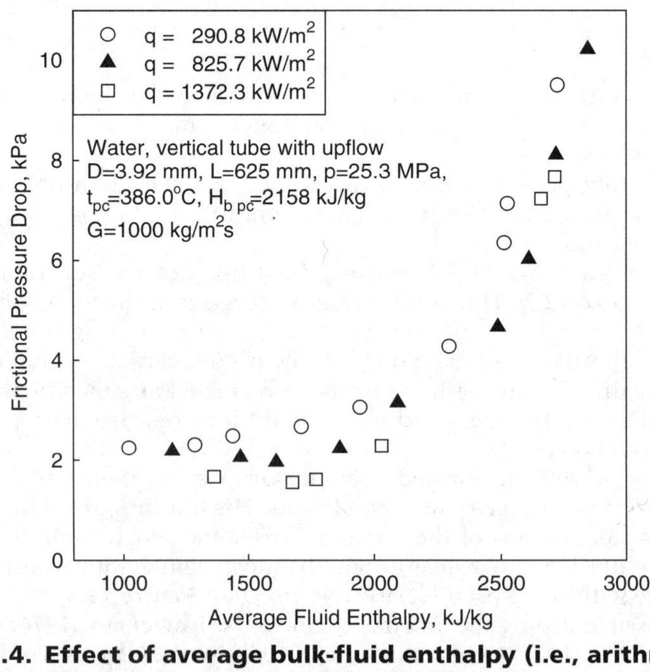

Figure 12.4. Effect of average bulk-fluid enthalpy (i.e., arithmetic average between inlet and outlet enthalpies) on frictional pressure drop (Ishigai et al. 1976).

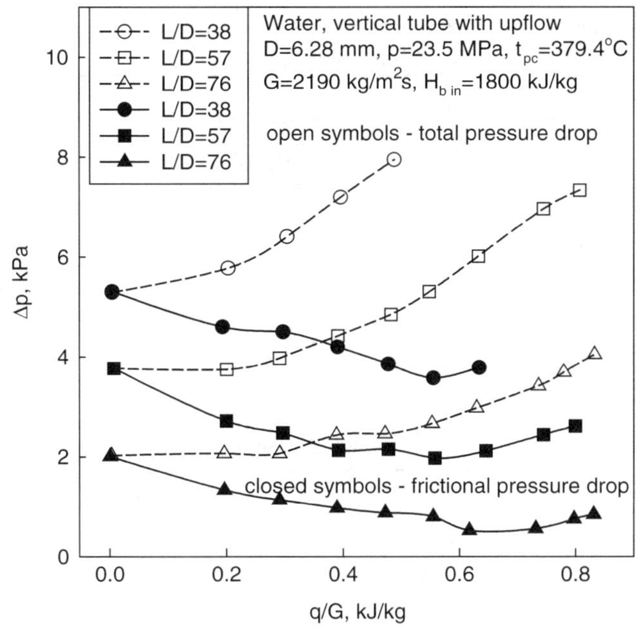

Figure 12.5. Total and frictional pressure drops vs. *q*/*G* ratio (Razumovskii et al. 1984).

(pressure drop due to acceleration of flow was accounted for) were lower than those calculated according to subcritical pressure-drop correlations.

Razumovskii et al. (1984) conducted experiments in vertical tubes within a wide range of flow conditions. Their findings are shown in Figure 12.5. In their calculations of the total pressure drop, they accounted for pressure drop due to acceleration of flow. They showed that for large values of q/G the pressure drop due to flow acceleration is a significant part of the total pressure drop.

12.3 EXPERIMENTS ON HYDRAULIC RESISTANCE OF CARBON DIOXIDE AT SUPERCRITICAL PRESSURES

All references with experimental data for hydraulic resistance of carbon dioxide are listed in Table 12.2.

Popov (1967) found (Figure 12.6) that his experimental data on frictional resistance coefficient are lower than calculated values according to the smooth-tube correlation for frictional resistance coefficient in isothermal flow and, therefore, a correction factor for non-isothermality of flow is needed.

Petukhov et al. (1980) obtained data for the frictional resistance in the case of flow within the region of maximum values of specific heat (i.e., pseudocritical region), and also for the local and average frictional and acceleration pressure drops in a heated horizontal tube at the normal and deteriorated

Table 12.2. Range of investigated parameters for hydraulic resistance experiments with carbon dioxide at supercritical pressures.

Reference	*P*, MPa	*t*, °C	*q*, kW/m²	*G*, kg/m²s	Flow Geometry
		Tubes (Vertical)			
Petukhov et al. 1983	7.7; 8.9	t_b = 0–80	384–1053	1000–4100	Vertical and horizontal SS tubes (*D* = 8 mm, *L* = 1.67 m), upward, downward and horizontal flows
Kurganov et al. 1986	9	t_{in} = 33–36	170–440	2100	SS tube, (*D* = 22.7 mm, *L* = 5.2 m), downward flow
		Horizontal Tubes			
Kuraeva Protopopov 1974	8; 10	19–88; t_w up to 500	up to 2500	1140–7400	SS tube (*D* = 4.1 mm, *L* = 0.21 m)
Petukhov et al. 1980	7.5–7.8; 8; 9	t_{in} = 18–20	870–1480	3270; 4130; 5230	SS tube (*D* = 8 mm, *L* = 1.8 m)
Petukhov et al. 1983	7.7; 8.9	t_{in} = 0–80	384–1053	1000–4100	Horizontal and vertical SS tubes (*D* = 8 mm, *L* = 1.67 m) , upward, downward and horizontal flows

heat-transfer regimes. It was shown that, in the deteriorated heat-transfer region, the acceleration pressure drop differed significantly from that determined on a basis of the one-dimensional flow model.

Later Petukhov et al. (1983) described a new procedure for measuring the hydraulic characteristics of fluid flows with variable properties. This procedure was used to obtain comprehensive experimental data on flow resistance associated with the heating of carbon dioxide at supercritical pressures in horizontal and vertical tubes, at the normal and deteriorated heat-transfer regimes. The experimental data concerning frictional and acceleration resistance values during the heating of CO_2 were generalized by empirical correlations.

Kurganov et al. (1986) proposed a method for the experimental investigation of heat transfer, friction, velocity and temperature fields in the heating of a turbulent flow of carbon dioxide at supercritical pressures in a vertical tube (*D* = 22.7 mm). Probe measurements were performed in flow with Pitot microtubes and microthermocouples in several sections along the heated length. The experimental data from thermal, hydraulic, and probe measurements were used together with the integral equations of motion, energy, and continuity, in order to determine drag and frictional resistance coefficients and the distributions of shear stresses, radial heat flux, and radial mass flux in

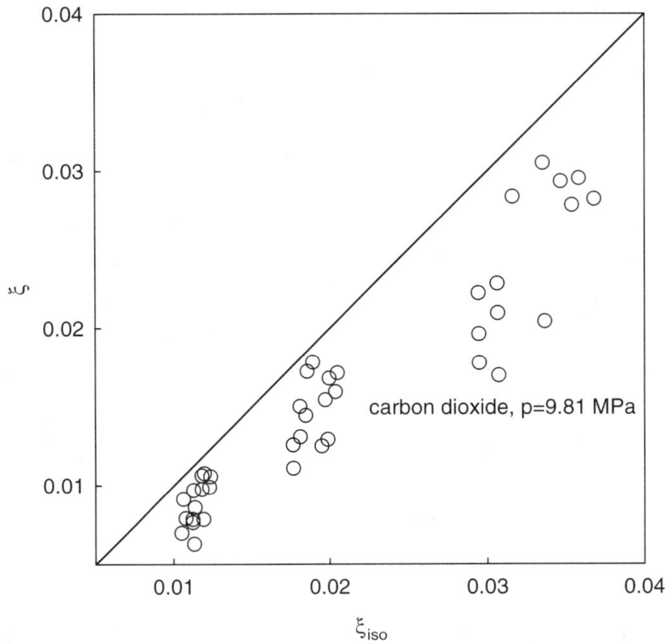

Figure 12.6. Results of calculation of frictional resistance coefficient (Popov 1967): Solid line – $\xi = \xi_{iso} f(\mathrm{Re})$.

different sections of the tube. The investigation was conducted for an upward flow of CO_2 with a mass flux of about 2100 kg/m²s, at the normal and deteriorated heat-transfer regimes. The relationship between the structure of the averaged flow and heat transfer was also discussed.

Kurganov et al. (1989) analyzed the experimental data of the total and frictional pressure drops associated with upward / downward flows of carbon dioxide at supercritical pressure in a heated tube. The momentum, kinetic energy, and density factors of the fluid flow—which were needed to determine the pressure drops along the tube length—were determined from the results of probe measurements of the velocity and temperature profiles.

At CRL, experiments (Pioro et al. 2004, 2003) were performed on measuring a pressure drop in a vertical smooth tube (for details, see Section 10.3). Selected results are shown in Figure 12.7 just for reference purposes. In these experiments, the local pressure drop due to obstructions along the heated length was 0 because of the smooth test section. Therefore, the measured pressure drop in the current experiments consists of three components (for details, see Figure 12.7):

$$\Delta p_{meas} = \Delta p_{fr} + \Delta p_{ac} + \Delta p_{g}. \tag{12.8}$$

The total pressure drop of the test section (PDT-9) was measured together with four local pressure drops (PDT-110 – PDT-113) along the heated length.

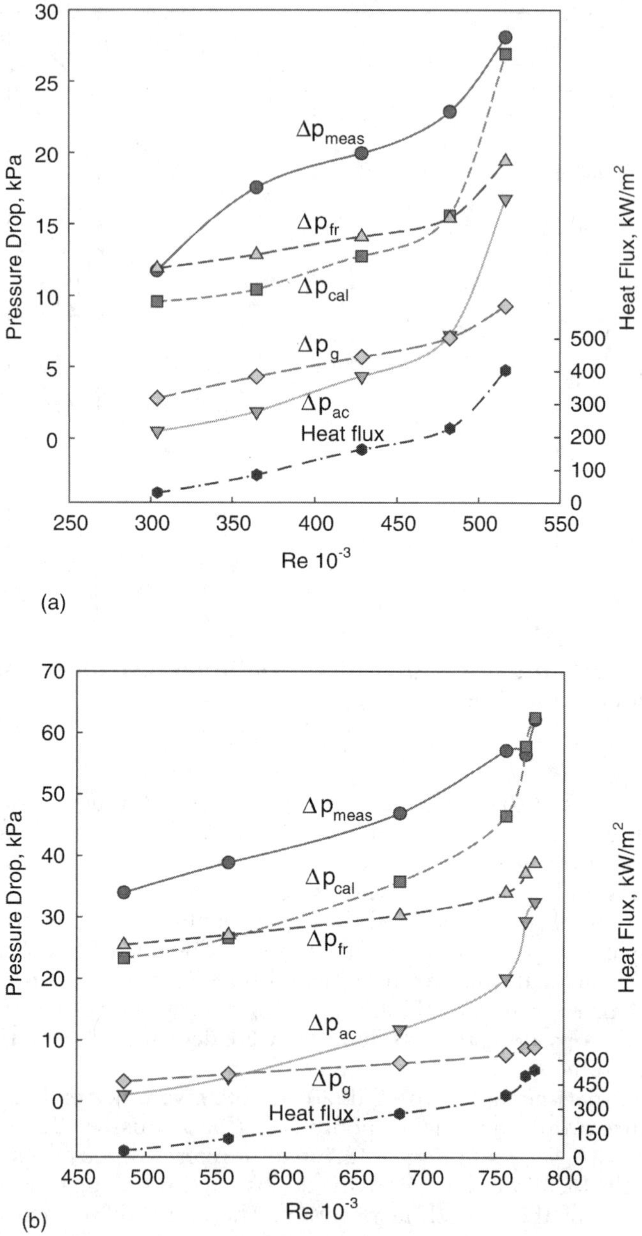

Figure 12.7. Effect of Reynolds number on total pressure drop (measured and calculated) and its components (calculated values) in supercritical carbon dioxide flowing in vertical circular tube: p_{out}=8.8 MPa; (a) G=2040 kg/m²s, t_{in}=32°C; and (b) G=3040 kg/m²s, t_{in}=31°C.

12.4 PRACTICAL PREDICTION METHODS FOR HYDRAULIC RESISTANCE AT SUPERCRITICAL PRESSURES

12.4.1 Tubes

Mikheev (1956) found that frictional resistance coefficient of non-isothermal flow of water and other fluids can be calculated as the following:

$$\xi_{fr} = \left(\frac{1}{\left(1.82 \log_{10} \mathbf{Re}_{\mathrm{b}} - 1.64\right)^2} \left(\frac{\mathbf{Pr}_{\mathrm{w}}}{\mathbf{Pr}_{\mathrm{b}}} \right)^{\frac{1}{3}} \right). \tag{12.9}$$

Equation (12.9) is valid for smooth tubes within the range of **Re** > 4000 (turbulent stabilized flow).

Chakrygin (1964, 1965) proposed a method to calculate pressure drop in heated tubes cooled with supercritical water (p = 23.3 – 35.5 MPa). The main idea of the proposed method was to use integral characteristics of thermophysical properties. Later, Chakrygin (1967) obtained correlations to estimate pressure losses with non-uniform heating at supercritical pressures.

Popov (1967) proposed to calculate frictional resistance coefficient for carbon dioxide with the following correlation:

$$\left(\frac{\xi}{\xi_{iso}} \right)_{fr} = \left(\frac{\rho_{b-w}^{ave}}{\rho_b} \right)^{0.74}, \tag{12.10}$$

where ρ_{b-w}^{ave} is the average density within a temperature range from t_w to t_b. Equation (12.10) has uncertainty of ±5%. With a higher degree of uncertainty (±10%) his data can be approximated with the following correlation:

$$\left(\frac{\xi}{\xi_{iso}} \right)_{fr} = \left(\frac{\rho_w}{\rho_b} \right)^{0.74}, \tag{12.11}$$

However, the range of validity for Equations (12.10) and (12.11) was not clearly specified in the paper.

Tarasova and Leont'ev (1968) found that

$$\left(\frac{\xi}{\xi_{iso}} \right)_{fr} = \left(\frac{\mu_w}{\mu_b} \right)^{0.22}. \tag{12.12}$$

Equation (11) was obtained with a deviation of ±5% between the experimental points and fitting.

Kondrat'ev (1969) found that his pressure-drop data could be approximated with a correlation for isothermal flow:

$$\xi_{fr} = 0.188\, \mathbf{Re}^{-0.22}. \tag{12.13}$$

Equation (12.13) has an uncertainty of ±10% outside of the pseudocritical region, where this uncertainty is much greater (see Figure 12.3).

Razumovskii et al. (1984) proposed to calculate the non-dimensional frictional resistance coefficient $(\tilde{\xi} = \xi / \xi_{iso})_{fr}$, as follows.

(A) Within a zone of viscous flow of water, bounded by low bulk-fluid enthalpies ($H_b \le 650$ kJ/kg):

$$\tilde{\xi}_{fr} = 1 - 0.5\,(1+M)^n\, log_{10}(1+M), \tag{12.14}$$

where $M = (\mu_b/(\mu_w - 1))\,(\mu_b/\mu_w)^{0.17}$ and $n = 0.17 - 2\cdot 10^{-6}\,\mathbf{Re}_b + 1800/\mathbf{Re}_b$ Equation (12.14) has an uncertainty of ±10% (Razumovskiy 2003).

(B) Within a zone of viscous-inertial flow of water, including higher bulk-fluid enthalpies of the "liquid" phase, i.e., high-density fluid, ($H_b \le 1500$ kJ/kg) and the region of bulk-fluid enthalpies of the "gaseous" phase, i.e., low-density fluid, ($H_b \ge 2250$ kJ/kg) and at $q/G > 0.8 - 1.0$ kJ/kg:

$$\tilde{\xi}_{fr} = \left(\frac{\mu_w}{\mu_b}\frac{\rho_w}{\rho_b}\right)^{0.15}, \tag{12.15}$$

Equation (12.15) has an uncertainty of ±15%.

Kirillov et al. (1990) stated that the frictional resistance coefficient for an isothermal stabilized turbulent flow of fluid at the near-critical state follows the same trends as the frictional resistance coefficient for turbulent fluid flow at subcritical pressures in smooth tubes. They proposed to calculate the frictional resistance coefficient for isothermal stabilized turbulent flow of fluid at the near-critical state using Equation (12.3). They stated that Equation (12.3) is valid for a reduced pressure, $\pi = 1.016 - 1.22$ and $\mathbf{Re} = 8\cdot 10^4 - 1.5\cdot 10^6$. Within the same range of parameters, the frictional resistance coefficient for a heated tube at the normal and deteriorated heat transfer regimes can be calculated using

$$\left(\frac{\xi}{\xi_{iso}}\right)_{fr} = \left(\frac{\rho_w}{\rho_b}\right)^{0.4} \tag{12.16}$$

In this equation, the density has to be determined using the *p-V-H* diagram, ξ_{iso} using Equation (12.3), and ρ_w evaluated at the wall temperature.

12.4.2 Helically Finned Bundles

Dyadyakin and Popov (1977) performed experiments with a tight-lattice 7-rod bundle with helical fins cooled with supercritical water (Table 12.3). They generalized their data with the following correlation for the hydraulic resistance:

$$\xi_{fr\,x} = \left(\frac{0.55}{\log_{10}\dfrac{\mathbf{Re}_x}{8}}\right)^2 \left(\frac{\rho_w}{\rho_b}\right)_x^{0.2} \left(\frac{\mu_b}{\mu_{in}}\right)_x^{0.2} \left(\frac{\rho_b}{\rho_{in}}\right)_x^{0.1} \tag{12.17}$$

where x is the axial location along the heated length, m, and D_{hy} is the characteristic dimension, m. One hundred and seventy experimental points or 94% of the data have deviation of ±20% from the fitting curve.

Table 12.3. Range of investigated parameters for experiments with water flowing in bundles at supercritical pressures.

Reference	*P*, MPa	*t*, °C (*H* in kJ/kg)	*q*, MW/m²	*G*, kg/m²s	Flow Geometry
Dyadyakin and Popov 1977	24.5	t_b = 90–570; H_b = 400–3400	<4.7	500–4000	Tight bundle (7 rods (6+1), D_{rod} = 5.2 mm, L = 0.5 m), each rod has 4 helical fins (fin height 0.6 mm, thickness 1mm, helical pitch 400 mm), pressure tube hexagonal in cross section

12.5 CONCLUSIONS

The following is a summary of the main findings:

- The majority of the experimental studies deal with heat transfer and relatively few with hydraulic resistance of fluids, mainly water and carbon dioxide, in circular tubes. Limited studies were devoted to pressure drop in other flow geometries. Only one publication deals with the pressure drop in tight helically finned bundles typical in nuclear reactors.
- There are some analytical and numerical approaches for estimating hydraulic resistance at near-critical and supercritical pressures. However, satisfactory analytical and numerical methods have not yet been developed due to the difficulty in dealing with the steep property variations, especially in turbulent flows and at high heat fluxes.
- In general, the pressure drop of supercritical fluid flow (similar to subcritical fluid flow) consists of four components: frictional resistance, flow obstruction, flow acceleration and gravity. The total pressure drop at supercritical pressures can be estimated based on general correlations for pressure drop at subcritical pressures with the correction factors for the effect of significant thermophysical properties variations and high heat flux.
- There is no one correlation suitable for hydraulic resistance calculations in water at supercritical pressures flowing in reactor bundles. The Dyadyakin-Popov (1977) correlation was obtained in water at supercritical pressures flowing in a short tight finned bundle and is expressed in terms of the inlet flow conditions and hence is valid only for the specific experimental configuration. More work is needed for developing prediction methods for hydraulic resistance in reactor bundles at supercritical pressures.

Chapter 13

ANALYTICAL APPROACHES FOR ESTIMATING HEAT TRANSFER AND HYDRAULIC RESISTANCE AT NEAR-CRITICAL AND SUPERCRITICAL PRESSURES

Unfortunately, satisfactory analytical methods have not yet been developed due to the difficulty in dealing with the steep property variations, especially in turbulent flows and at high heat fluxes. However, for completeness of the monograph it was decided to summarize the latest findings in analytical and numerical approaches in this chapter.

13.1 GENERAL

According to Polyakov (1991), heat transfer at high heat fluxes in a single-phase flow near a wall is subjected to very large variations in the fluid physical properties with temperature. The principal focus for analytical approaches is on fluctuations in flow about the mean (quasi-stationary in turbulent fluctuation scales). Momentum and heat transport are essentially a coupled heat-transfer problem. The mathematical form of the *steady-state conservation equations* is the following system written in cylindrical coordinates (r is the radius) as an approximation for a boundary layer:

Energy

$$\rho u \frac{\partial H}{\partial x} + \rho v \frac{\partial H}{\partial r} = \frac{1}{r}\frac{\partial}{\partial r}\left[r \left(\frac{k}{c_p}\frac{\partial H}{\partial r} - \overline{\rho v' H'} \right) \right]; \tag{13.1}$$

Momentum

$$\rho u \frac{\partial u}{\partial x} + \rho v \frac{\partial u}{\partial r} = -\frac{dp}{dx} \pm \rho g + \frac{1}{r}\frac{\partial}{\partial r}\left[r \left(\mu \frac{\partial u}{\partial r} - \overline{\rho v' u'} \right) \right] \tag{13.2}$$

and Mass

$$\frac{\partial(\rho u)}{\partial x}+\frac{1}{r}\frac{\partial(r\rho v)}{\partial r}=0, \tag{13.3}$$

where $(\overline{v'H'})$ represents the turbulent heat transport and $(\overline{v'u'})$ the turbulent momentum transport. The positive sign in front of $\rho \cdot g$ (Equation (13.2)) conventionally refers to upward flow in heated tubes; and consequently the negative sign to downward flow.

The energy, momentum, and mass equations are written without taking into account the physical properties fluctuations: that is, their variations are supposed to be in compliance with changes in the mean temperature (enthalpy), and their instantaneous variations caused by the fluctuating temperature are neglected. This is an averaging approximation.

The main difficulties in solving Equations (13.1) and (13.2), apart from numerical considerations, involve the search for the most reliable approximations for correlations characterizing turbulent heat $(\overline{v'H'})$ and momentum $(\overline{v'u'})$ transport.

One of the most important factors affecting supercritical heat transfer is the very large change in fluid density. In the first place, an occurrence of regimes with a sharp local wall temperature maximum ("peak") may be considered specific to supercritical flow. These regimes were conventionally referred by Petukhov (1970) as "degraded heat-transfer regimes," contrary to "normal regimes" without the "peak" in the wall temperature distribution. It is noted that, in Russian literature, such regimes of unusually low heat transfer are also called or termed "deteriorated heat transfer" (Ankudinov and Kurganov 1981), "worsened heat transfer" (Petukhov and Polyakov 1974) or "degenerated heat transfer" (Kurganov et al. 1986). However, the term "deteriorated heat transfer" is used in the current monograph instead of all other similar terms.

Others (Hall and Jackson 1978; Tanaka et al. 1973) relate this local deterioration of turbulent heat transfer to a free-convection effect, when wall temperature peaks are obtained experimentally in vertical heated tube upward flows. However, the peaks are absent in downward flows at the same conditions.

The mechanism of the buoyancy and acceleration effects, as well as quantitative correlations between the development of these effects and heat transfer changes were not explained for a long time (Polyakov 1991). In 1975, Polyakov proposed, apparently for the first time, to take into account the influence of buoyancy and acceleration effects for the analysis of heat transfer at supercritical pressures, connecting them with density fluctuations by means of a turbulent energy balance equation in the following form:

$$\overline{\rho u' v'}\frac{\partial u}{\partial y}+\overline{\rho' u'}\left(\pm g+u\frac{\partial u}{\partial x}\right)+\varepsilon=0 \tag{13.4}$$

The first two terms in Equation (13.4) take into account the density fluctuations. The term ($\pm g$) accounts for the acceleration due to gravity (as before, the positive sign refers to upward flow in heated tubes, the negative sign refers to downward flow). The term, ($u\partial u/\partial x$), is related to the individual particle acceleration in averaged motion and is written as superimposed on the presence of a mean fluctuating mass flux only along the tube. The first and the last terms in

Equation (13.4) describe turbulence production due to mean velocity gradients and the dissipation of turbulence, respectively. This formulation provides a basis for explanation of the heat-transfer peculiarities mentioned above. The formulation can be used for the further development of analysis, generalization, and numerical modeling.

The calculations with Equations (13.1) – (13.3) make it possible to follow local heat transfer development immediately from the start of fluid heating (Polyakov 1991).

13.2 CONVECTION HEAT TRANSFER

For completeness, but at the same time not to overload it with lengthy numerical and analytical solutions, only selected analytical and numerical expressions in a final form are given. However, the relevant references are listed in full below.

13.2.1 Laminar Flow

In general, the study of heat transfer in turbulent flow is more important for practical purposes. However, for a complete understanding of heat transfer at supercritical pressures, it is useful to consider the influence of variable physical properties when only molecular momentum and heat transport affect heat transfer from the wall (Polyakov 1991).

Koppel and Smith (1962) obtained a solution for heat transfer from a circular tube to a fluid with variable properties in fully developed laminar flow inside a tube. The major assumption employed was that the radial velocity component could be neglected. Their method used the example of supercritical carbon dioxide with the boundary condition of constant wall heat flux.

Hasegawa and Yoshioka (1966) conducted an analysis of laminar free convection from a heated vertical plate with uniform surface temperature for supercritical fluids. The variations of thermophysical properties were evaluated from the enthalpy using a perturbation method.

Nowak and Konanur (1970) investigated analytically heat transfer to supercritical water (at 23.4 MPa and within the pseudocritical region) assuming stable laminar free convection from an isothermal, vertical plate. Fair agreement was found between the analytical values and existing experimental data.

Shenoy et al. (1975) obtained the numerical solution of Equations (13.1 – 13.3) for laminar flow (setting $\overline{u'v'} = \overline{H'v'} = 0$) without taking into account buoyancy forces ($g = 0$), by imposing the following boundary conditions on velocity and temperature:

$$\begin{array}{l} u = v = 0, \quad T = T_w \quad \text{for} \quad r = r_o \\ u = const, \quad v = 0, \quad T = T_{in} = const; \\ \text{for} \quad x = 0, \quad 0 \le r \le r_o \end{array} \tag{13.5}$$

$$\left(\frac{\partial u}{\partial r}\right) = 0, \qquad v = 0, \left(\frac{\partial T}{\partial r}\right) = 0 \quad \text{for} \quad x > 0, r = 0.$$

The results of calculations for the hydrodynamic entry region, that is, without a preliminary developed velocity profile for flow over a heated surface ($T_w \geq T_{pc} > T_{in}$), demonstrate a large increase in the HTC over downstream heat transfer.

The numerical solution was also carried out for upward flow in vertical heated tubes, taking into account buoyancy forces, that is, the term (ρ g) in Equation (13.2) was presented in a dimensionless form as $(\mathbf{Ga} / \mathbf{Re}^2)_{in}(\rho/\rho_{in}) = gD\rho_{in}^2/G$. By neglecting buoyancy effects ($\mathbf{Ga}$ = 0), it was found that the heat-transfer rate decreases as q_w increases at $t_{in} < t_{pc}$.

Popov and Yan'kov (1979) performed an analytical study of laminar natural convection of supercritical carbon dioxide and helium near a vertical plate. They made allowance for variability of thermophysical properties. Popov and Yan'kov compared their results with the experimental data and found that they were in satisfactory agreement.

Seetharam and Sharma (1979) developed numerical predictions for laminar free convective heat transfer to fluids in the near-critical region from a vertical flat plate with uniform heat flux. The governing equations were integrated using the Patankar-Spalding implicit finite difference scheme. Computations were made for carbon dioxide in the near-critical region within a wide range of Rayleigh numbers ($5{\cdot}10^6 - 5{\cdot}10^{10}$).

Ghajar and Parker (1981) developed a reference temperature method for heat transfer in the supercritical region with variable property conditions in laminar free convection on a vertical plate.

Popov and Yankov (1982) calculated heat transfer in a laminar natural convection near a vertical plate for water, carbon dioxide, and nitrogen in the supercritical region for boundary conditions $t_w = const$ and $q_w = const$. It was shown that a consideration of the thermal conductivity peak had a significant effect on the results of heat-transfer calculations. An interpolation formula was selected that gave the Nusselt numbers for the considered fluids, for both types of boundary conditions and for previously obtained data for helium.

Stephan et al. (1985) investigated convective heat transfer to carbon dioxide near its critical point. The boundary layer equations were solved with variable properties for a vertical plate of constant temperature. The calculated HTCs were compared with experimental results.

Valueva and Popov (1985) performed numerical modeling of mixed laminar and turbulent convection in water at subcritical and supercritical pressures. They found that their method of calculation enabled them to reproduce the heat transfer observed in the experiments with upward and downward flows at conditions of strong free-convection effects.

Comparison of the results of Valueva and Popov (1985) with the data shows a different character for the heat transfer at different hydrodynamic and heat boundary conditions and different temperature ranges, even in the simplest case of viscous flow. The case of mixed laminar convection is more complicated, with buoyancy effects being coupled with varying physical property effects. At low heat flux and at $t_{in} < t_{pc}$, these effects lead to an increase in heat transfer of 30% – 40%, as compared with the case of constant physical properties (Polyakov 1991). As heat flux increases, buoyancy leads to increased heat transfer at constant properties.

In this case, however, the effect of a significant decrease of fluid thermal conductivity in the wall region (compared to the flow core) dominates other effects, manifesting itself in a reduction of the Nusselt number. As fluid heats at $t_b \geq t_{pc}$, its physical properties vary with temperature in a fashion similar to changes in gas properties. In this case, the increase of thermal conductivity near the wall and the effect of the buoyancy forces lead to increasing heat transfer (Polyakov 1991).

13.2.2 Turbulent Flow

Deissler and Taylor (1953) and Deissler (1954) found that the effects of variable fluid properties on the Nusselt number and friction factor correlations can be accounted for by evaluating the properties in the **Nu** and **Re** numbers at a reference temperature that is a function of both the wall temperature and the ratio of wall to bulk temperatures.

Goldmann (1954) developed a new analysis method to predict heat transfer and pressure drop for fluids with temperature-dependent properties in fully developed turbulent flow. The proposed method is a further extension of the Reynolds analogy between turbulent momentum transfer and heat transfer.

Hsu and Smith (1961) derived equations to calculate HTCs in turbulent flow with significant changes of density across the tube (in the critical region). Their results were compared with experimental data of carbon dioxide and showed good agreement.

Popov (1967; 1966) conducted theoretical calculations of heat transfer and friction resistance for supercritical carbon dioxide based on the analytical expression for the Nusselt number for steady-state axisymmetric turbulent flow of incompressible liquid in a tube with variable physical properties.

Graham (1969) modified the traditional steady-state model of turbulent convection in a thermal boundary layer to include a non-steady penetration component of heat transfer. He assumed that penetration mechanism results from appreciable changes in the specific volume of local agglomerates of fluid near the wall under heating conditions. Moreover, it was found that in some respects the penetration mechanism is similar to boiling. With this model, some success was achieved for accounting for the differences between the experimental data and conventional turbulent heat-transfer correlations for variable property fluids.

Leontiev (1969) considered some problems of deterioration in heat transfer at supercritical pressures at forced flow of fluid in vertical channels. The analysis carried out showed similarity between the processes of the laminarization of the turbulent boundary layer under the influence of buoyancy forces and of the negative pressure gradient.

Shlykov et al. (1971a,b) carried out a calculation of the temperature of a tube wall cooled with water at supercritical conditions. Their results appeared to be in qualitative agreement with experimental data. Their calculations confirmed a possibility of the existence of deteriorated heat transfer.

Kamenetskii (1973) considered conditions at which free convection had a substantial effect on heat transfer in turbulent flow in vertical tubes with variable fluid properties. To estimate this effect, he obtained a dimensionless number that characterized the laminar boundary layer under the action of buoyancy forces.

By assuming that the turbulent boundary layer was constructed by the superposition of locally developed layers, Tanaka et al. (1973) proposed an approximate theory to calculate temperature and velocity profiles under the large effects of buoyancy and acceleration. Based on their theory, a criterion of the reverse transition from turbulent to laminar flow was proposed.

Kakarala and Thomas (1974) developed the surface renewal and penetration model, which provided a useful new approach to the analysis of turbulent convective heat transfer to supercritical fluids.

Khabenskii et al. (1974) solved for the temperature profile in the tube wall with non-uniform radial heat flux at the outer surface and heat transfer to the medium at supercritical pressures at the inner surface.

Polyakov (1975) examined the effect of thermo-gravitational forces and local acceleration of flow due to a change in density on turbulent momentum transfer at supercritical coolant parameters based on analytic estimates. He determined the limits on the causal origin of these effects on heat transfer. It was shown that local deterioration of heat transfer in heated tubes (in the case of upward flow) was associated with the effect of thermo-gravitation and "thermal acceleration" on turbulent momentum transfer; in the case of downward flow, it was associated with the effect of "thermal acceleration." Available experimental data on the local deterioration of heat transfer in the case of the flow of water, carbon dioxide, and helium supported this conclusion.

Watson (1977) found analytically that for a tube or wire heated electrically and cooled by convection with the HTC decreases with temperature increase, non-uniform steady-state axial temperature distributions can occur.

Popov and Valueva (1988) supplemented a method of numerical modeling of turbulent mixed convection described previously by them in 1986 by an approximate means of calculation for turbulent viscosity with flow at low Reynolds numbers. This approximation explained features observed in the experiments of the temperature regimes for subcritical fluids and fluids in the supercritical regions, with a strong effect of free convection and within a wide range of Reynolds numbers.

As was shown by Polyakov (1991), even for laminar flow, a large change in physical properties at subcritical fluid conditions results in a unique heat-transfer characteristic compared with heat transfer for constant-property fluids. However, finding a mathematical solution to this problem is a very difficult task, because it involves finding a solution for three-dimensional, non-linear equations with sharply varying coefficients.

In the case of turbulent flow, major difficulties are related to the determination of averaged expressions for turbulent momentum and heat transport (Polyakov 1991). The regimes with deteriorated local heat transfer cause significant difficulties in practice. At present, it is known that, in addition to physical property variability causing heat transfer decrease in some cases, buoyancy and thermal acceleration cause significant deterioration of heat transfer. All three effects are to be taken into account in Equations (13.1) to (13.3) for mean values and for the mathematical description of turbulent momentum and heat transfer.

According to Polyakov (1991), the manifestation of buoyancy forces and thermal acceleration is coupled with a density change that becomes more intense with the increasing of heat flux, and is naturally accompanied by an increase in the variability of other physical properties. He started the analysis by presenting results for a rather low heat flux, which corresponded to the small temperature difference case ($t_w - t_b$), when the effect of thermal acceleration and buoyancy forces can be neglected. The investigation of heat transfer in turbulent flow based on the system of Equations (13.1) to (13.3) require the specification of relations for the terms $(\overline{u'v'})$ and $(\overline{H'v'})$. The traditional approximations by the Boussinesq relation leads to:

$$\overline{u'\,v'} = -\nu_T\,\frac{\partial u}{\partial y} \tag{13.6}$$

and by a similar relation,

$$\overline{u'\,v'} = -\nu_T\,\mathbf{Pr_T}\,\frac{\partial H}{\partial y} \tag{13.6a}$$

These approximations are widely used for the prediction of supercritical heat transfer.

Polyakov (1975) and Petukhov and Polyakov (1988) proposed, on the basis of Equation (13.4), the following estimation of the boundaries, below which it is possible to neglect heat-transfer changes in vertical tubes due to variations of turbulent momentum transport induced by buoyancy and thermal acceleration effects:

$$\left|\pm\mathbf{Gr_q}\ \pm\mathbf{J}\right| < 4\cdot10^{-4}\,\mathbf{Re_b^{2.8}}\,\overline{\mathbf{Pr}} = \mathbf{B}_{\text{th}}, \tag{13.7}$$

where:

$$\mathbf{J} = 4\,\frac{\mathbf{Re_b}}{\mathbf{Pr}}\left(\frac{\rho_b-\rho_w}{t_w-t_b}\,\frac{q_w\,D\,\rho_b}{k_b\,\rho_f^2}\right)^2, \tag{13.8}$$

$$\mathbf{Gr_q} = \frac{g\,(\rho_b-\rho_w)\,q_w\,D^4}{(t_w-t_b)\,\nu_b^2\,k_b}, \tag{13.9}$$

and ρ_f is evaluated at $\frac{t_w+t_b}{2}$.

The positive sign in front of the Grashof number in Equation (13.7) is as usual for upward flow in heated vertical tubes; while the negative sign is for downward flow in cooled vertical tubes. The positive sign in front of the parameter **J** is for the case of fluid heating ($t_w > t_b$), and the negative sign is for fluid cooling ($t_w < t_b$).

Sastry and Schnurr (1975) developed a numerical solution for heat transfer to fluids near the critical point for turbulent flow in a circular tube with constant wall heat flux. They used an adaptation of the Patankar-Spalding implicit finite difference marching procedure. The results were compared to the experimental data of water and carbon dioxide and showed good agreement.

Popov (1977) obtained a relationship between the coefficient of turbulent momentum transfer for variable physical properties and the coefficient of turbulent heat transfer for constant properties. A method of calculating heat transfer and hydraulic resistance was proposed for turbulent flow in a circular tube of a compressible fluid, with arbitrarily varying physical properties. By this method, calculations in the tube cross-section can be made solely on the basis of the hydrodynamic and thermal conditions in the cross-section. In addition, the effect of the previous hydrodynamic and thermal development of the flow can be approximately taken into account. As an example, results were given for the frictional drag and the temperature recovery coefficient in the case of air at Mach numbers in the range of 0 – 2.6.

Popov et al. (1977) carried out calculations for heat transfer under conditions of turbulent flow of liquids in a circular tube, with various types of dependencies of physical properties on temperature (water, air, and nitrogen (nitrogen at supercritical pressure)) and under strong variability of physical properties during heating and cooling. The results showed that the use of a one-dimensional flow model to determine the local values of the frictional resistance coefficient from experimental data for a liquid with supercritical parameters could lead to serious errors.

Protopopov (1977) analyzed the experimental heat-transfer data for water and carbon dioxide in a heated tube with upward flow, and proposed the following criteria for an estimation of the boundaries of the absence of buoyancy effects:

$$\left(\frac{\mathbf{Gr}_\rho}{\mathbf{Re}^2}\right) < 0.01 \text{ or } \left(1-\frac{\rho_w}{\rho_b}\right)\left(\frac{\mathbf{Gr}_\rho}{\mathbf{Re}^2}\right) < 0.01 \tag{13.10}$$

where $\mathbf{Gr}_\rho = \dfrac{g\,(\rho_b-\rho_w)\,D^3}{\rho_b\,\nu_b^2}$.

Petukhov et al. (1977) conducted the numerical investigation of heat transfer for a turbulent flow of fluid with strongly temperature-dependent physical properties in a circular tube, based on equations of energy, motion, and continuity in the boundary layer form. The problem was solved for the case when the heat flux at the wall was constant along a tube length. In their calculations, nine different models were used to describe the turbulent transfer. The system of equations was solved with a numerical method using a two-layer, six-point implicit difference scheme. The finite-difference equations were solved with successive approximation. The results were compared with the experimental data obtained for the flow of water and carbon dioxide at near-critical states in a tube. They demonstrated that the use of relations for turbulent transport coefficients proposed for forced flow with constant properties, without taking into account variable physical properties, buoyancy forces and thermal acceleration, does not allow for a correct description of the heat transfer behavior of a single-phase fluid with parameters near the critical point.

Grigor'ev et al. (1977) demonstrated the possibility of using Duhamel integral-type relations (superposition principle) for calculating the heat transfer for a turbulent flow in a tube at supercritical pressures. Experimentally determined Nusselt numbers and values of the wall temperature obtained under conditions of linear increase or decrease of heat flux along the length of the tube were compared with values from the superposition method. In performing the calculations, the properties of heat transfer for a constant heat flux were assumed to be known.

Popov et al. (1978) investigated the effect of the flow history using Equations (13.1) to (13.3). They used helium at the near-critical point (p = 0.25 – 2 MPa, **Re** = 5·10³ – 10⁶, T_{in} = 4 – 6 K, T_w = 25 K). The calculations showed that, with an increase in heat flux in the region of the tube where the mean-mass temperature is close to the pseudocritical temperature, the change in the wall temperature over the length has a peak (deterioration in heat transfer). The results of heat-transfer calculations are in good agreement (within the limits of experimental error) with the known experimental data. Data on the local drag coefficient indicated that inertial forces made a considerable contribution to the hydraulic drag, and that calculation of this contribution using a one-dimensional model may lead to large errors.

Investigating the effect of upstream flow history on the HTCs, Popov et al. (1978) found that the largest difference (up to 50%) between the HTCs was at small values of x/D. For $x/D > 50$, the value of the HTC varied by ±15%. This variation decreased with an increasing value for x/D. Thus, a key conclusion is that the heat transfer in a turbulent flow can be considered to be fully developed for $x/D > 50$.

Popov et al. (1979) presented results of heat-transfer calculations in turbulent flow of supercritical helium in a circular tube. They compared their results with the data of Giarratano and Jones (1975) and found the agreement between them to be acceptable.

Popov et al. (1978) presented the results of numerical calculations of the heat transfer in turbulent flow of helium in a heated circular tube at supercritical pressures. The calculations assume that thermogravitation had no effect on the pronounced variability of the physical properties over the tube cross-section (corresponding to a ratio of up to 0.1 between the densities at the wall temperature and at the bulk-fluid temperature). The ranges of the parameters are as follows: p = 0.25 – 2 MPa, **Re** = 5·10³ – 10⁶, T_{in} = 4 – 6 K, and $T_w \leq 25$ K. The calculation showed that, with an increase in heat flux in the region of the tube where the bulk-fluid temperature is close to the pseudocritical temperature, the change in the wall temperature over the length had a peak (deterioration in heat transfer). The results of heat-transfer calculations were in good agreement with known experimental data, within the limits of experimental error.

Sevast'yanov et al. (1979) carried out a theoretical and experimental study of heat transfer in a turbulent flow of liquid at supercritical pressure under conditions of high-frequency oscillations. By numerically solving a system of differential equations, it was possible to find the local and average HTCs as functions of the amplitude-frequency characteristics of the oscillations.

Ivlev (1979) examined the results of calculation of heat transfer in turbulent tube flow of supercritical helium. The calculations were performed by a technique suggested by Melik-Pashaev (1966). The behavior of the calculated HTC was in qualitative agreement with available experimental data.

Yaskin (1980) showed that the temperature on the top part of the wall of a horizontal tube with supercritical water could be computed from the dimensionless equation for heat transfer proposed by the author. This equation was corrected for the buoyancy effects.

Yeroshenko et al. (1980) developed a method for calculating heat transfer in supercritical helium flowing through circular tubes with correction for the effect of variable properties, density fluctuations and thermal acceleration of the flow. The results were found to be in satisfactory agreement with the experimental data.

Popov et al. (1980) presented results of an analytical calculation of turbulent flow of supercritical helium in a circular tube at conditions of significant variability of critical properties and free convection (in upwards and downwards flows). The analytic results were in satisfactory agreement with the experimental results by other researches.

Ivlev et al. (1980) analytically investigated the appearance of the improved and deteriorated heat transfer regimes for forced convection of helium.

Adelt and Mikielewicz (1981) conducted a theoretical analysis of the convective heat transfer at supercritical pressures in a channel. Their analysis is based on the division of flow into two zones with average properties, and with the interface between them being the surface of the pseudocritical temperature. The theoretical results were compared with CO_2 data showing a fairly good agreement.

Kurganov (1982) calculated the heat transfer in smooth tubes with turbulent flow of gaseous working fluids (mixtures of helium and hydrogen) with constant and variable physical properties. He proposed equations for heat transfer to gases and vapors in the heated tubes with turbulent flow, and boundary conditions of the first and second kind.

Bellmore and Reid (1983) developed a method for numerical prediction of wall temperatures for near-critical para-hydrogen in turbulent upflow inside vertical tubes, which includes density variations in the equations of turbulent transport.

Renz and Bellinghausen (1986) determined from a numerical solution of the turbulent conservation equations that the effects, similar to film boiling, are due to the influence of gravity on the velocity profile and the turbulence structure in the near wall region of the flow. The calculated results were compared with experimental data and showed good agreement.

Popov and Valueva (1986) calculated heat transfer and turbulent flow of water at supercritical conditions in a vertical tube, with a significant effect of free convection based on a system of differential equations of motion, continuity, and energy. They claimed that due to the method of calculating turbulent heat transfer it became possible to reproduce the regimes with deteriorated heat transfer in upward flow for various heat fluxes and flows.

Petrov and Popov (1988) used the numerical method previously verified with carbon dioxide and helium for calculating heat transfer and hydraulic resistance with turbulent flow of water in a tube at supercritical pressure. They found that water, carbon dioxide, and helium are dissimilar with respect to type of dependence of thermophysical properties on temperature.

Koshizuka et al. (1995) and Koshizuka and Oka (2000) analyzed numerically the deteriorated heat transfer in supercritical water cooled in a vertical tube. They found that heat transfer to supercritical water can be analyzed by a numerical calculation using a k-ε turbulence model. Their numerical results agreed with the experimental data.

Zhou and Krishnan (1995) studied the integration and incorporation of models for transport properties of fluids in the supercritical regime into a general purpose Computational Fluid Dynamics (CFD) code for an analysis of flow and heat transfer in aircraft fuel systems. They used supercritical carbon dioxide for a preliminary validation and application of the code to laminar and turbulent flow.

Lee and Howell (1997) carried out a numerical modeling to investigate the characteristics of convective heat transfer in turbulent developing flow near the critical point in a tube with and without buoyancy effects at constant wall temperature. The numerical modeling results showed heat-transfer and fluid-flow characteristics, which included velocity profiles, HTC and the friction factor along the tube. They found that steep variation of density near the pseudocritical temperature resulted in high buoyancy forces. Close to the critical pressure, fluid near the wall undergoes more acceleration and this increases the HTC. With increasing wall temperature for the same inlet fluid conditions, the HTC and friction factor reach a minimum at some distance from the entrance. The minimum is closer to the entrance for the friction factor than for the HTC.

Li et al. (1999) performed a numerical modeling of the developing turbulent flow and heat-transfer characteristics of water near the critical point in a curved tube. Based on the results of their research, the velocity, temperature, HTC, friction-factor distribution, and effective viscosity were presented graphically and were analyzed.

Kitoh et al. (1999) carried out a safety analysis for a high-temperature core reactor with supercritical water. A new formula for the heat-transfer correlation was proposed based on numerical simulation.

Scalabrin and Piazza (2002) applied the neural networks method for deriving a heat-transfer correlation for supercritical carbon dioxide flowing inside tubes. They found that the best correlation architecture is the one that takes into account the property variations along the radial coordinate.

Dumaz and Antoni (2003) have modified the recent version of the CATHARE2 code for simulation transients at supercritical and subcritical regimes. Their preliminary assessment of this model for supercritical water showed that it can be used for various types of one-dimensional calculations including LOCAs.

Yoon and Bae (2003) have developed a computer code for the safety analysis of a SCWR with a gravity-driven, passive-safety system. This code employs one-dimensional governing equations for the coolant mass, energy, and momentum. The main objective of this study was investigation whether the passive heat-removal system can supply enough coolant flow to the reactor core at various transient conditions.

Dreitser and Lobanov (2004) developed a theoretical model based on a four-layer scheme of turbulent flow for calculating heat and hydraulic resistance of the enhanced turbulent flow of a supercritical jet-propulsion propellant flowing in tubes. This model enables to predict heat transfer and hydraulic resistance within wider ranges than the existing experimental data.

He et al. (2004) performed computational simulations of turbulent mixed-convection heat transfer to supercritical carbon dioxide by solving the Reynolds averaged transport equations using an elliptic formulation. To some extend, the simulations were able to reproduce the effects of very strong influences of buoyancy forces on the heat transfer. However, the models were not able to predict exactly the onset of such effects.

Kim et al. (2004) simulated numerically a vertical upward flow of water in the heated tube at supercritical pressure using a commercially available CFD code. They tested several turbulence models. Their conclusion is that a modification of the turbulence transport equation may be required to improve predictions of the wall temperature.

Cheng et al. (2005) investigated numerically heat transfer of supercritical water in various flow channels using the CFD code CFX 5.6. The intent of this work was to obtain a basic knowledge of heat-transfer behavior and gathering the first experience in the application of CFD codes to heat transfer in supercritical fluids. They investigated effects of mesh structures, turbulence models, and flow-channel configurations.

Cho et al. (2005) studied numerically heat transfer to supercritical carbon dioxide flowing upward in a heated vertical tube (4.4 mm ID and heated length of 3 m) using the FLUENT code. They found that within the normal heat-transfer condition, the low-Reynolds model (ABID) predicted the experimental data better than the high-Reynolds model (RNG). However, in the deteriorated heat transfer, the difference between this model and the experiment was remarkable.

He et al. (2005) conducted computational simulations of experiments on turbulent-convection heat transfer of supercritical carbon dioxide in a vertical tube of 0.948 mm inside diameter. These computational simulations were carried out using low-Reynolds number eddy viscosity turbulence models. In general, the simulations were able to reproduce the features exhibited in the experiments. However, in some cases, the results were very diffirent.

Kitou et al. (2005) evaluated heat transfer in a circular tube cooled with supercritical R-22 and water, and in single-rod experiments with R-22, using unified analysis model. They found that deteriorated heat transfer can be also simulated using this model.

Seo et al. (2005) used the commercially available CFD code FLUENT, which solves the Navier-Stokes and energy equations with the standard k-ε model and the standard wall function for various fluid-dynamics and heat-transfer applications, to predict heat transfer to supercritical water. Two sets of the experimental data were used in this study: Yamagata et al. (1971) and Shitsman (1963). The simulations showed a good agreement with the experimental data within high mass-flux conditions, where the buoyancy effects are minor. However, the FLUENT code had difficulties in predicting the deteriorated heat transfer.

Vasic and Khartabil (2005) used the CATHENA (Canadian Algorithm for THErmalhydraulics Network Analysis (Hanna 1998)) code simulations to optimize the performance of the insulated CANDU fuel channel under decay heat generation conditions and variable thermophysical characteristics of the insulating layer. Their study showed that the advanced CANDU fuel-channel design is promising and can prevent overheating of the fuel even during very severe accident scenarios. The final simulation results showed also that this channel design combined with the passive moderator heat rejection has the potential to reduce significantly or even eliminate the possibility of core damage.

13.3 HYDRAULIC RESISTANCE

Tanaka et al. (1973) considered turbulent heat and momentum transfer for a fluid flowing in a vertical tube. They studied the shear-stress distribution in a tube, by taking the buoyancy forces and the inertia force due to acceleration into consideration. It was shown that the effects of both forces operated quite similarly and resulted in a very sharp decrease of the shear stress near the wall. By considering how the velocity profile depends on the shear-stress gradient at the wall, the authors deduced the criteria for the prominent effects of buoyancy and acceleration. By assuming that the turbulent boundary layer was constructed by the superposition of the locally developed layers, they proposed an approximate theory to calculate velocity and temperature profiles under the large effects of buoyancy and acceleration. Based on their theory, a criterion of the reverse transition from turbulent to laminar flow was proposed.

Popov (1977) proposed a method for calculating the hydrodynamic resistance and recovery coefficient for turbulent flow in a circular tube (far from inlet, with closed boundary layer) of a compressible fluid, with arbitrarily varying physical properties.

Popov et al. (1977) carried out calculations for the hydraulic resistance at conditions of turbulent flow in a circular tube, with various types of dependencies of physical properties on temperature (water, air, and nitrogen (nitrogen at supercritical pressure)) and under strong variability of physical properties during heating and cooling. The results showed that the use of a one-dimensional flow model in the experimental determinations of the local values of the frictional resistance coefficient for a liquid with supercritical parameters could lead to serious errors.

Popov et al. (1978) presented the results of numerical calculations of the hydraulic drag in a turbulent flow of helium in a heated circular tube at supercritical pressures. The calculations assume that thermogravitation had no effect on the pronounced variability of the physical properties over the tube cross-section (corresponding to a ratio of up to 0.1 between the densities at the wall temperature and at the bulk-fluid temperature). The ranges of the parameters were as follows: $p = 0.25 - 2$ MPa, $\mathbf{Re} = 5{\cdot}10^3 - 10^6$, $T_{in} = 4 - 6$ K, and $T_w \le 25$ K. Data on local drag coefficients indicated that inertial forces made a considerable contribution to the hydraulic drag, and that the calculation of this contribution using a one-dimensional model may lead to large errors.

Petukhov and Medvetskaya (1978, 1979) proposed a computational model. This model used the simplified equation of turbulent kinetic energy balance similar to Equation (13.4) to find a turbulent momentum transport coefficient. The coefficient $\mathbf{Pr_T} \cdot \nu_T$ was obtained in accordance with the simplified enthalpy balance equation. Also, this model included some approximations borrowed from the general theory of turbulence. Adopted approximations and constant values were verified to be acceptable through comparisons with the experimental data obtained for turbulent flow of water and air in tubes under significant influence of a gravity field.

Sinitsyn (1980) suggested a linearized system of equations describing the distribution of pressure waves in a channel, taking account of the friction and thermal exchange with the walls. It was shown that the presence of a liquid boundary layer in which sound velocity is low, leads to oscillatory enhancement of the flow parameters.

Popov (1983) proposed to use model equations for turbulent stresses and heat fluxes for deriving expressions for the turbulent viscosity under conditions of free convection.

Popov and Petrov (1985) presented the results of a numerical solution of the flow and heat transfer in the turbulent flow of supercritical carbon dioxide in a tube at cooling conditions.

Jiang et al. (1995) investigated numerically forced- and mixed-convection heat and mass transfer in water containing metallic corrosion products in a heated or cooled vertical tube with variable thermophysical properties at supercritical pressures. Their paper shows the fouling mechanisms and models.

Howell and Lee (1999) investigated numerically the turbulent convective heat transfer for developing flow of water near the thermodynamic critical point in a constant wall temperature vertical tube with and without buoyancy force. Also, they looked into wall temperature effects on momentum and heat transfer, velocity profiles, property variations, HTC, and friction-factor distributions close to the inlet. They found that flow acceleration near the wall increases near the critical pressure.

Bae et al. (2003) investigated numerically heat transfer to carbon dioxide at supercritical pressure in a vertical tube. In this simulation, no turbulence was adopted. Based on their direct numerical simulation (DNS), they explained the basic mechanism of the local deterioration for turbulent mixed convection.

Bae et al. (2005) investigated numerically an influence of strongly varying properties of supercritical-pressure fluids on turbulent convective heat transfer. They considered thermally developing upward flows in a vertical annular channel where the inner wall was heated with a constant heat flux and the outer wall was insulated. Carbon dioxide was chosen as the working fluid at a pressure of 8 MPa and the inlet Reynolds number was about 8900. They found that the streamwise turbulent heat flux showed a very peculiar transitional behavior due to the buoyancy effect.

Bogoslovskaia et al. (2005) presented results of the analysis of the thermalhydraulic characteristics of fuel assemblies cooled with supercritical water based on subchannel analysis. They used a modified subchannel code MIF – MIF-SCD developed by the IPPE (Obninsk, Russia). This modified code permits calculation of coolant temperature and velocity distributions in fuel assemblies.

Chapter 14

FLOW STABILITY AT NEAR-CRITICAL AND SUPERCRITICAL PRESSURES

For near-critical and supercritical pressures single-phase heat transfer, as well as for subcritical pressures two-phase heat transfer, flow oscillations can occur. These flow oscillations can significantly narrow the region with stable flow and affect heat transfer and reliability of heat-transfer equipment.

For two-phase flows at subcritical pressures, the appearance of flow pulsations and thermo-acoustic oscillations are described in book by Gerliga and Scalozubov (1992). Also, Rohatgi and Duffey (1998) using homogenous equilibrium theory obtained a close-form solution for the stability region for parallel channels and derived the critical subcooling number, which is the lowest for unstable flow. Some conclusions from these two sources can be applicable for supercritical flows.

Papers related to flow instability at supercritical pressures are listed below in the chronological order.

Hines and Wolf (1962) performed experiments with RP-1 and di-ethyl-cyclohexane (DECH) flowing inside circular tube at supercritical pressures and temperatures. They found that at these conditions vibration of the test section occurred which lead to heat-transfer increase.

Harden and Boggs (1964) investigated experimentally and analytically the transient behavior of a closed natural-circulation loop with Freon-114 as the working fluid. They found that stable operation of the loop could be maintained as long as the working fluid was not in the thermodynamic region characterized by a maximum in the density-enthalpy product versus temperature plot. When the working fluid in the loop approached this region from the low temperature side, pressure and flow fluctuations were encountered with the frequency range of 10 to 20 Hz. An approach from the high temperature side resulted in fluctuations within the range of 0.1–0.5 Hz.

Krasyakova and Glusker (1965) investigated flow stability in parallel plain-tube coils. They investigated three types of coils (U-type, ∩-type and N-type) and found that these coils can work in the normal regime, i.e., without flow stagnation. Flow stagnation is possible at mass fluxes below 300 kg/m^{2}s, heat fluxes below 80 kW/m^2 and subcooled enthalpies more than 420 kJ/kg.

Cornelius and Parker (1965) investigated flow instabilities in a loop with natural and forced convection at supercritical pressures. They identified two types of instabilities during natural circulation.

Zuber (1966) analyzed thermally induced flow oscillations in the near-critical and supercritical regions. In his comprehensive analysis, three mechanisms responsible for inducing thermohydraulic oscillations were distinguished and discussed. He found that low-frequency oscillations were most prevalent in supercritical pressure systems. In his work, the conditions leading to aperiodic and periodic flow oscillations were investigated and stability maps and stability criteria were proposed.

Walker and Harden (1967) used the "density effect" model formulated by Boure to predict the flow instability threshold in a natural-circulation loop operating near the critical point. They found that this model had an excellent agreement with the experimental data.

Kafengauz and Fedorov (1968) investigated heat transfer at surface boiling ($p < p_{cr}$) and pseudo-boiling ($p > p_{cr}$) regimes. They found that these regimes are related with the corresponding natural oscillations. An increase in the frequency of these oscillations prevented a rise in the temperature of the cooled surface.

Kaplan and Tolchinskaya (1969) experimentally investigated high-frequency pressure pulsations developing under heat transfer to n-heptane at different mass fluxes and pressures.

Treshchev et al. (1971) investigated flow oscillations in water flowing in a heated channel at supercritical pressures (p = 30.4 MPa, t up to 600°C). They found that within the investigated range flow throttling at the inlet of test section did not significantly affect heat flux value at which auto-oscillations started. However, within the same range of operational parameters, increase in unheated length downstream of the test section outlet narrows the stable flow region.

Johannes (1972) conducted forced-convection experiments with supercritical helium at pressures of 0.3–0.6 MPa and inlet temperature of 4.2 K. He found that stable flow conditions existed without heat input. However, with heat input and regardless of tube diameter, heat flux and flow rate, the wall temperatures started to oscillate with a frequency of 20 Hz and amplitude of 2 K. These oscillations were damped before reaching the pressure transducers, therefore, no pressure oscillations were observed.

Stewart et al. (1973) conducted heat-transfer measurements in supercritical-pressure water (p = 25 MPa) flowing through horizontal tubes (D = 1.524 and 3.1 mm, L = 0.203 and 0.61 m, respectively). They investigated high-frequency oscillations, which occurred spontaneously in water at supercritical pressures. These oscillations were measured, and it was shown that they were associated with pressure oscillations in the test section resulting from a standing pressure wave between the entry and the exit. Several modes of standing waves were identified.

Chakrygin et al. (1974) performed experiments for defining the limits of hydrodynamically unsteady flow regimes in heated tubes (for investigated ranges, see Table 5.1). Their results showed that within the investigated range the experimental data on limits of aperiodic instability are in agreement with the calculated values.

Kaplan and Tolchinskaya (1974a) recorded an anomalous increase in hydraulic resistance at heat transfer to a relatively cold n-heptane in the velocity range of 3–6 m/s when wall temperature exceeds pseudocritical temperature. Strong oscillations were encountered with a frequency of 2560–3220 Hz.

Dashkiyev and Rozhalin (1975) examined the stability of operation of a system of parallel steam-generating tubes in the presence of thermohydraulic stratification. They proposed an analytical method, which is based on two equations: (1) the equation of compatibility and (2) the equation of state. Their generalized results were presented in the form of nomograms.

Shvarts and Glusker (1976) proposed a method for determining minimum permissible flows with respect to conditions of stability in ∩- and U-shaped elements at supercritical pressures. This method is also based on solving several equations such as an empirical equation of state and pressure drop correlations. The final product is nomograms for ∩- and U-shaped elements.

Vetrov et al. (1977) investigated thermoacoustic oscillations in supercritical water. They found that these oscillations are quite common within a wide range of operating parameters. Therefore, it was proposed to account for these effects using criteria for pulsating flows (Galitseyskiy et al. 1977).

Sevast'yanov et al. (1980) conducted a theoretical and experimental study of heat transfer in a turbulent fluid flow at supercritical pressure under conditions of high-frequency oscillations. Equations were obtained for the secondary dynamic and thermal flows in a standing pressure wave, allowing for variability of the flow parameters and thermophysical properties of the fluid along the channel. By numerically solving a system of differential equations, it was possible to find the local and mean HTCs as functions of the amplitude, i.e., the frequency characteristics of the oscillations. The experimental results showed satisfactory agreement with the theory.

Sinitsyn (1980) suggested a linearized system of equations to describe the distribution of pressure waves in a channel, with account taken of friction and heat transfer at the walls. It was shown that the presence of a liquid boundary layer in which the sound velocity is low leads to oscillation enhancement of the flow parameters.

Labuntsov and Mirzoyan (1983) analyzed the boundaries of flow stability of helium at supercritical pressures in heated channels. Later on, in 1986, they analyzed the flow stability of helium at supercritical pressure with a non-uniform distribution of heat flux along the length of the channel.

Kafengauz and Borovitskii (1985) established experimentally that solid carbon deposits formed during heat transfer to kerosene in small diameter tubes induce self-excited thermoacoustic oscillations.

Labuntsov and Mirzoyan (1986) investigated the stability of helium flow at supercritical pressures with a non-uniform distribution of heat flux over the length of a channel. It was shown that the non-uniformity of axial heat flux had an effect on the stability boundary.

Bogachev et al. (1986, 1988) investigated the conditions for the offset of thermally induced oscillations and their effect on heat transfer in low-temperature helium in forced and mixed convection.

Vetrov (1990) analyzed frequencies of thermoacoustic oscillations and their dependence on problem parameters, on a basis of the wave equation. The calculated results were compared with experimental data.

Chatoorgoon (2001) examined supercritical flow stability in a single-channel, natural-convection loop using a non-linear numerical code. A theoretical stability criterion was developed to verify the numerical prediction. The numerical results showed good agreement with the analytical results.

Lomperski et al. (2004) conducted experiments to study closed-loop natural circulation of supercritical carbon dioxide. Tests were conducted with a 2-m high loop (with the maximum flow area and calibrated orifice plate in hot leg), input power of up to 15 kW and mass flux of 500 kg/m^{2}s. No flow instabilities were observed in these experiments.

Chatoorgoon and Upadhye (2005) performed a linear stability analysis of supercritical flow in a natural-convection loop to verify the predictions of the non-linear SPORTS code. Three fluids were used for this verification. They found that the linear stability predictions were within 95% agreement to the non-linear predictions in all cases. Also, it was found that supercritical carbon dioxide behaved similarly to water from a stability point of view.

Zhao et al. (2005) applied a non-homogeneous, i.e., drift-flux, non-equilibrium two-phase flow model for investigation of a stability of SCWR during steady-state and sliding pressure start-up. The SCWR was the US design of pressure-vessel nuclear reactor. They developed a sliding pressure SCWR start-up strategy to avoid thermal-hydraulic flow instabilities.

Chapter 15

OTHER PROBLEMS RELATED TO SUPERCRITICAL PRESSURES

Some general problems in the design and reliability of supercritical "steam" generators were discussed in papers by Styrikovich et al. (1967) and Rudyka et al. (1971).

15.1 DEPOSITS FORMED INSIDE TUBES IN SUPERCRITICAL "STEAM" GENERATORS

One of the first analytical works devoted to the problem of feed water impurity behaviour in supercritical "steam" generation was published in 1966 by Styrikovich et al. In this paper, the authors presented the results of the theoretical analysis of the solubility and distribution of feed water impurities in "steam" generators operating at 25 and 29.4 MPa.

Martynova and Rogatskin (1969) investigated the formation of calcium sulphate deposits in "steam" generators operating at 24 MPa and 560°C. The feed water was treated with a 50 – 100 μg/kg solution of hydrazine, which was injected after the deaerators. It was found that low thermally conductive, loose calcium sulphate deposits on the heating surfaces of the "steam" generators were considerably more dangerous for tube failure than the corrosion product deposits, which have a dense structure and good thermal conductivity.

Tret'yakov (1971) noted that temperature conditions of "steam" generating channels were governed by heat transfer and depositions of salt on the inside surface of the channel. Normally, these processes are considered separately, without allowance for their interaction. Usually, more attention was paid to the investigation of heat transfer, and the conditions of impurity deposition on a channel surface were studied to a less extent. The main series of experiments by Tre'yakov was conducted with the addition of 400 – 500 g/kg of sodium sulphate. The investigated range was p = 24, 30, and 34 MPa, t_b = 173 – 445°C, t_w = 213 – 630°C, G = 970 – 2320 kg/m^{2}s, and q = 174 – 513 kW/m^2. The effect of the maximum rate of increase in wall temperature and deposit thickness on mass flux and heat flux was presented.

Klochkov (1975) evaluated the corrosivity of water in the condensate-feed loop of high-pressure and supercritical pressure power generating units. The operating temperatures were from 291°C to 473°C. He found that, for a further reduction in the intensity of corrosion of high-pressure heating zones, it was desirable to reduce, by the maximum amount, the concentration of CO_2 in the feed loop.

Vasilenko et al. (1975) determined the allowable concentration of aluminium in the feed water of a supercritical power-generating unit equipped with a Heller system air-condensing plant, which incorporated an aluminium-cooling tower. It was found that, to ensure scale-free operation of a supercritical "steam" generator, the concentration of aluminium compounds in the feed water must not exceed 10 μg/kg of aluminium.

Belyakov (1976) investigated the temperature conditions of tubes in supercritical "steam" generators, where an iron oxide deposit was created. He made a comparison of the temperature conditions of a clean tube and that of a tube with the outer layer of deposits removed. Also, the effect of mass flux on the thermal resistance of the porous layer of iron oxide deposits was presented.

Glebov et al. (1978), and Vasilenko and Sutotskii (1980) reported on the formation of deposits inside tubes in supercritical "steam" generators, when an ammonia-hydrazine treatment of the feed water was used.

Glebov et al. (1978) presented the results of an experimental investigation of the overall thermal resistance of deposits in tubes cooled by "steam" at supercritical pressure. Together with the determination of the thermal resistance, they conducted a structural, quantitative, and chemical analysis.

Vasilenko and Sutotskii (1980) presented several graphs, which showed the change in concentration of iron compounds over the circuit of a supercritical "steam"-generating unit. One of the iron compounds is Fe_3O_4, which is a product of the high temperature decomposition of $Fe(OH)_2$. Fe_3O_4 (or magnetite) is the main deposit, which forms on the inner heating surfaces of a supercritical "steam" generator.

In 1983, Glebov et al. published the book "Deposits in Tubes of Supercritical Pressure Boilers" in Russian, in which they summarized existing industrial experience gathered during the operation and servicing of supercritical "steam" generators in Russia. The book contains four chapters: (1) formation of inside-tube deposits during lengthy operation of supercritical "steam" generators, (2) structural and physico-chemical characteristics of inside-tube ferro-oxide deposits, (3) thermophysical properties of inside-tube deposits, and (4) ferro-oxide deposits and reliability of supercritical "steam"-generator operation.

Sutotskii et al. (1989) analyzed data for damage to tubes from 73 supercritical "steam" generators. They found that the tubes were damaged in 43 "steam" generators (about 60% of all generators) in 1987.

15.2 MATERIAL PROBLEMS IN SUPERCRITICAL WATER

The latest review paper (169 references) devoted to the problems of corrosion in high-temperature and supercritical water and aqueous solutions was prepared by Kritzer (2001). According to Kritzer, corrosion in these systems up to supercritical temperatures is determined by several solution properties (density, temperature, pH value, electrochemical potential and "aggressiveness" of attacking anions) and material factors (surface condition and material purity).

Pressure-channel reactors of the Russian named by I.V. Kurchatov Beloyarsk NPP (Grigor'yants et al. 1979; Baturov et al. 1978; Samoilov et al. 1976; Aleshchenkov et al. 1971; Dollezhal' et al. 1974, 1971, 1958) had superheating zirconium channels with stainless steel fuel elements (for steel content, see Table 15.1),

Table 15.1. Content of stainless steels used in Russian superheating steam fuel elements (Sorokin et al. 1989; Samoilov 1985; Arsen'ev and Koledov 1976).

Steel	Element Content, %										
	C	Cr	Ni	Nb	Mo	V	Si	Mn	S	P	N
ЭИ-847	≤0.06	16–18	15–16	0.65	3.0	–	≤0.6	≤0.8	–	–	–
ЭП-753	≤0.01	17.5–19	39–41	0.25–0.6	4.5–5.0	0.05–0.02	≤0.01	1–2	0.015	<0.01	<0.015

which were successfully operated for several years. Steam parameters were: pressure of 7 – 10 MPa and temperature of 400°C – 550°C.

The importance of water density on corrosion and oxidation was pointed out by Watanabe et al. (2001).

Latanision and Mitton (2001) considered stress-corrosion cracking in supercritical water systems.

Scientists from Japan (Suzuki 2001; Sekimura et al. 2001) investigated irradiation-assisted corrosion cracking in supercritical systems.

Konobeev and Birzhevoi (2004) published a review paper on a possible application of the high-nickel alloys in SCWRs. They concluded that behaviour of these alloys is not fully investigated at SCWR conditions. However, some of these alloys showed a relatively low swelling rate and their radiation creep rate and long-time strength are comparable to those of austenitic stainless steels.

Sridharan et al. (2004) studied the corrosion performance of the candidate alloys such as 316 austenitic stainless steel, Inconel 718 and Zircaloy-2, which were exposed to supercritical water at 300°C – 500°C in a corrosion loop at the University of Wisconsin. Their study also included an examination of the austenitic steel samples from a component that was exposed to supercritical water for about 30 years at a fossil-fired thermal power plant.

Allen et al. (2005a) investigated corrosion and radiation response of advanced ferritic-martensitic steels, which are considered candidates materials for core internals, cladding and pressure vessel of the SCWRs. Two materials were used in this study: HCM12A (Grade 122) and 9Cr Oxide Dispersion Strengthened (ODS) steel.

In another investigation, Allen et al. (2005b) looked into behavior of ferritic-martensitic alloys in pure supercritical water to get a better understanding of the kinetics and thermodynamics of the oxidation mechanism.

Bojinov et al. (2005) investigated oxidation kinetics and morphologies for steels exposed to supercritical water conditions. Results of this investigation are reported in their paper.

Hwang et al. (2005) investigated corrosion and stress corrosion cracking (SCC) tests of ferritic-martensitic (F/M) steels and Ni alloys in deaerated supercritical water. SCC tests were performed with slow strain rate tester (SSRT) and U-bent specimens at a pressure of 25 MPa. No SCC was observed on the F/M alloy T91 specimens in the supercritical water within a temperature range of 500°C – 600°C. The ultimate tensile strength of the alloy T91 at 600°C was less than at 500°C. However, the elongations were similar about 20%. In general, F/M alloys T91, 92, and 122 showed higher corrosion rates than the high Ni alloys at 500°C and 550°C. The corrosion rates of the F/M steels at 550°C were three times larger than those at 500°C.

Jang et al. (2005) tested several 9Cr F/M steel specimens for corrosion behaviour in non-aerated and deaerated supercritical water at 25 MPa pressure and within the range of temperatures from 350°C to 627°C for up to 500 h. They found that all steel specimens showed oxide layers on their surfaces.

Kaneda et al. (2005) analyzed corrosion performance from the viewpoints of thickness, morphology and chemical compositions of the oxide films formed by supercritical water exposure. The corrosion supercritical water tests were performed with austenitic and ferritic steels, Ni-based alloys and Ti-based alloys within the following ranges: temperatures—290°C, 380°C, and 550°C; pressure —25 MPa; dissolved oxygen—8 ppm; conductivity—less than 0.1 μS/cm; and test period—500 h.

Was and Allen (2005) studied the corrosion behavior of austenitic and ferritic-martensitic alloys in supercritical water to get better understanding of kinetics and thermodynamics of the oxidation mechanism. These alloys together with Ni-base alloys were exposed in supercritical water within the range of temperatures from 400°C to 550°C for periods of up to 1026 h. They found that weight gains, which vary roughly according to alloy class, were dependent on both the temperature and oxygen concentration in supercritical water. In general, Ni-based alloys showed the smallest weight gain compared to that of other tested alloys.

15.3 EFFECT OF DISSOLVED GAS ON HEAT TRANSFER

Petukhov et al. (1985) conducted an experimental study of heat transfer to a turbulent flow of carbon dioxide at supercritical pressure in a heated tube with different concentrations of a nitrogen impurity. The data obtained were used to determine the character and scale of the effect of the gas impurity on heat transfer.

Chapter 16

SUMMARY

The following is a summary of the main findings of this study:

- There is a plethora of data, empirical correlations and simple models for heat transfer and pressure drop published in the literature on the use of many supercritical fluids, mainly for simplified test sections. We have compiled and reviewed as much of the information as possible (in general, about 650 literature sources), and have provided all the key references and inter-compared the experimental and theoretical approaches. The data largely cover all the ranges of interest, but of course for mainly commercial reasons some design specific information is missing from the open papers and reports.
- There are hundreds of fossil power plants in the world using supercritical conditions (thermal parameters: water pressure of up to 25 – 30 MPa, turbine inlet temperatures of up to 625°C (but mainly lower than 600°C) and power output of up to 1400 MW_e), which have been successfully operated for many years. Their main advantage is high thermal efficiency of up to 45% – 53%. The demonstrated experience in their design and operation is very helpful for current developments in fossil-fired units and in nuclear-powered reactors cooled with supercritical water and provides useful benchmark data.
- After a 30-year hiatus, because of the fossil experience and the need to improve the overall cycle thermal efficiency, the idea of developing nuclear reactors cooled with supercritical water became attractive again, and several countries (Canada, Germany, Japan, Russia, and the USA) have started to work in that direction. However, none of these concepts is expected to be implemented in practice before 2015 – 2020.
- The major limits in designing supercritical heat-transfer systems seems to be with the materials reliability and corrosion rates at high temperatures, pressures, and for nuclear systems with neutron fluxes, within a highly aggressive medium such as supercritical water. The combined effect of these parameters is yet to be fully defined and investigated.

- Heat transfer at supercritical pressures is strongly influenced by the significant and rapid changes in thermophysical properties at these conditions near the critical point. For many working fluids that are used at supercritical conditions, their physical and thermophysical properties are well established and available via the ASME tables and NIST computerized tabulations. All thermophysical properties undergo significant changes near the critical and pseudocritical points. In the vicinity of pseudocritical points with an increase in pressure, these changes become less pronounced. In general, density and dynamic viscosity undergo a significant drop within a very narrow temperature range, while specific enthalpy and kinematic viscosity undergo a sharp increase. Volume expansivity, specific heat, thermal conductivity, and Prandtl number have a peak near both the critical and pseudocritical points. The magnitudes of these peaks decrease very quickly with an increase in pressure. The heat transfer and pressure drop show corresponding variations. However, satisfactory analytical methods have not yet been developed due to difficulty in dealing with the steep property variations, especially in turbulent flows and at high heat fluxes.
- The majority of the experimental studies deal with heat transfer and relatively few with hydraulic resistance of working fluids, mainly water, carbon dioxide, and helium, in circular tubes. A limited number of studies were devoted to heat transfer and pressure drop in annuli, rectangular-shaped channels and bundles (just two such data sets have been found so far). In general, experiments at supercritical pressures are very expensive and require sophisticated equipment and measuring techniques. Therefore, some studies (for example, heat transfer in bundles) are proprietary and hence remain unknown or are not published in the open literature.
- In general, experiments showed that there are three modes of heat transfer somewhat loosely defined in fluids at supercritical pressures: (1) normal heat transfer, (2) so-called deteriorated heat transfer with lower values of the HTC (and hence higher values of wall temperature) than those for (1) within some part of a test section; and (3) relatively increased or improved heat transfer with higher values of the HTC within some part of a test section. We give a more precise definition based on the relative magnitudes of the HTCs. The deteriorated heat transfer is of limited extent, usually appears at high heat fluxes and low mass fluxes in simple tubes, and is generally considered to be due to buoyancy forces dominating the formation and behavior of the heat transfer boundary layer near the heated wall. Importantly, this decreasing HTC phenomenon can be entirely suppressed or significantly offset by increasing the turbulence level with flow obstructions and other heat-transfer enhancing devices.
- In consequence of the above, the limited region of deteriorated heat transfer has not been detected in bundles cooled with supercritical water, as based on the only two available references.
- There are many heat-transfer correlations (empirical fits to data) obtained at various supercritical conditions, which describe heat

transfer mainly in circular tubes and similar simple flow geometries. A comparison of these correlations showed that several of them can be used for preliminary estimations of HTC. However, no one correlation is presently able to completely describe deteriorated heat transfer.

- There exists a single correlation suitable for heat-transfer calculations in water at supercritical pressures flowing in reactor bundles. The Dyadyakin-Popov (1977) correlation was obtained in water at supercritical pressures flowing in a short tight finned bundle and hence is not suitable for other types of bundles.
- While useful progress has been reported on scaling heat transfer between different fluids using dimensionless groups. Scaling parameters should be selected and used with caution. In general, they can be used for scaling operating conditions from one fluid to another just for comparative reference purposes. Due to scaling parameters simplicity, the special behavior of thermophysical properties at supercritical pressures and complexity of the processes involved, causes some discontinuities to exist.
- There are considerably fewer publications related to hydraulic resistance at supercritical pressures than on HTC (about 30 papers). According to some of the cited literature sources, the hydraulic resistance of an isothermal turbulent flow of fluid at the near-critical state follows the same trends as that at subcritical pressures in smooth tubes.
- There is no one correlation suitable for hydraulic-resistance calculations in water at supercritical pressures flowing in heated bundles. The Dyadyakin-Popov (1977) correlation was obtained in water at supercritical pressures flowing in a short tight-finned bundle.
- Because supercritical fluids are thermally expandable, and the flow and pressure-drop multiple values for a given heating profile, the flow can be unstable in certain regions. Hence, the heat transfer and hydraulic resistance at supercritical pressures can be accompanied by flow oscillations and other instabilities at some operating conditions. However, experimental data on these aspects remains very limited.

Appendix A

BOOKS AND REVIEW PAPERS

There are a number of books and review papers devoted to the problem of heat transfer and hydraulic resistance of fluids at near-critical and supercritical pressures. These literature sources are listed below in chronological order for completeness.

1961–1970

Possibly the first review (48 references, including four Russian publications) of heat transfer and fluid flow of water in the supercritical region with forced convection was prepared by A.A. Bishop, L.E. Efferding, and L.S. Tong (Atomic Power Department, Westinghouse Electric Corporation, USA) in 1962.

S.S. Kutateladze and A.I. Leont'ev (1964) (Russian scientists, USSR) discussed heat-transfer basics in a channel at critical and supercritical pressures in their book "Turbulent Boundary Layers in Compressible Gases." They provided just several early correlations on supercritical heat transfer.

In 1968, W.B. Hall, J.D. Jackson and A. Watson (University of Manchester, UK) published a review paper (41 references, including seven Russian publications) on forced-convective heat transfer to fluids at supercritical pressures. Their brief survey of experimental data sets and empirical correlations was supplemented with a discussion of the main semi-empirical theories that have been proposed. It was concluded that the correlations and theories were in a good agreement with experimental data only within very limited ranges and more experimental and theoretical studies were needed.

In 1969, R.V. Smith (Cryogenics Division, National Bureau of Standards, Boulder, CO, USA) published a paper (37 references, including two Russian publications), which reviewed heat transfer to helium I, including heat transfer at supercritical conditions.

In 1970, R.C. Hendricks, R.J. Simoneau, and R.V. Smith (1970a,b) (1970b seems to be a short version of the same report) (Lewis Research Centre, NASA, USA) published an extensive survey of heat transfer to near-critical fluids (217 references, including 24 Russian publications). Their survey covers such topics as:

- near-critical fluid properties—thermodynamics of the critical point, p–ρ–T data – equations of state, transport properties, and pseudo-critical properties;

- heat-transfer regions—region I – gas-fluid, region II – liquids, region III – two phase, and boundaries of region IV – near-critical region;
- near-critical heat-transfer region—peculiarities of the near-critical region, heat transfer in free-convection systems, heat transfer in loops – natural-convection systems, heat transfer in forced-convection systems (heated-tube experiments, detailed investigations into mechanisms), near-critical heat transfer in relation to conventional geometric effects (curved tubes, twisted tapes and rifle boring, body-force orientation, entrance effects), theoretical considerations in forced convection (mixing length analyses, acceleration—strain rates, penetration model), oscillations (general remarks, thermal-acoustic oscillation, system oscillations), choking phenomenon, and zero-gravity operation; and
- summary of results.

Several reviews were prepared by a well-known scientist in the area of heat transfer, B.S. Petukhov, from the Institute of High Temperatures, Russian Academy of Sciences in 1968 and 1970. His first survey, "Heat transfer in a single-phase medium under supercritical conditions" (39 references, including 19 Western publications), covered the following topics:

- results of theoretical analysis; and
- results of experimental investigations (normal regimes, regimes with deteriorated heat transfer, and regimes with improved heat transfer).

In his second review, "Heat transfer and friction in turbulent pipe flow with variable physical properties" (97 references, including 60 Western publications), B.S. Petukhov (1970) considered such topics as:

- analytical methods—basic equations, eddy diffusivities of heat and momentum, analytical expressions for temperature and velocity profiles, heat transfer, and skin friction;
- heat transfer with constant physical properties—analytical results and experimental data;
- heat transfer and skin friction for liquids with variable viscosity—theoretical results, experimental data, and empirical equations;
- heat transfer and skin friction for gases with variable physical properties-analytical results, experimental data, and empirical equations; and
- heat transfer and skin friction for single-phase fluids at subcritical states—analytical results, experimental data, and empirical equations for normal heat-transfer regimes; experimental data for regimes with deteriorated and improved heat transfer.

1971–1980

In 1971, W.B. Hall (University of Manchester, UK) published a review paper on heat transfer at the near-critical point (57 references, including 12 Russian publications). In this literature survey, the following topics were covered:

- physical properties near the critical point (thermodynamic properties, molecular structure near the critical point, transport properties and the implications of physical property variation on heat transfer);
- the equations of motion and energy (boundary-layer flow, channel flow, turbulent shear stress and heat flux);
- forced convection (methods of presenting data, experimental data, correlation of experimental data, and semi-empirical theories);
- free convection (experimental results, theoretical methods and correlations);
- combined forced and free convection (experimental results and proposed mechanism for heat-transfer deterioration); and
- boiling (nucleate boiling, film boiling and pseudo-boiling).

G.V. Alekseev and A.M. Smirnov (1976) (Institute of Physics and Power Engineering, Obninsk, Russia) prepared an analytical review (206 papers, including 55 Western publications) of the literature devoted to the heat transfer and hydraulic resistance of fluids at supercritical pressures. The analytical review consisted of the following parts:

- physical properties of water at supercritical pressures;
- results of experimental investigation of heat transfer at supercritical pressures of fluids (normal regimes, regimes with deteriorated heat transfer and regimes with improved heat transfer);
- experimental investigation of hydraulic resistance of friction at turbulent flow of supercritical fluids in tubes;
- experimental investigation of non-isothermal flow structure of supercritical fluids;
- theoretical analysis of heat transfer and friction resistance; and
- conclusions and tasks for future investigations.

In 1978, W.B. Hall and J.D. Jackson (University of Manchester, UK) presented the upgraded review, "Heat transfer at the near-critical point" (71 references, including 22 Russian publications). The review covered such topics as:

- physical properties at the near-critical point;
- supercritical forced convection (low, intermediate and high heat fluxes, heat transfer accompanied by pressure oscillations and the criterion for buoyancy affected flow);
- forced convection in the absence of buoyancy;
- mixed convection in tubes (vertical and horizontal tubes);
- theoretical studies of convection heat transfer (models based on the "universal" velocity distribution and direct solution of the momentum and energy equations);
- free convection; and
- boiling (nucleation at high sub-critical pressures, pool boiling and flow boiling).

In 1979, J.D. Jackson and W.B. Hall (1979a) (University of Manchester, UK) published their review devoted to forced-convection heat transfer to fluids at supercritical pressures (111 references, including 33 Russian publications). In their review, the following topics were considered:

- special features of heat transfer to fluids at supercritical pressures (improved heat transfer, effect of increased heat flux—deterioration of heat transfer, acceleration due to heating—a mechanism for heat-transfer deterioration, effect of buoyancy, effect of wall conduction (conjugated effect) and nonuniformity of heat generation under conditions of deteriorated heat transfer, and thermoacoustic oscillations in supercritical convection);
- correlation of data for supercritical pressure forced convection (dimensionless form of the basic equations for variable properties heat transfer, evaluation of forced convection correlations and similarity considerations); and
- theoretical studies of forced convection to supercritical pressure fluids (governing equations, turbulence models, and comparison of theoretical models).

In addition to the previous review, J.D. Jackson and W.B. Hall (1979b) (University of Manchester, UK) analyzed the effect of buoyancy on heat transfer to fluids flowing in vertical tubes under turbulent conditions and published a review on this topic (52 references, including 15 Russian publications).

In 1980, A.P. Ornatskiy, Yu.G. Dashkiev and V.G. Perkov (Polytechnic Institute, Kiev, Ukraine) published a book on "steam" generators operating at supercritical pressures. The book (288 pages, 129 references including one Western paper, 165 figures, and 11 tables) covers such topics as:

- peculiarities of the heat transfer and internal deposit creation in supercritical "steam" generators;
- hydrodynamics of the heating surfaces of supercritical "steam" generators;
- peculiarities of the processes in combustion chambers and their designs;
- design of supercritical "steam" generators;
- supercritical "steam" generators for powerful electrical units;
- control of supercritical "steam" generators;
- reliability of supercritical "steam" generators and methods for its improvement; and
- perspectives and future developments in supercritical "steam" generators.

1981–1990

Y.Y. Hsu and R.W. Graham (1986) (USA) published a book on transport processes in boiling and two-phase systems in which Chapter 15 "Heat transfer to near-critical fluid" devoted to heat transfer at supercritical pressures. In this chapter, the following problems were covered:

- peculiarities of near-critical region;
- laminar free convection;
- natural-convection loops;
- laminar flow boundary-layer analysis;
- laminar fully developed pipe flow; turbulent forced-convection (channel flow, correlation of forced-convection data, pseudo-phase model, boundary-layer approach, body force effects in forced convection, acceleration effects, external boundary-layer flow);
- comparison of boiling two-phase flow and heated supercritical fluids; and
- summary.

B.S. Petukhov, L.G. Genin, and S.A. Kovalev (1986) (Moscow, Russia) published a book on heat transfer in nuclear power reactors in which Section 8.5 was devoted to heat transfer in a single-phase near-critical region.

S. Kakaç, R.K. Shah, and W. Aung (1987) (USA) reviewed heat transfer at supercritical pressures in their handbook on single-phase heat transfer (see Sections 18.3.1 and 18.3.2). They listed 18 supercritical heat-transfer correlations.

M. Malandrone, B. Panella, G. Pedrelli, and G. Sobrero (1987a,b) (Turin Polytechnic Institute and ENEL, Pisa, Italy) published two review papers: Paper (a) contained 28 selected references, including 12 Russian publications and paper (b) contained 25 selected references, including 12 Russian publications (some references are identical in both papers) related to the deteriorated heat transfer to supercritical fluids and boundaries of this phenomenon.

Later, in 1988, B.S. Petukhov and A.F. Polyakov (Institute of High Temperatures, Russian Academy of Sciences, Moscow, Russia) published a book "Heat Transfer in Turbulent Mixed Convection" devoted to heat transfer in turbulent mixed convection, in which, Chapter VII "Gravitational effects on heat transfer in a single-phase fluid near the critical point" dealt with heat transfer at supercritical pressures (16 selected references, including three Western publications). The topics covered in this chapter were:

- heat transfer at supercritical pressures in vertical channels; and
- heat transfer at supercritical pressures in horizontal channels.

In 1989, D. Kasao and T. Ito (Kyushu University, Japan) reviewed existing experimental findings on forced-convection heat transfer to supercritical helium. In their paper (21 selected references, including eight Russian publications), the deterioration of heat transfer, the effect of buoyancy forces, and heat-transfer correlations for supercritical helium were discussed.

Several Russian books (or chapters in these books) were devoted to mass transfer and corrosion processes in water at supercritical pressures: Handbook on Thermal and Atomic Power Station (1988); Margulova and Martynova (1987); Glebov (1983); Antikain (1977); Mankina (1977); and Akolzin et al. (1972).

In 1990, P.L. Kirillov, Yu.S. Yur'ev, and V.P. Bobkov (Institute of Physics and Power Engineering, Obninsk, Russia) published a second edition of their handbook "Handbook for Thermal-Hydraulic Calculations (Nuclear Reactors, Heat Exchangers, Steam Generators)," in which two Sections, 3.2 and 8.4, contained parts devoted to the hydraulic resistance of working fluids at near-critical parameters and the heat transfer at near-critical and supercritical pressures.

1991–2000

In 1991, A.F. Polyakov (Institute of High Temperatures, Russian Academy of Sciences, Moscow, Russia) prepared a literature review of 83 references, including 25 Western publications. In this review, the following problems of heat transfer at supercritical pressures were given special attention:

I. General description of the problem
   - thermophysical properties of fluids; and
   - approaches to problem solving.

II. Heat transfer at forced convection in circular tubes
   - laminar flow and turbulent flow without substantial influence of the gravity field;
   - turbulent mixed convection (vertical and horizontal tubes);
   - turbulent heat transfer at non-uniform axial heat flux; and
   - turbulent heat transfer under cooling.

III. Free convection
   - vertical surfaces (laminar and turbulent flow); and
   - horizontal wires.

IV. Special problems
   - data on transient heat transfer (transient determined by external conditions and thermo-acoustic oscillations); and
   - heat-transfer enhancement at conditions corresponding to the deteriorated heat transfer.

In 1993, B.S. Petukhov (Institute of High Temperatures, Russian Academy of Sciences, Moscow, Russia) published a book entitled "Heat Transfer in Flowing Single-Phase Medium" (Editor A.F. Polyakov). In this book, Section 9.3, "Free convection near vertical plate in the medium at near-critical state parameters" was devoted to heat transfer under free convection at near-critical pressures.

In 1998, V.A. Kurganov (1998a,b) (Institute of High Temperatures, Russian Academy of Sciences, Moscow, Russia) published a summary paper in two parts: Part 1—29 selected references, including eight papers by the author and 5 Western publications and Part 2—contained 32 selected references, including 14 papers by the author and 4 Western publications (some references are identical in both parts). This paper covered the following areas of heat transfer and pressure drop at supercritical pressures:

- variability of the fluid properties, heat transfer, and hydrodynamics in the supercritical region;
- heat transfer and pressure drop in the regimes of normal heat transfer;
- heat transfer and pressure drop at high heat fluxes, regimes of deteriorated heat transfer; and
- heat transfer in the liquid-state region, the effects of additional factors on heat transfer at high heat fluxes and enhancement of deteriorated heat transfer.

One of the latest (by date of publication, but not by material) concise reviews was prepared by J.D. Jackson (University of Manchester, UK) and published in the "International Encyclopedia of Heat & Mass Transfer" in 1998. According to the author, his review was mainly based on the publication by Hall and Jackson (1978).

An extensive literature review was prepared by I.L. Pioro and S.C. Cheng (University of Ottawa, Canada) in 1998. They reviewed 150 publications, including 134 Russian publications and 16 selected Western papers, devoted to the heat transfer and hydraulic resistance of fluids at near-critical and supercritical pressures. In this review, the following topics were given special attention:

- concepts of nuclear reactors for supercritical pressures;
- physical properties of fluids and the HTC at the near-critical point;
- analytical approaches for estimating heat transfer and hydraulic resistance at near-critical and supercritical pressures;
- heat transfer and hydraulic resistance of water at supercritical pressures;
- heat transfer and hydraulic resistance of CO_2 at supercritical pressures;
- heat transfer and hydraulic resistance of helium at supercritical pressures;
- heat transfer and hydraulic resistance of other fluids at supercritical pressures;
- practical prediction methods for heat transfer and hydraulic resistance at supercritical pressures;
- flow stability at near-critical and supercritical pressures; and
- some problems related to supercritical pressures.

E.K. Kalinin, G.A. Dreitser, I.Z. Kopp and A.S. Myakochin (Kalinin et al. (1998; see also, Dreitser (1993) and Dreitser et al. (1993)) (State Moscow Aviation Institute, Russia) published a book devoted to single- and two-phase flow heat-transfer enhancement, in which Section 2.6.8 described methods for heat-transfer enhancement in hydrocarbons flowing in circular tubes at supercritical pressures.

S.S. Pitla, D.M. Robinson, E.A. Groll and S. Ramadhyani (School of Mechanical Engineering, Purdue University, Indiana, USA) published their review devoted to the heat transfer of supercritical carbon dioxide in tube flow in 1998. They reviewed 75 publications including 33 Russian papers.

Recently, P.L. Kirillov (2000) (Institute of Physics and Power Engineering, Obninsk, Russia) published a short review of Russian studies devoted to heat and mass transfer at supercritical pressures.

Two review papers by R.V. Smith (1999) (USA) and Yo. Oka (2002, 2000) (University of Tokyo, Japan) dealt with supercritical power-plant "steam" generators and modern concepts of nuclear reactors at supercritical pressures, respectively.

In 2000, A.M. Smirnov (Institute of Physics and Power Engineering, Obninsk, Russia) published a bibliography on hydrodynamics and heat transfer at supercritical pressures, which contained titles of the published works since 1954 (several hundred Russian and Western publications).

In 2000, P. Kritzer (Freudenberg Vliesstoffe KG, Germany) published a review paper (169 references) on corrosion in high-temperature supercritical water and aqueous solutions.

S. Yoshida and H. Mori (2000) (Kyushu University, Japan) published a concise overview of the current knowledge of heat transfer in fluids at supercritical pressures (20 selected references, including six Russian publications).

In 2000, the 1st International Symposium on Supercritical Water-Cooled Reactor Design and Technology (SCR-2000) was held in Tokyo (Japan). Thirty-four papers were presented at the symposium on the following topics:

- conceptual design study and development program of SCWRs (4 papers);
- thermal-hydraulics (2 papers);
- experience of supercritical fossil-fired power plants (2 papers);
- physics and chemistry of supercritical water (2 papers);
- material issues and water chemistry (5 papers);
- reactor design (2 papers);
- thermo-hydrodynamics (2 papers);
- radiation chemistry of supercritical water (2 papers);
- radiation-induced reactions in supercritical fluids (4 papers);
- corrosion and high-temperature materials (four papers); and
- damage-formation mechanisms in dielectric materials (five papers).

In addition to the symposium mentioned above, the International Symposium on Supercritical Fluids had been convening regularly for some time. The latest, 5th symposium, was held in Atlanta (USA) in April of 2000. This symposium was mainly devoted to chemical and pharmaceutical applications of supercritical fluids, such as catalysis, separation, extraction, and reactions in supercritical fluids and other fluids.

2001–PRESENT

In 2001, X. Cheng and T. Schulenberg (Forschungszentrum Karlsruhe, Germany) prepared a literature review of selected papers (54 references, including 16 Russian publications) with relevance to the development of the HPLWR. The literature survey covered the following topics: general features of heat transfer at supercritical pressure, experimental and numerical studies, prediction methods, deterioration of heat transfer, friction pressure drop, and application to HPLWR.

J.D. Jackson (2001) (University of Manchester, UK) presented a keynote paper at the International Conference on Energy Conservation and Application (ICECA 2001) held in Wuhan (China). His paper lists some striking features of heat transfer with fluids at pressures and temperatures near the critical point. The paper references 28 open literature sources.

In 2002, a book on supercritical fluids molecular interactions, physical properties and new applications in chemical and pharmaceutical industries (Supercritical Fluids 2002) edited by Ya. Arai, T. Sako, and Yo. Takebayashi was published by Springer-Verlag Publishing House. This monograph contains the following chapters:

1. Solution structure in supercritical fluids.
2. Phase equilibria and static properties.
3. Transport properties of supercritical fluids.
4. Extraction and separation using supercritical fluids.
5. Material processing using supercritical fluids.
6. Reactions in supercritical fluids.

In 2003, I.L. Pioro and R.B. Duffey (Pioro and Duffey 2003b) (Chalk River Laboratories, AECL, Canada) published the latest and the most extensive literature survey. This survey consists of 430 references, including 269 Russian publications and 161 Western publications, devoted to the problems of heat transfer and hydraulic resistance of a fluid at near-critical and supercritical pressures. The objective of the literature survey is to compile and summarize findings in the area of heat transfer and hydraulic resistance at supercritical pressures for various fluids for the last 50 years published in the open Russian and Western literature. The analysis of the publications showed that the majority of the papers were devoted to the heat transfer of fluids at near-critical and supercritical pressures flowing inside a circular tube. Three major working fluids are involved: water, carbon dioxide, and helium. The main objective of these studies was the development and design of supercritical "steam" generators for power stations (utilizing water as a working fluid) in the 1950s, 1960s, and 1970s. Carbon dioxide was usually used as the modeling fluid due to lower values of the critical parameters. Helium, and sometimes carbon dioxide, was considered as possible working fluids in some special designs of nuclear reactors. Later, several chapters of this report were updated and published as separate papers (see Duffey et al. 2006, 2005, 2003; Pioro and Duffey 2005 2003a; Duffey and Pioro 2006, 2005a,b, 2004; Pioro et al. 2004a,b; 2003).

R. Kurihara, K. Watanabe, and S. Konishi (2003) (Japanese Atomic Energy Research Institute (JAERI), Japan) published a literature survey (in Japanese) devoted to supercritical water application in fusion demo reactor. In this survey, they evaluated blowdown behaviour of supercritical water, LOCA, etc.

T. Chen (2004) (Xi'an University, Xi'an, China) published a book in Chinese on two-phase flow and heat transfer in which several chapters are devoted to heat transfer in supercritical "steam" generators.

In general, the following major international regular meetings are partially or fully devoted to heat transfer and pressure drop at supercritical pressures and to SCWR concepts:

- International Conference on Nuclear Engineering (ICONE)
- International Congress on Advances in Nuclear Power Plants (ICAPP)
- International Heat Transfer Conference
- International Symposium on Supercritical Water-Cooled Reactor Design and Technology (SCR)
- International Conference GLOBAL
- International Topical Meeting on Nuclear Reactor Thermal Hydraulics (NURETH)
- Pacific Basin Nuclear Conference
- American Nuclear Society (ANS) International Meeting
- Joint International Conference Global Environment and Nuclear Energy Systems/Advanced Nuclear Power Plants (GENES4/ANP2003)

Appendix B

THERMOPHYSICAL PROPERTIES OF CARBON DIOXIDE, R-134A AND HELIUM NEAR CRITICAL AND PSEUDOCRITICAL POINTS

The following figures of carbon dioxide, R-134a, and helium show thermophysical properties variation near the critical point for each substance (solid lines) and the pseudocritical point at a pressure (dashed lines), which is equivalent to the water pressure of 25 MPa (p/p_{cr} = 1.133) (for details, see Table B1). Data in these figures were obtained with temperature increments of 0.01°C (for helium – 0.05°C). It should be noted that height of the peaks in specific heat, thermal conductivity, volume expansivity, and Prandtl number in the critical point and pseudocritical points near the critical point might vary with a temperature increment value.

Table B1. Critical and pseudocritical (at p/p_{cr}=1.133) parameters of water, carbon dioxide, R-134a, and helium (based on NIST (2002)).

Fluid	Type of Parameters	*p*, MPa	*t*, °C	H_b, kJ/kg
Water	Critical	22.064	373.95	2146.6
	Pseudocritical	25	384.90	2152.2
Carbon dioxide	Critical	7.3773	30.98	342.39
	Pseudocritical	8.36	36.60	340.75
R-134a	Critical	4.0593	101.06	385.61
	Pseudocritical	4.6	107.41	397.75
Helium	Critical	0.2275	–267.95	10.239
	Pseudocritical	0.2577	–267.79	12.070

B1. Thermophysical Properties of Carbon Dioxide Near Critical and Pseudocritical Points (NIST 2002)

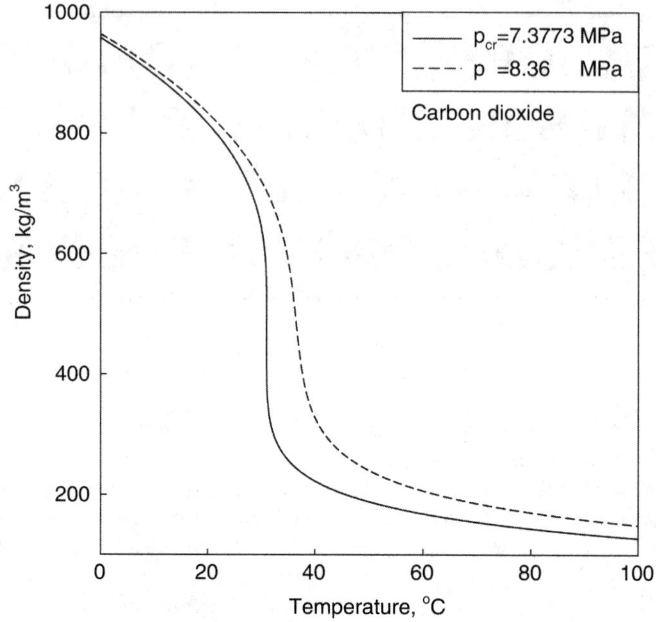

Figure B1.1. Density vs. temperature.

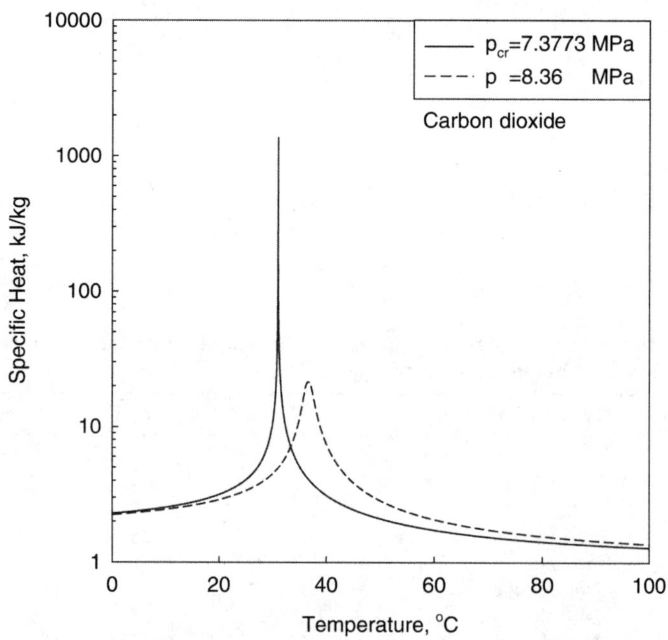

Figure B1.2. Specific heat vs. temperature.

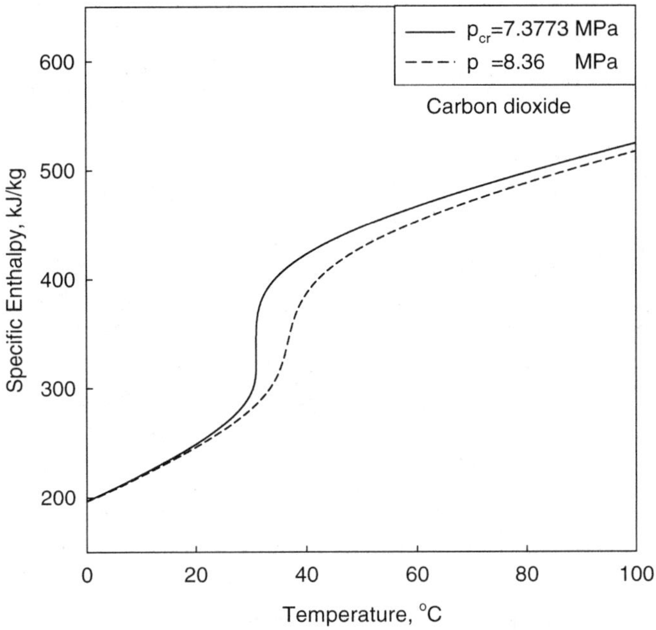

Figure B1.3. Specific enthalpy vs. temperature.

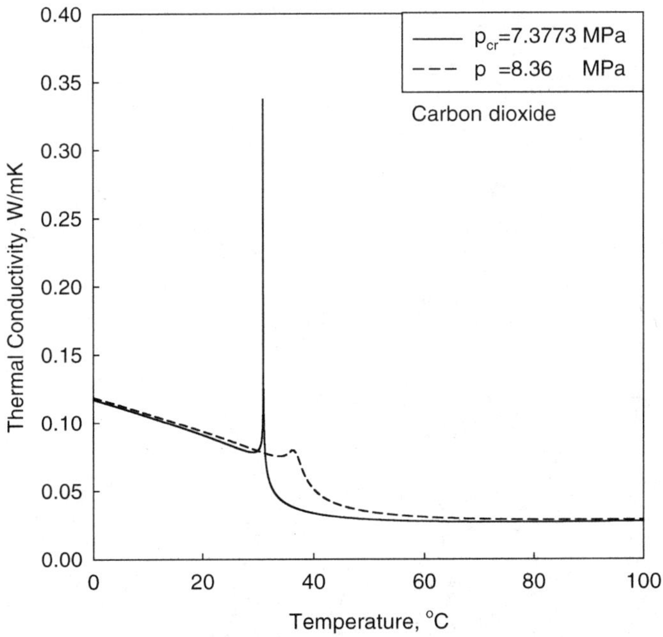

Figure B1.4. Thermal conductivity vs. temperature.

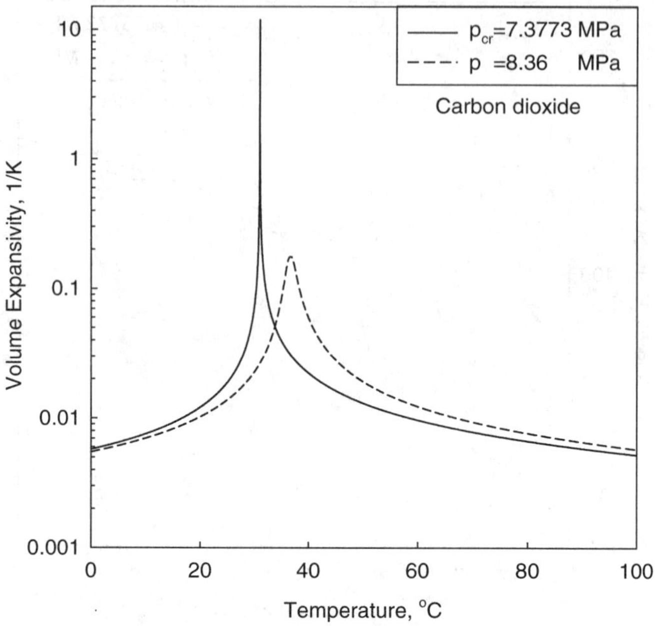

Figure B1.5. Volume expansivity vs. temperature.

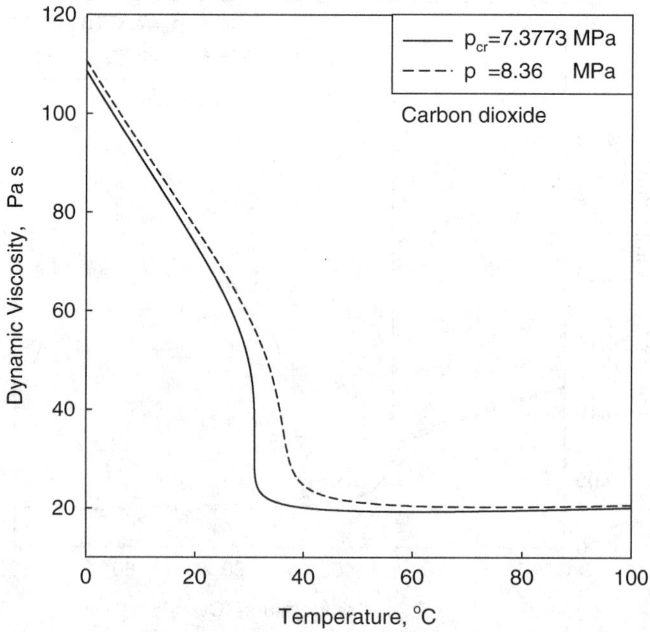

Figure B1.6. Dynamic viscosity vs. temperature.

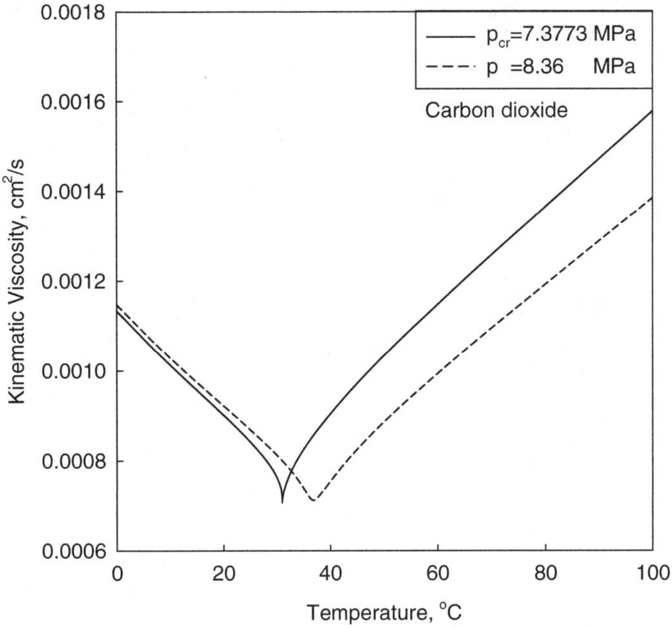

Figure B1.7. Kinematic viscosity vs. temperature.

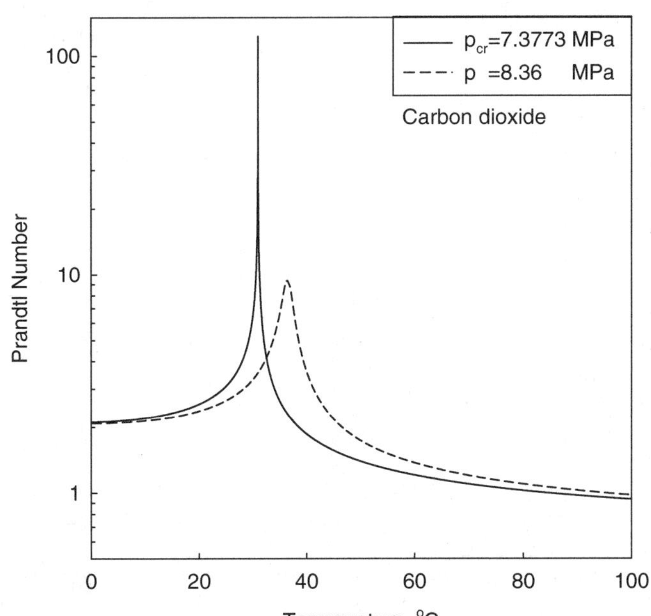

Figure B1.8. Prandtl number vs. temperature.

B2. Thermophysical Properties of R-134a Near-Critical and Pseudo-critical Points (NIST 2002)

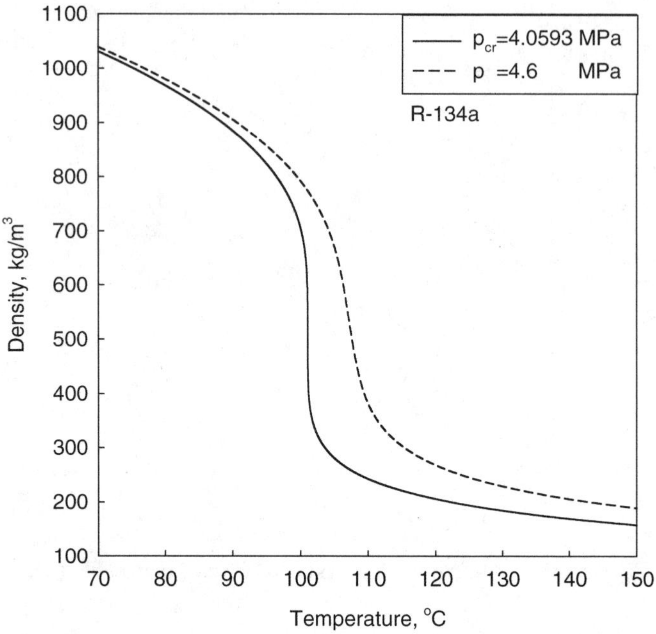

Figure B2.1. Density vs. temperature.

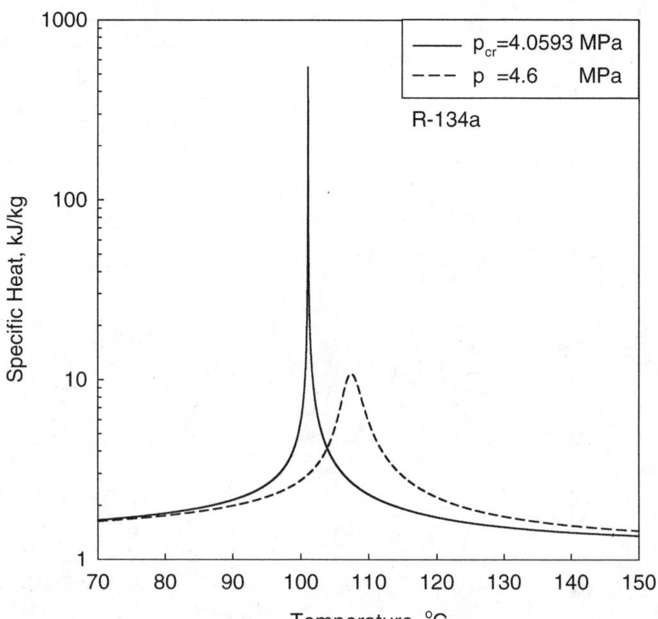

Figure B2.2. Specific heat vs. temperature.

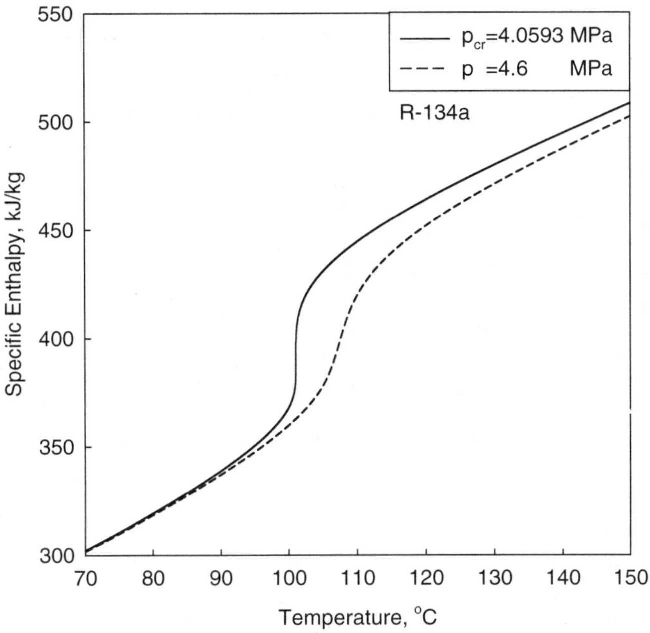

Figure B2.3. Specific enthalpy vs. temperature.

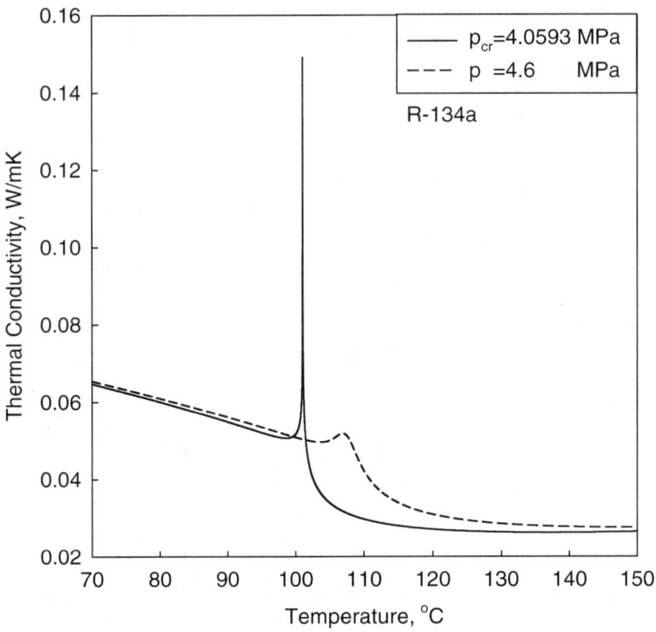

Figure B2.4. Thermal conductivity vs. temperature.

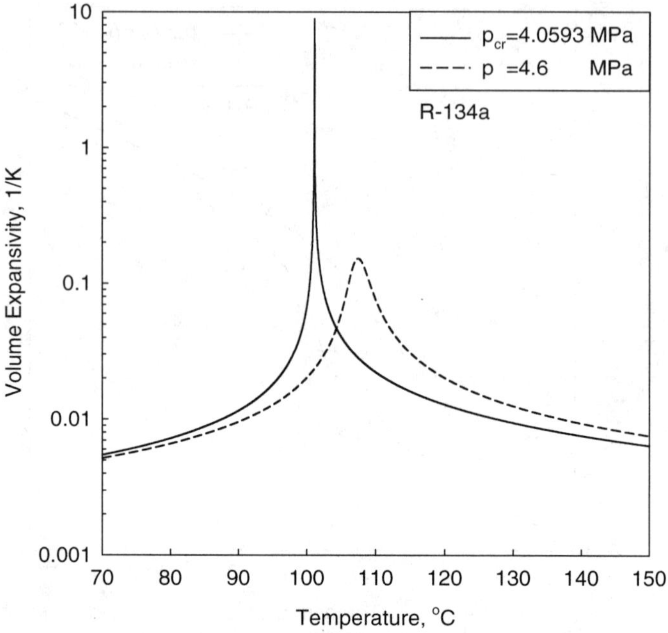

Figure B2.5. Volume expansivity vs. temperature.

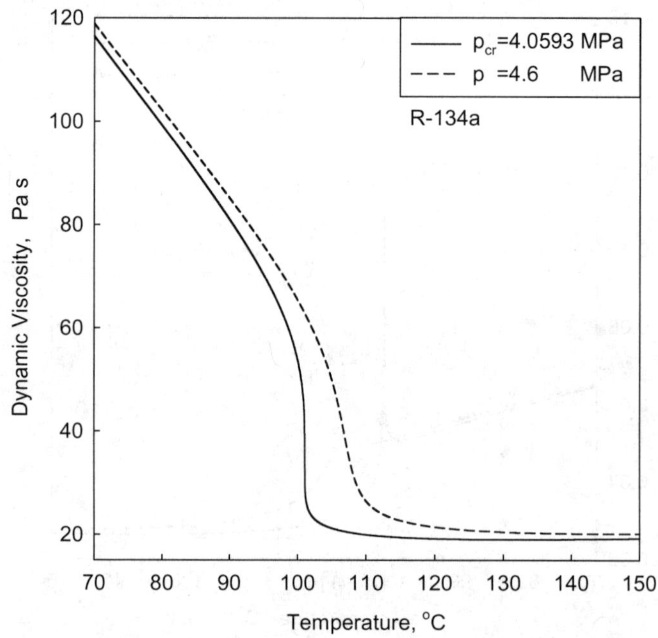

Figure B2.6. Dynamic viscosity vs. temperature.

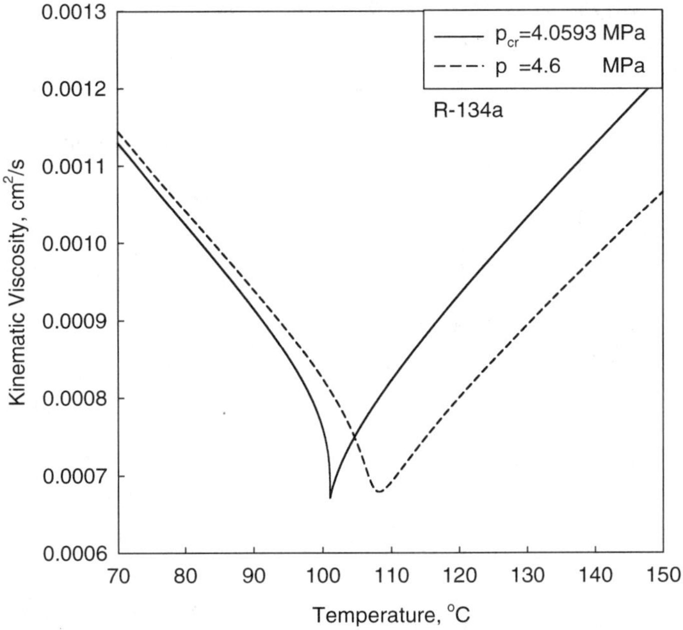

Figure B2.7. Kinematic viscosity vs. temperature.

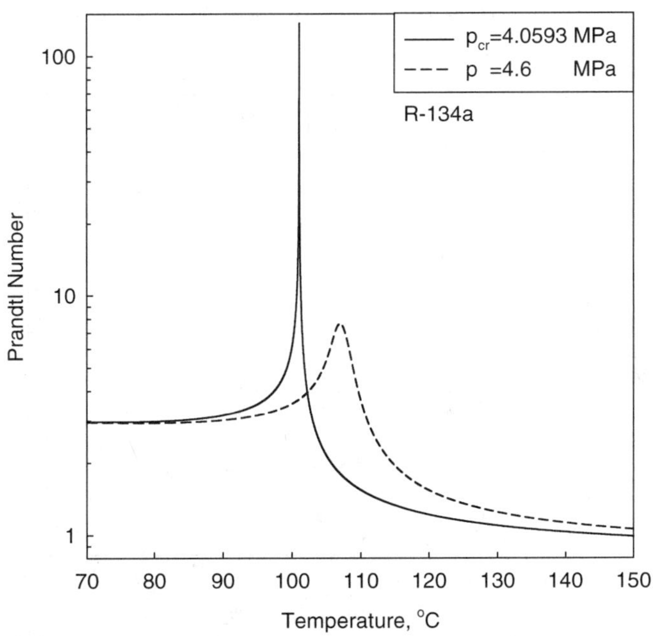

Figure B2.8. Prandtl number vs. temperature.

B3. Thermophysical Properties of Helium Near Critical and Pseudo-critical Points (NIST 2002)

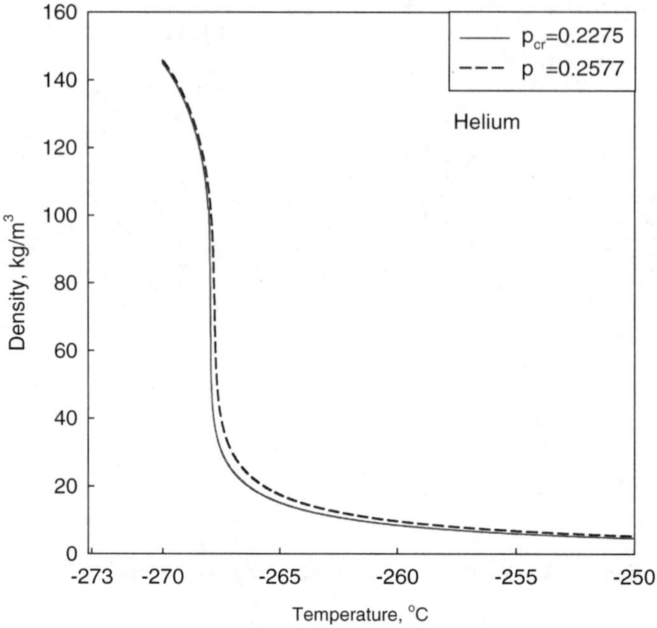

Figure B3.1. Density vs. temperature.

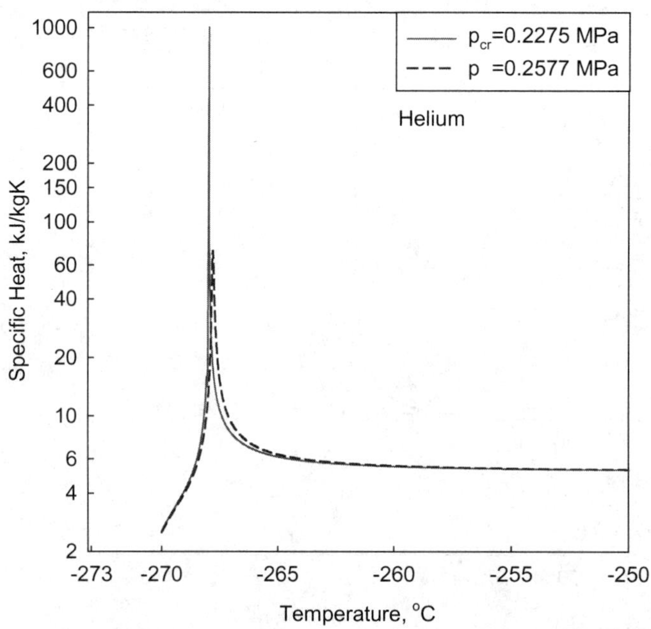

Figure B3.2. Specific heat vs. temperature.

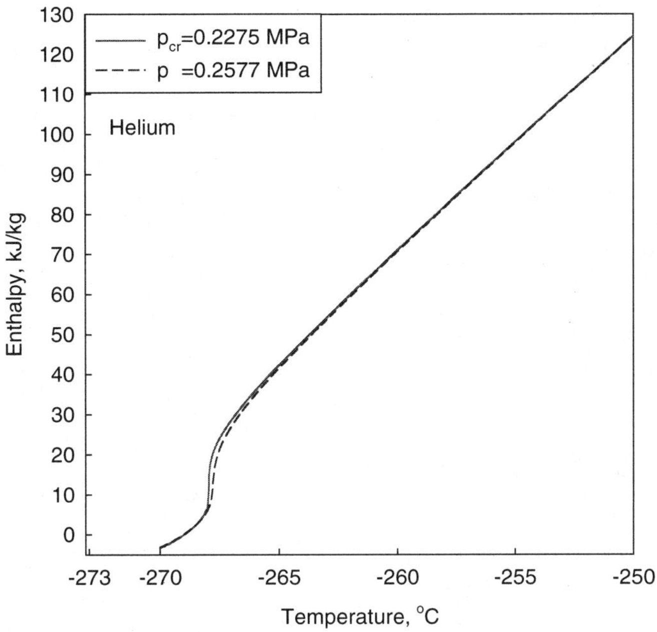

Figure B3.3. Specific enthalpy vs. temperature.

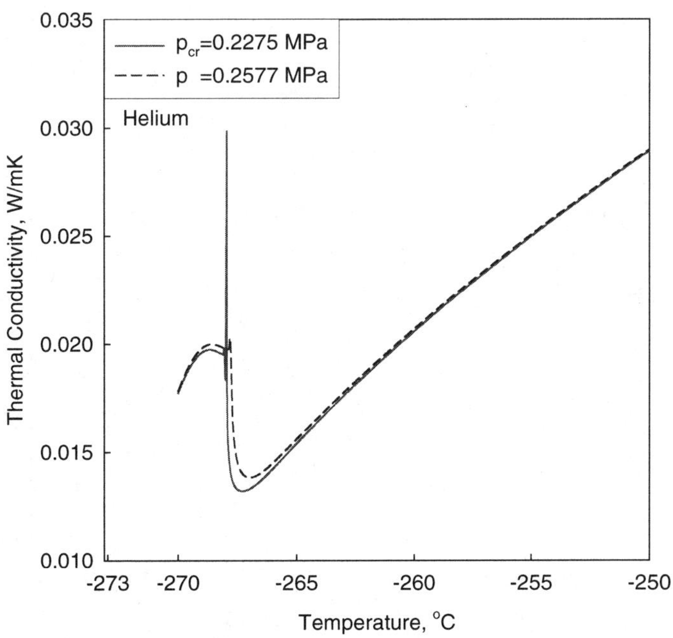

Figure B3.4. Thermal conductivity vs. temperature.

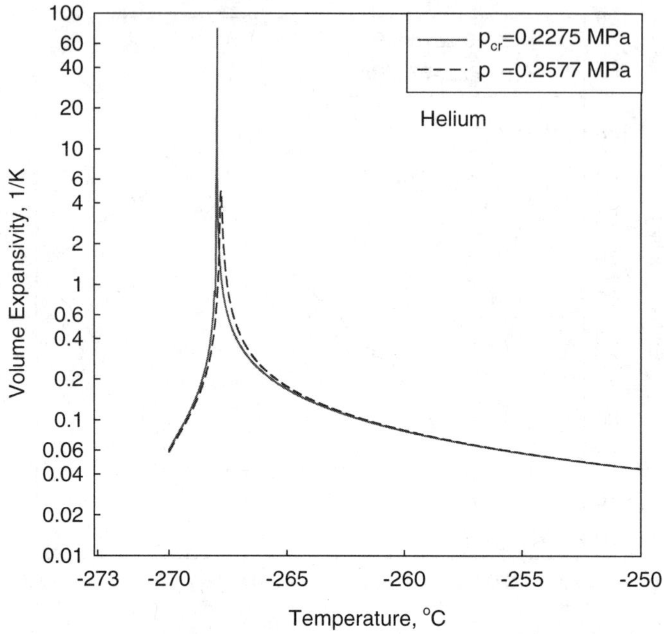

Figure B3.5. Volume expansivity vs. temperature.

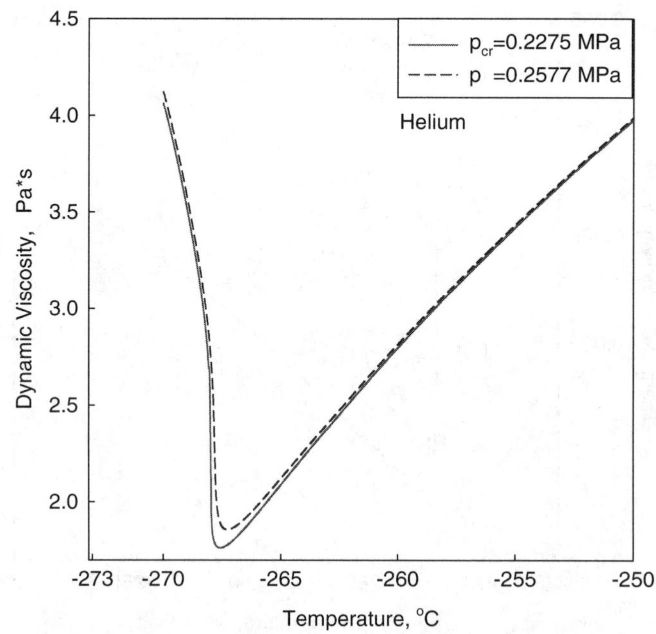

Figure B3.6. Dynamic viscosity vs. temperature.

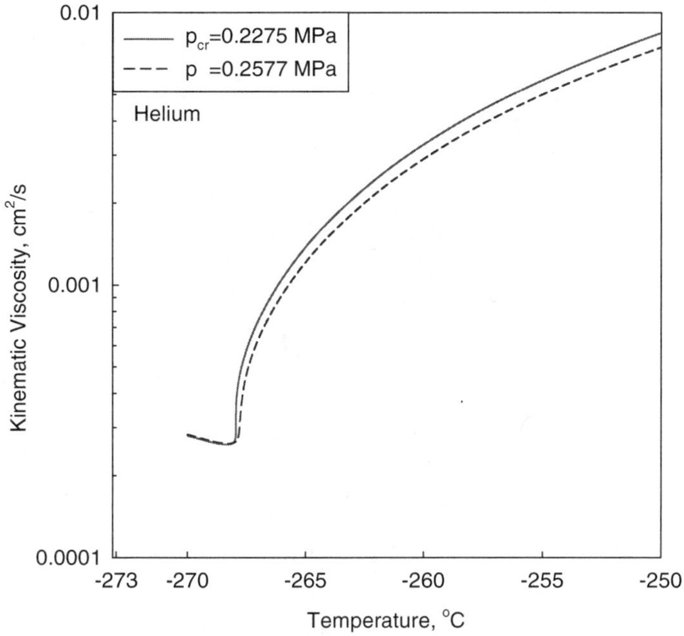

Figure B3.7. Kinematic viscosity vs. temperature.

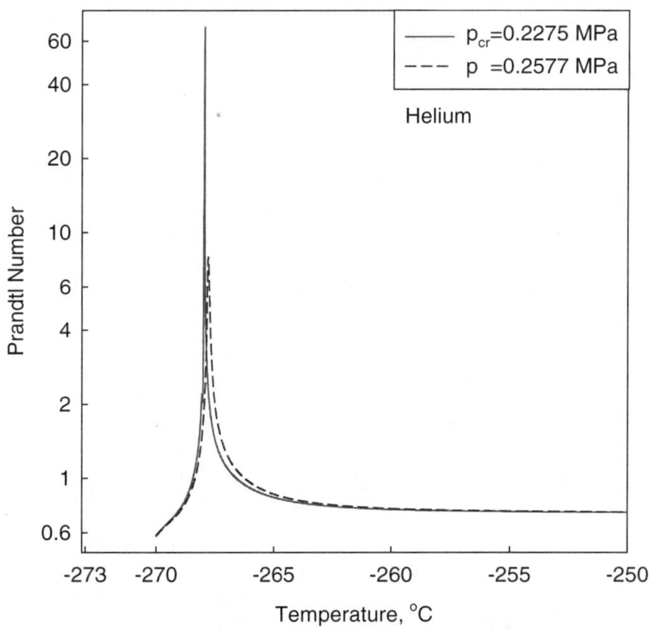

Figure B3.8. Prandtl number vs. temperature.

Appendix C

ADDITIONAL INFORMATION ON CRL SUPERCRITICAL CO_2 TEST FACILITY

C.1 CHEMICAL ANALYSIS OF TEST-SECTION MATERIAL

The test section was manufactured from Inconel 600 seamless tubing. The tube was cold drawn, bright annealed, and passivated (specification ASTM-B-167-00). The heat number of the tube is 769570. The tube was hydro tested at 6.9 MPa (1000 psi) and was subjected to an eddy current test. The tube has a tensile strength of 623.7 MPa (90,460 psi) and yield strength of 316.5 MPa (45,906 psi). The chemical content of the tube material is listed in Table C1.

Table C1. Chemical content of tube material (Inconel 600) in %.

C	Mn	S	Si	Ni	Cr	Mo	Fe	N	Cu	Co
0.021	0.81	0.001	0.36	72.80	16.54	0.01	8.85	0.015	0.010	0.015

The above-mentioned information is important for exact knowledge of tube material properties.

C.2 PRECISE MEASUREMENTS OF TEST-SECTION INSIDE DIAMETER

Measured values of the test-section ID are listed in Table C2. The measurement was performed using a digital bore micrometer with an accuracy of ±0.001 mm.

The above-mentioned information together with knowledge of the tube OD is important for accurate calculation of the volumetric heat flux. Also, information on ID and OD will show how perfect is a tube shape.

C.3 TEST-SECTION SURFACE ROUGHNESS PARAMETERS

It is well known (Incropera and DeWitt 2002; Tanaka et al. 1971) that for turbulent flow, the HTC increases with wall roughness. Therefore, the surface-roughness parameters of the heating surface may be important at some flow conditions.

Table C2. Precise measurements of test-section ID (nominal values are D = 8 mm and D_{ext} = 10 mm).

Distance from tube end inside tube, mm	Inside diameter, mm	
	Measurement along *x*-axis	Measurement along *y*-axis
	Tube end #1	
38	8.039	8.034
89	8.054	8.046
114	8.051	8.074
140	8.077	8.069
Average value	8.055	8.056
	Tube end #2	
38	8.036	8.046
89	8.074	8.067
114	8.067	8.067
140	8.061	8.061
Average value	8.060	8.060
Average inside diameter	**8.058±0.02 mm**	

Usually average or root-mean-square (RMS) surface roughness was used to characterize the heating surface. However, these two parameters cannot fully describe surface microstructure (see below). Therefore, it is a good practice (Pioro and Duffey 2003b) to list the surface-roughness parameters of the test section.

The current test-section surface-roughness parameters are listed below. All these roughness and microstructure parameters will be used to ensure that future test sections are close enough by these parameters to the existing test section.

A laser profilometer was used to determine the surface-roughness parameters that are listed in Table C3.

Table C3. Surface-roughness parameters of test section (circular tube, ID 8 mm, OD 10 mm, Inconel 600).

Surface-Roughness Parameters			
		Values on tube surface	
Symbol	**Unit**	**Internal**	**External**
R_a	μm	0.99	0.65
R_q	μm	1.25	0.91
R_p	μm	4.07	3.55
R_t	μm	8.51	10.50
R_{pm}	μm	3.68	2.94
R_z	μm	7.38	7.22
R_{z3}	μm	7.07	5.72
S_m	–	44.7	42.5
Δ_a	Degree	0.138	0.088
L_o	mm	9.69	9.62
R_{sk}	–	−0.07	−1.00

Explanations to Table C3.

Simple-Roughness-Amplitude Parameters

Mean parameters

R_a – average roughness. The average roughness is the most commonly used parameter in surface-finish measurements.

R_q – RMS roughness: the average roughness parameter calculated as $R_q = \sqrt{\frac{1}{n}\sum_{x=0}^{n}[s(x)-\overline{s(x)}]^2}$, where *s(x)* is the surface height at point *x* in the surface profile and $\overline{s(x)}$ is the average height of the surface profile. The RMS roughness was a commonly used parameter some time ago; however, nowadays it has been replaced with R_a in metal-machining specifications. Usually (but not necessarily), R_q is 1.1–1.3 times larger than R_a.

Extremes parameters

R_p – peak roughness (height of highest peak in the roughness profile over the actual profile length).

R_t – total roughness (vertical distance from the deepest valley to the highest peak), $R_t = R_p + R_v$.

Mean-extremes parameters

R_{pm} – mean-peak roughness (average peak roughness over the actual profile length).

R_z – mean-total roughness (average value of the five highest peaks plus the five deepest valleys over the actual profile length).

R_{z3} – mean-total roughness of third extremes parameters (average vertical distance from the third deepest valley to the third highest peak).

Mean-extremes parameters are less sensitive to single unusual features such as artificial scratches, gouges, burrs, etc.

Mean-roughness-spacing parameters

S_m – mean spacing between peaks (peaks cross above the mean line and then go back below it).

Roughness-hybrid parameters

Δ_a – average of absolute slope of the roughness profile over the actual profile length.

L_o – actual profile length (in all measurements, this was about 10 mm).

Statistical parameters

R_{sk} – skewness (it represents the symmetry of the profile variation over its mean line). Surfaces with $R_{sk} < 0$ have fairly deep valleys in a smoother plateau. Surfaces with $R_{sk} > 0$ have fairly high spikes, which protrude above a flatter average.

C.4 TEST-SECTION BURST PRESSURE

The test-section burst pressure (see Table C4) was calculated using the following correlation (Thermal Power Engineering... 1987):

$$p_{burst} = \frac{2\,S\,\delta_w}{D_{ext}}, \tag{C1}$$

where p_{burst} is the burst pressure in MPa, S is the maximum allowable stress in tension in MPa, δ_w is the tube wall thickness in m ($\delta_w = 0.001$ m), and D_{ext} is the outside tube diameter in m ($D_{ext} = 0.01$ m). Equation (C1) is valid for thin-walled tubes/pipes and cylindrical shells: at $\delta_w / D_{ext} \leq 0.3$ for tubes and pipes with $D_{ext} < 200$ mm and at $\delta_w / D_{ext} \leq 0.1$ for cylindrical shells with $D_{ext} \geq 200$ mm.

Table C4. Test-section burst pressure (calculated values).

Temperature	Tensile stress	Burst pressure
°C	MPa	MPa
38	137.9	27.58
93	137.9	27.58
204	137.9	27.58
316	137.9	27.58
343	136.5	27.30
371	135.1	27.02
399	133.7	26.74
427	131.6	26.32
454	128.9	25.78
482	110.3	22.06
510	73.0	14.60
538	48.2	9.64
566	31.0	6.20
593	20.6	4.12
621	15.1	3.02
649	13.7	2.74

The mentioned above information is important for a safe testing.

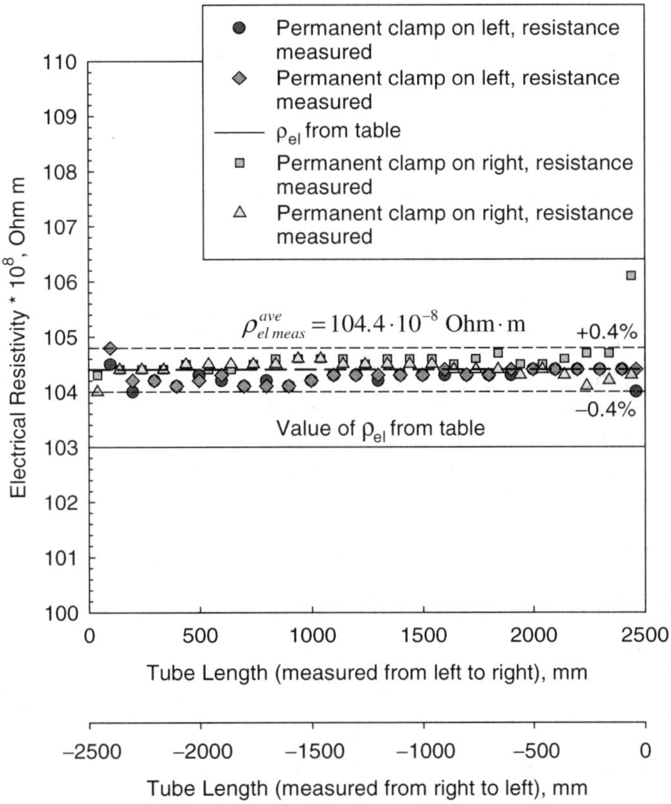

Figure C1. Electrical resistivity of Inconel-600 test section (D=8.058 mm, D_{ext}=10 mm, value of electrical resistivity from Special Metals Corp. datasheet is ρ_{el}[25]=103·10^{-8} Ohm·m).

C.5 TEST-SECTION ELECTRICAL RESISTANCE (MEASURED VALUES)

The test-section electrical resistance was measured using a micro-ohmmeter in a 4-wire configuration. This setup removes the measuring wire resistance from the measured value. The objectives are (*a*) to measure accurately the electrical resistance of the test section, and hence, to calculate a value of the electrical resistivity of Inconel 600, which will be later used in local heat flux calculations, and (*b*) to check wall thickness uniformity along the tube through the values of measured electrical resistance (see Table C5 and Figure C1). Electrical resistance was measured starting from both ends (negative values of length correspond to one direction of measurement, and positive values

[25] The value of ρ_{el} was obtained from the Inconel alloy 600 data information sheet of the Special Metals Corporation (http://www.specialmetals.com/products/data_600.htm), 2002.

Table C5. Electrical resistivity of Inconel 600.

t	°C	20	100	200	300	400	500	600	700	800	900
ρ_{el}	μΩ·m	1.03	1.04	1.05	1.07	1.09	1.12	1.13	1.13	1.13	1.15
$\rho_{el\,cal}$	μΩ·m	1.03	1.04	1.05	1.07	1.09	1.11	1.13	1.13	1.13	–

of length correspond to the opposite direction of measurement). The electrical resistivity was calculated as follows:

$$\rho_{el} = \frac{R_{el}\,A_{cr\,sect}}{L}, \tag{C2}$$

where $A_{cr\,sect}$ is the wall cross-sectional area $(\pi/4(D_{ext}^2 - D^2))$ in m² ($D_{ext} = 10$ mm and $D = 8.058$ mm), and L is the heated length in m.

The measurements showed that the test section had uniform electrical resistivity (see Table C5 and Figure C1), and therefore, wall thickness.

C.6 EFFECT OF TEMPERATURE ON INCONEL-600 ELECTRICAL RESISTIVITY[26]

The electrical resistivity of Inconel 600 (see Table C5) is less affected by temperature than other metals (see Figure 10.7) and its value is relatively large. Therefore, this material is more suitable for use in directly-heated test sections.

The electrical resistivity of Inconel 600 can be calculated using the following correlation:

$$\begin{aligned}\rho_{el\,cal} = (&103.1289703317 - 5.4963164982\cdot10^{-4}\cdot t_w^{ave} + 6.4711351326\cdot10^{-5}\cdot(t_w^{ave})^2 \\ &- 6.111698975\cdot10^{-8}\cdot(t_w^{ave})^3)\cdot10^{-8},\ \text{Ohm}\cdot\text{m}\end{aligned} \tag{C3}$$

where $t_w^{ave} = (t_w^{int} + t_w^{ext})/2$ is the average wall temperature in °C. Equation (C3) is valid within the temperature range of 20°C – 800°C and deviates from the fitting curve within a range of ±0.6%. The correlation coefficient (r^2) is 0.9931.

The above-mentioned information is important for a local heat flux calculation.

C.7. EFFECT OF TEMPERATURE ON INCONEL 600 THERMAL CONDUCTIVITY[27]

The values of thermal conductivity of Inconel 600 at different temperatures are listed in Table C6.

[26] Values for ρ_{el} were obtained from the Inconel alloy 600 data information sheet of the Special Metals Corporation (http://www.specialmetals.com/products/data_600.htm), 2002.

[27] Values for k were obtained from the Inconel alloy 600 data information sheet of the Special Metals Corporation (http://www.specialmetals.com/products/data_600.htm), 2002.

Table C6. Thermal conductivity of Inconel 600.

t	°C	20	100	200	300	400	500	600	700	800
k	W/m·K	14.9	15.9	17.3	19.0	20.5	22.1	23.9	25.7	27.5
k_{cal}	W/m·K	14.5	15.8	17.5	19.1	20.7	22.3	24.0	25.6	27.2

The thermal conductivity of Inconel 600 can be calculated using the following correlation:

$$k = 14.2214329176 + 0.0162450563 \cdot t_w^{ave}, \tag{C4}$$

where $t_w^{ave} = (t_w^{\text{int}} + t_w^{ext})/2$ is the average wall temperature in °C, and k is in W/m·K. Equation (C4) is valid within a temperature range of 20°C – 800°C and deviates from the fitting curve within a range of ±2.5%. The correlation coefficient (r^2) is 0.9976.

The above-mentioned information is important for a local HTC calculation.

Appendix D

SAMPLE OF UNCERTAINTY ANALYSIS

The proposed uncertainty analysis[28] is based on our current experience with heat-transfer and pressure-drop experiments in supercritical water (Kirillov et al. 2005; Pis'menny et al. 2005) and carbon dioxide (Pioro and Khartabil 2005) and on our long-term experience in conducting heat-transfer experiments at subcritical pressures (Guo et al. 2006; Bezrodny et al. 2005; Leung et al. 2003; Pioro et al. 2002a,b, 2001, 2000; Pioro 1999, 1992; Pioro and Pioro 1997; Kichigin and Pioro 1992; Pioro and Kalashnikov 1988; Pioro 1982). Also, basic principles of the theory of thermophysical experiments and their uncertainties were applied (Coleman and Steel 1999; Hardy et al. 1999; Guide... 1995; Holman 1994; Moffat 1988; Gortyshov et al. 1985; Topping 1971).

In general, an uncertainty analysis is quite complicated process in which some uncertainties[29] (for example, uncertainties of thermophysical properties (for details, see NIST (2002), uncertainties of constants, etc.) may not be known or may not be exactly calculated. Therefore, applying the engineering judgement is the only choice in some uncertainty calculations.

This section summarizes instrument calibrations and uncertainty calculations for the measured parameters such as temperature, pressure, pressure drop, mass-flow rate, power, tube dimensions, etc., and for the calculated parameters such as mass flux, heat flux, etc., in supercritical heat-transfer and pressure-drop tests. Uncertainties for these parameters are based on the RMS of component uncertainties. All uncertainty values are at the 2σ level, unless otherwise specified.

Calibration of the instruments used in the tests was performed either *in situ*, e.g., power measurements, test-section thermocouples, etc., or at an instrumentation shop, e.g., pressure transducers and bulk-fluid temperature thermocouples. In general, instruments were tested against a corresponding calibration standard.

[28] The authors of the current monograph express their appreciation to D. Bullock and Y. Lachance (CRL AECL) for their help in preparation of this uncertainty analysis.

[29] Uncertainty refers to the accuracy of measurement standards and equals the sum of the errors that are at work to make the measured value different from the true value. The accuracy of an instrument is the closeness with which its reading approaches the true value of the variable being measured. Accuracy is commonly expressed as a percentage of a measurement span, measurement value or full-span value. Span is the difference between the full-scale and the zero-scale value (Mark's Standard Handbook for Mechanical Engineers 1996).

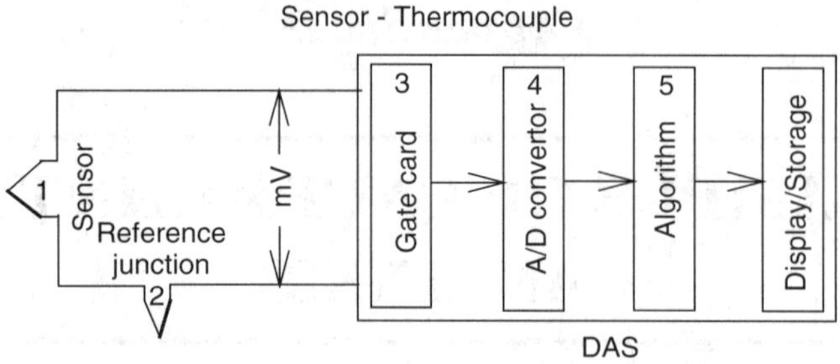

Figure D1. Schematic of signal processing for temperature (based on thermocouple) measurements. Numbers in figure identify uncertainty of particular device in measuring circuit: 1—sensor uncertainty, 2—reference junction uncertainty, 3—Analog Input (A/I) uncertainty, 4—Analog-to-Digital (A/D) conversion uncertainty, and 5—DAS algorithm uncertainty.

When the same calibration standard is used for serial instruments, the calibration standard uncertainty is treated as a systematic uncertainty. In general, high-accuracy calibrators were used, hence systematic errors for calibrated instruments are considered to be negligible. All other uncertainties are assumed to be random. Also, errors correspond to the normal distribution. Usually, the uncertainties have to be evaluated for three values of the corresponding parameter: minimum, mean and maximum value within the investigated range.

Uncertainties are presented below for instruments, which are commonly used in heat-transfer and pressure-drop experiments. It is important to know the exact schematics for sensor signal processing. Some commonly used cases, which are mainly based on a DAS recording, are shown in Figure D1 for thermocouples and in Figure D2 for RTDs, pressure cells and differential pressure cells.

Also, absolute and relative errors for commonly used functions are listed in Table D1 for reference purposes.

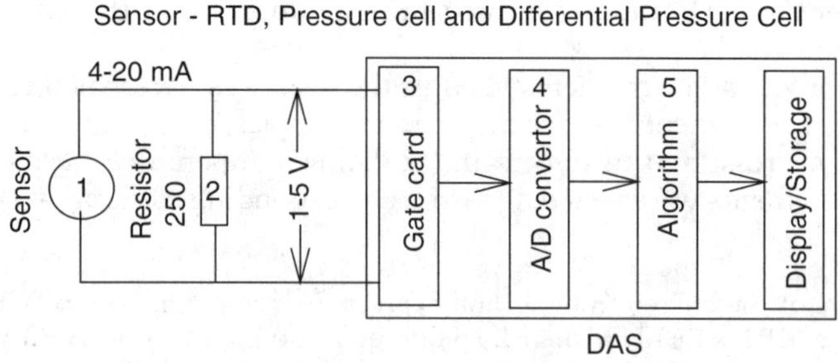

Figure D2. Schematic of signal processing for temperature (based on RTD), absolute pressure and differential pressure. Numbers in figure identify uncertainty of particular device in measuring circuit: 1—sensor uncertainty, 2—uncertainty due to temperature effect, 3—A/I uncertainty, 4—A/D conversion uncertainty, and 5—DAS algorithm uncertainty; for RTD and both types pressure cells—DAS algorithm uncertainty is usually 0 due to linear fit.

Table D1. Absolute and relative errors for commonly used functions (based on Gortyshov et al. (1985)).

Function	Absolute Error	Relative Error
$Y = X_1 + X_2 + \cdots + X_n$	$\pm\sqrt{\Delta X_1^2 + \Delta X_2^2 + \cdots + \Delta X_n^2}$	$\pm\dfrac{\sqrt{\Delta X_1^2 + \Delta X_2^2 + \cdots + \Delta X_n^2}}{X_1 + X_2 + \cdots + X_n}$
$Y = \dfrac{X_1 + X_2 + \cdots + X_n}{n}$	$\pm\dfrac{\sqrt{\Delta X_1^2 + \Delta X_2^2 + \cdots + \Delta X_n^2}}{\sqrt{n}}$	$\pm\dfrac{\sqrt{n}\cdot\sqrt{\Delta X_1^2 + \Delta X_2^2 + \cdots + \Delta X_n^2}}{X_1 + X_2 + \cdots + X_n}$
$Y = X_1 - X_2$	$\pm\sqrt{\Delta X_1^2 + \Delta X_2^2}$	$\pm\dfrac{\sqrt{\Delta X_1^2 + \Delta X_2^2}}{X_1 - X_2}$
$Y = X_1 \cdot X_2$	$\pm\sqrt{(X_1 \cdot \Delta X_2)^2 + (X_2 \cdot \Delta X_1)^2}$	$\pm\sqrt{\left(\dfrac{\Delta X_1}{X_1}\right)^2 + \left(\dfrac{\Delta X_2}{X_2}\right)^2}$
$Y = a \cdot X$	$\pm a \cdot \Delta X$	$\pm\dfrac{\Delta X}{X}$
$Y = X^n$	$\pm n \cdot X^{n-1} \Delta X$	$\pm\dfrac{n \cdot \Delta X}{X}$
$Y = \sin X$	$\pm \cos X \cdot \Delta X$	$\pm \mathrm{ctg} X \cdot \Delta X$
$Y = \cos X$	$\pm \sin X \cdot \Delta X$	$\pm \mathrm{tg} X \cdot \Delta X$
$Y = \mathrm{tg} X$	$\pm\dfrac{\Delta X}{\cos^2 X}$	$\pm\dfrac{2 \cdot \Delta X}{\sin(2 \cdot X)}$
$Y = \mathrm{ctg} X$	$\pm\dfrac{\Delta X}{\sin^2 X}$	$\pm\dfrac{2 \cdot \Delta X}{\sin(2 \cdot X)}$

D.1 TEMPERATURE

For the calibrated thermocouples, the following linear characteristics were found:

$$V_{act} = a \cdot V_{meas} + b, \tag{D1}$$

where V_{act} is the "actual" value[30] of the given parameter, V_{meas} is the value measured by the given instrument, and a and b are the calibration coefficients.

D.1.1 Measured Bulk-Fluid Temperature

The test section (see Figures 10.5 and 10.6) has three thermocouples to measure the inlet and outlet bulk-fluid temperatures. Also, the temperature at the flow-meter is monitored by a thermocouple for fluid-density calculations.

[30] The value obtained from the calibration standard.

Table D2. Linear coefficients for inlet and outlet temperature thermocouples (from instrument calibration records).

TC	Coefficient		Uncertainty, °C	Number of Points
–	*a*	*b*	Maximum (2σ)	–
TE-1	1.000	-0.1798	0.12	5
TE-2	0.9980	0.1502	0.12	5
TE-3	0.9985	0.0980	0.12	5

The test-section inlet and outlet bulk-fluid temperatures were measured with sheathed K-type thermocouples (for thermocouple signal processing, see Figure D1). These thermocouples were calibrated against the temperature standard RTD over the temperature range from 0°C to 100°C. For the reference RTD, the maximum error was ±0.3°C. The maximum uncertainty of a data fit for inlet and outlet bulk-fluid temperature measurements is listed in Table D2.

The inlet and outlet bulk-fluid temperature measurement uncertainties[31] are as follows:

Calibration system uncertainty	±0.3°C
Thermocouple sensor accuracy after linear fit	± 0.12°C
A/I accuracy	± 0.06°C, i.e., ±0.025% of f.s.; $\left(=\pm\dfrac{0.00025\cdot 10\,\text{mV}}{0.045\,\dfrac{\text{mV}}{°\text{C}}}\right)$; where f.s. is the full scale
A/D resolution accuracy (minimum 1 bit)	±0.03°C $\left(=\pm\dfrac{10\,\text{mV (f.s.)}}{8192\,\text{counts}\cdot 0.045\,\dfrac{\text{mV}}{°\text{C}}}\right)$, where 0.045 mV/°C is the conversion rate, i.e., 4.509 mV for 100°C (The Temperature Handbook 2000)
Reference junction accuracy	±0.4°C

For a given test-section inlet or outlet temperature *T*, the uncertainty Δt is given by:

$$\frac{\Delta t}{t}=\sqrt{\left(\frac{0.3}{t}\right)^2+\left(\frac{0.12}{t}\right)^2+\left(\frac{0.03}{t}\right)^2}. \tag{D2}$$

The first term is the maximum error of the calibration system (±0.3°C). The second term is the maximum error for the sheathed thermocouple (≤100°C), obtained from the calibration. The third term is the uncertainty introduced by the DAS, i.e., the A/D resolution uncertainty (±0.03°C). Note that since the calibration was done *in situ* using the DAS as the measuring system for the RTD

[31] All inputs are from instrument calibration records and device manuals unless otherwise specified.

and for the calibrated thermocouples, the uncertainty introduced by the reference junction and the A/I accuracy was included in calibration curves.

All bulk-fluid temperature thermocouples were calibrated *in situ*, only within the range of 0°C – 100°C. Therefore, individual correction factors were implemented for each thermocouple within the range of 0°C – 100°C (see Table D2). For this range of temperatures, the uncertainty Δt is

- for t_{min} = 20°C $\Delta t = \pm 0.32$°C (or ±1.62%), and
- for t = 100°C $\Delta t = \pm 0.32$°C (or ±0.32%).

Beyond this range, thermocouple uncertainties were taken as per The Temperature Handbook (2000), i.e., ±2.2°C.

Thermocouple installed near the flowmeter was calibrated using another calibrating system and procedure. All inputs below are from instrument calibration record and device manuals unless otherwise specified.

Calibration system uncertainty:

±0.5°C, i.e., $\left(= \pm\sqrt{0.06^2 + 0.5^2 + 0.041^2}\right.$, where the first term is the accuracy of standard RTD, the second term is the accuracy of thermocouple signal measuring device and the third term is the accuracy of RTD signal measuring device (all uncertainties are in °C).

TC maximum calibration accuracy[32] (>2σ) within 0.0°C – 45.0°C	±0.53°C
A/I accuracy	±0.06°C, i.e., ±0.025% of f.s. $\left(= \pm \frac{0.00025 \cdot 10\,\text{mV}}{0.045\,\frac{\text{mV}}{°\text{C}}}\right)$
A/D resolution accuracy (minimum 1 bit)	±0.03°C $\left(= \pm \frac{10\,\text{mV (f.s.)}}{8192\,\text{counts} \cdot 0.045\,\frac{\text{mV}}{°\text{C}}}\right)$ where 0.045 mV/°C is the conversion rate, i.e., 4.509 mV for 100°C (The Temperature Handbook 2000)
Reference junction accuracy	±0.02°C

For a given flowmeter bulk-fluid temperature t_{fm}, the uncertainty Δt_{fm} is given by:

$$\left(\frac{\Delta t_{fm}}{t_{fm}}\right)_{TC} = \left(\pm\sqrt{\left(\frac{0.5}{t_{fm}}\right)^2 + \left(\frac{0.53}{t_{fm}}\right)^2 + \left(\frac{0.06}{t_{out}}\right)^2 + \left(\frac{0.03}{t_{out}}\right)^2 + \left(\frac{0.02}{t_{out}}\right)^2}\right)_{TC}. \quad \text{(D3)}$$

[32] The TC calibration accuracy is the maximum difference in °C between what the calibration standard measured and what TC indicated.

Therefore, the flowmeter bulk-fluid temperature uncertainty is:

- for $t_{fm\ min} = 19°$ C $\Delta t_{fm} = \pm 0.74°C$ (or ±3.9%), and
- for $t_{fm\ max} = 35°C$ $\Delta t_{fm} = \pm 0.74°C$ (or ±2.1%).

Additional uncertainties due to thermocouple installation and possible electrical pickup have been minimized by using good engineering practices.

If a bulk-fluid temperature is measured with an RTD, then the following will apply.

The bulk-fluid temperature measurement uncertainties at the 2σ level are characterized with the following for an RTD (for RTD signal processing, see Figure D2):

Calibration system uncertainty in °C (from the instrument calibration record):,

$$\text{Cal. Sys. Unc.} = \pm\sqrt{\left[\left(0.1\%\text{of Reading}\frac{16\text{ mA}}{100°\text{C}} + 0.015\%\text{ of f.s.}\right) + \frac{100°\text{C}}{16\text{ mA}}\right]^2 + (0.05°\text{C})^2}$$

$$\approx 0.06°\text{C}$$

where the first term is the accuracy of calibrator in which reading is in °C and f.s. is 30 mA and a conversion rate is 16 mA for 100°C; and the second term is the accuracy of standard RTD.

The RTD accuracy after linear fit, i.e., maximum deviation (from the instrument calibration record), is about 0.08°C;

A/I accuracy (from the device manual):
±0.032°C (±0.025% of f.s.), i.e., $\left(= \pm (0.00025 \cdot 5.12\text{ V (f.s.)})/0.04\text{ V/°C}\right)$

A/D conversion accuracy (minimum 1 bit accuracy) (from the device manual):

(0.016°C, i.e., $(= \pm (5.12\text{ V(f.s.)}/(8192\text{ counts} \cdot 0.04\text{ V/°C}))$, where 0.04 V/°C is the conversion rate, i.e., 4 V for 100°C (from the instrument calibration record). DAS algorithm uncertainty is 0 due to a linear fit.

Therefore, for a given test-section inlet temperature, its uncertainty (ΔT) is given by

$$\frac{\Delta t}{t} = \pm\sqrt{\left(\frac{\text{Cal. Sys. Unc.}}{t}\right)^2 + \left(\frac{0.08}{t}\right)^2 + \left(\frac{0.032}{t}\right)^2 + \left(\frac{0.016}{t}\right)^2}. \tag{D4}$$

The resulting uncertainties in the bulk-fluid temperature are

- For $t = 10°C$ $\Delta t = \pm 0.10°C$ (or (±1.2%); and
- For $t = 50°C$ $\Delta t = \pm 0.11°C$ (or (±0.2%).

If the bulk-fluid temperature is measured with several devices installed in a one cross-section (for example, two RTDs and one thermocouple), the following equation may apply:

$$\frac{\Delta \bar{t}}{\bar{t}} \cong \pm \frac{\sqrt{(\Delta t_{RTD1})^2 + (\Delta t_{RTD2})^2 + (\Delta t_{TC})^2}}{\sqrt{3}\,\bar{t}} \tag{D5}$$

In this case, the resulting uncertainty will be close to the larger uncertainty, i.e., the thermocouple uncertainty. Therefore, if several devices have to be used for measuring a non-uniform temperature or any other parameter, they have to be with a similar accuracy.

D.1.2 External Wall Temperature

Temperatures for the test-section external surface (see Figure 10.6) were measured using fast-response K-type thermocouples (see Figure D3). In general, thermocouple uncertainties for K-type thermocouples are ±2.2°C within a range of 0°C – 277°C (The Temperature Handbook 2000). However, all fast-response thermocouples were calibrated *in situ* within a range of 0°C – 100°C prior to use (for details, see below). Therefore, individual correction factors were implemented for each thermocouple within the range of 0°C – 100°C. Beyond this range, thermocouple uncertainties were taken as per The Temperature Handbook (2000), i.e., ±2.2°C.

All K-type thermocouples were calibrated against the temperature calibration standard (i.e., the reference RTD) over the temperature range from 0°C to 100°C. These thermocouple assemblies were immersed in a liquid bath thermostat

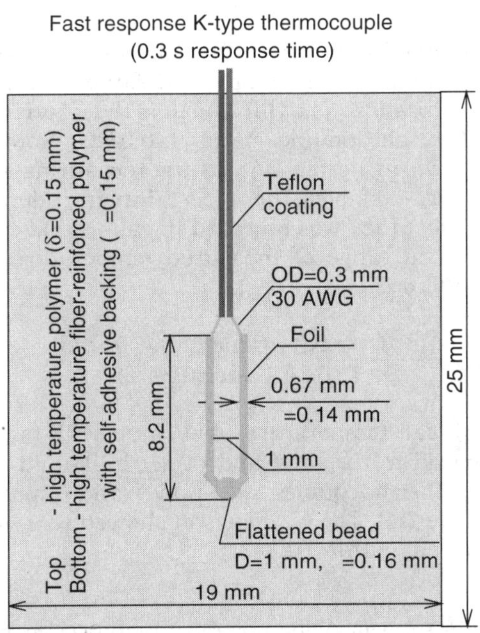

Figure D3. Sketch drawing of fast-response K-type thermocouple.

together with the RTD. For the reference RTD, the maximum uncertainty is ±0.3°C. The combined uncertainty[33] for wall temperature measurements is as follows:

Calibration system accuracy	±0.3°C
Thermocouple sensor accuracy after linear fit	±0.16°C max at values ≤100°C
A/I accuracy	±0.06°C, i.e., ±0.025% of f.s. $\left(= \pm \frac{0.00025 \cdot 10\,\text{mV}}{0.045\,\frac{\text{mV}}{^\circ\text{C}}}\right)$
A/D resolution accuracy (minimum 1 bit)	±0.03°C, $\left(= \pm \frac{10\,\text{mV (f.s.)}}{8192\,\text{counts} \cdot 0.045\,\frac{\text{mV}}{^\circ\text{C}}}\right)$ where 0.045 mV/°C is the conversion rate, i.e., 4.509 mV for 100°C (The Temperature Handbook 2000).

For a given test-section wall temperature *t*, the uncertainty Δt is given by:

$$\frac{\Delta t}{t} = \sqrt{\left(\frac{0.3}{t}\right)^2 + \left(\frac{0.16}{t}\right)^2 + \left(\frac{0.03}{t}\right)^2}. \tag{D6}$$

The first term is the maximum error of the calibration system (±0.3°C). The second term is the maximum error of the sheathed thermocouple (≤100°C), obtained from the calibration. The third term is the uncertainty introduced by the DAS, i.e., the A/D resolution uncertainty (±0.03°C). Note that since the calibration was done *in situ* using the DAS as the measuring system for the RTD and the calibrated thermocouples, the uncertainty introduced by the reference junction and the A/I accuracy was included in calibration curves.

Within the calibrated range of measured temperatures, i.e., from 0°C to 100°C, the uncertainty Δt is

- for t_{min} = 25°C $\Delta t = \pm 0.34$°C (or ±1.36%), and
- for t = 100°C $\Delta t = \pm 0.34$°C (or ±0.34%).

Also, the external wall temperatures measured with fast-response thermocouples were compared to the inlet and outlet bulk-fluid temperatures measured with sheathed thermocouples, at 0 power and 0 mass flux through the test section (see Figure D4). The comparison showed that, in general, all measured temperatures were within ±0.3°C.

[33] All inputs are from instrument calibration records and device manuals unless otherwise specified.

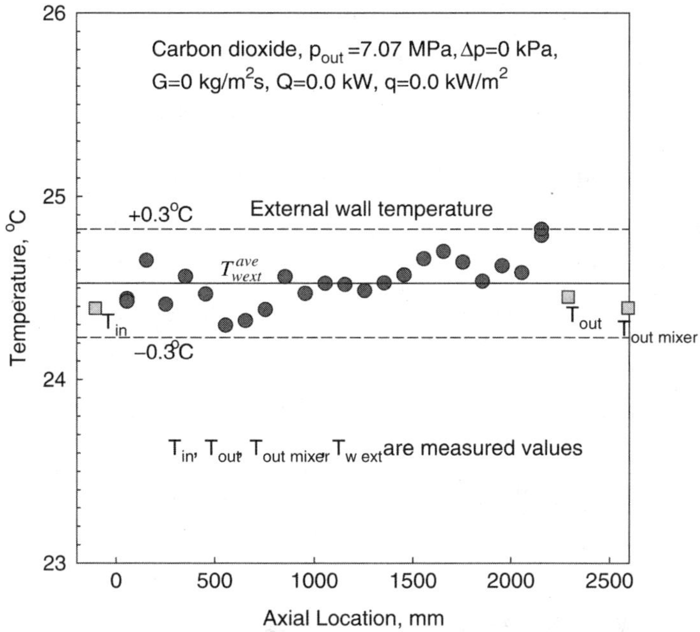

Figure D4. Temperature profile along test section at 0 power and 0 mass flux values.

D.2 ABSOLUTE PRESSURE

A high-accuracy gauge pressure cell with a range of 0 – 10,000 kPa (0 – 10 MPa) was used for the outlet-pressure measurements (for pressure signal processing, see Figure D2). A small correction (77.2 kPa) is applied in the DAS program for the elevation difference between the pressure tap and transmitter. The combined uncertainty for absolute pressure measurements is as follows.

Accuracy of gauge pressure cell (from the calibration record) is 0.1% of calibrated span (10,000 kPa), and this accuracy was verified during the calibration check.

Cal. Sys. Unc. =

$$\pm\sqrt{[(0.015\% \text{ of Reading } \frac{16 \text{ mA}}{10{,}000 \text{ kPa}} + 0.015\% \text{ of f.s.}) \frac{10000 \text{ kPa}}{16 \text{ mA}}]^2 + (0.1\% \text{ of Reading})^2},$$

Calibration system uncertainty in kPa (from the instrument calibration record), where the first term is the accuracy of calibrator in which reading is in kPa and f.s. is 30 mA and conversion rate is 16 mA for 10,000 kPa; and the second term is the accuracy of tester.

Uncertainty due to temperature effect in 250-Ω resistor:
±0.1%.

A/I accuracy (from the device manual):

±3.2 kPa, i.e., ±0.025% of f.s., i.e., 5.12 V $(=\pm(0.00025\cdot 5.12\text{V})/(0.0004\,\text{V/kPa}))$

A/D conversion accuracy (minimum 1 bit accuracy) (from the device manual): ±1.56 kPa, $(=\pm(5.12\,\text{V(f.s.)})/(8192\text{ counts}\cdot 0.0004\,\text{V/kPa}))$; where 0.0004 V/kPa is the conversion rate, i.e., 4 V for 10,000 kPa (from the instrument calibration record).

For a given test-section outlet pressure p, the uncertainty Δp is given by:

$$\frac{\Delta p}{p}=\pm\sqrt{\left(\frac{0.001\cdot 10000}{p}\right)^2+\left(\frac{0.1}{100}\right)^2+\left(\frac{\text{A/I}}{p}\right)^2+\left(\frac{\text{A/D}}{p}\right)^2}, \tag{D7}$$

For the range of p from 7.6 to 8.8 MPa, the uncertainty Δp is given by:

- for p_{min} = 7,600 kPa Δp = ±13.1 kPa (or ±0.17%),
- for p = 8,400 kPa Δp = ±13.5 kPa (or ±0.16%).
- for p_{max} = 8,800 kPa Δp = ±13.8 kPa (or ±0.16%).

D.3 DIFFERENTIAL-PRESSURE CELLS

Five differential-pressure transducers for measuring test-section pressure drops (for differential-pressure signal processing, see Figure D2) were connected to the corresponding pressure taps installed as shown in Figure 10.6. They were used for measuring the test-section axial pressure gradient and the overall pressure drop. Also, one differential-pressure transducer was used to measure a pressure drop across the flowmeter (see Figure 10.5). All these pressure drops were measured using pressure transmitters.

A calibrator and a pressure module were used for the calibration check of the differential-pressure transducers. Basic characteristics of the test-section and flowmeter differential-pressure cells are listed in Table D3.

Table D3. Basic characteristics of differential-pressure cells.

Instrument Name	Description	Output	Output kPa	Span kPa	Accuracy ±% of Span
PDT-1	Total test-section pressure drop	10–50 mV	0–300	300	0.5
PDT-2	Test-section pressure drop	1–5 V	0–50	50	0.5
PDT-3	Test-section pressure drop	1–5 V	0–50	50	0.5
PDT-4	Test-section pressure drop	1–5 V	0–50	50	0.5
PDT-5	Test-section pressure drop	1–5 V	0–50	50	0.5
PDT-FM-1	Orifice-plate pressure drop	10–50 mV	0–37	37	0.5

Accuracy, includes combined effects of linearity, hysteresis, and repeatability in % of a calibrated span are listed in Table D3.

Calibration system uncertainty in kPa (from instrument calibration records):

Cal. Sys. Unc. =

$$\pm\sqrt{\left[0.015\% \text{ of Reading}\frac{16\text{ mA}}{\text{Span kPa}}+0.015\% \text{ of f.s.}\frac{\text{Span kPa}}{16\text{ mA}}\right]^2+(0.05\% \text{ of f.s.})^2},$$

where the first term is the accuracy of process calibrator in which reading is in kPa, f.s. is 30 mA and conversion rate is 16 mA for span in kPa; and the second term is the accuracy of calibrator in which f.s. is 690 kPa (100 psig).

Uncertainty due to temperature effect in 250-Ω resistor:
±0.1%.

A/I accuracy (from a device manual):

±0.025% of f.s., i.e., $\left(=\pm\frac{0.00025\cdot 5.12\text{ V}}{\frac{4\text{ V}}{\text{Span kPa}}}\right)$.

A/D conversion accuracy (minimum 1 bit accuracy) (from a device manual):

$$\left(=\pm\frac{5.12\text{ V (f.s.)}}{8192\text{ counts}\frac{4\text{ V}}{\text{Span kPa}}}\right).$$

For a given pressure drop (Δp) for PDT-1, PDT-2 to PDT-5 and PDT-FM-1, the uncertainty $\Delta(\Delta p)$ is given by:

$$\frac{\Delta(\Delta p)}{\Delta p}=\pm\sqrt{\left(\frac{\%\text{ of span in kPa}}{\Delta p}\right)^2+\left(\frac{0.1}{100}\right)^2+\left(\frac{\text{A/I}}{\Delta p}\right)^2+\left(\frac{\text{A/D}}{\Delta p}\right)^2}. \quad \text{(D8)}$$

For the range of the total Δp from 5 to 70 kPa, the uncertainty $\Delta(\Delta p)$ for PDT-1 is given by:

- for Δp_{min} = 5 kPa $\Delta(\Delta p)$ = ±1.50 kPa (or ±30.1%), and
- for Δp_{max} = 70 kPa $\Delta(\Delta p)$ = ±1.51 kPa (or ±2.2%).

For the range of the local Δp from 5 to 30 kPa, the uncertainty $\Delta(\Delta p)$ for PDT-2 – PDT-5 is given by:

- for Δp_{min} = 5 kPa $\Delta(\Delta p)$ = ±0.25 kPa (or ±5.0%), and
- for Δp_{max} = 30 kPa $\Delta(\Delta p)$ = ±0.25 kPa (or ±0.84%).

For the local Δp equals to 37 kPa, the uncertainty $\Delta(\Delta p)$ for PDT-FM-1 is given by:

- for Δp_{min} = 1.5 kPa $\Delta(\Delta p)$ = ±0.19 kPa (or ±12.5%), and
- for Δp_{max} = 16.9 kPa $\Delta(\Delta p)$ = ±0.19 kPa (or ± 1.1%).

D.4 MASS-FLOW RATE

The loop mass-flow rate FM-1 (see Figure 10.5) is measured by a small orifice plate[34] with an orifice diameter of 0.308" (7.8232 mm), and monitored by a differential-pressure cell with the range of 0 – 37 kPa. This cell has a square root output, with an accuracy of ±0.5% of full scale. The square root output is converted in the program to obtain kPa for use in the following flow equation, for a mass-flow rate of 0 – 0.24 kg/s (see Figure D4):

$$m = C_{fl}\sqrt{\rho\,\Delta p}, \tag{D9}$$

where C_{fl} = 0.00130 is the constant (White 1994), ρ is the density at the orifice plate in kg/m³, and Δp is the pressure drop across the orifice plate in kPa. It is known that orifice-plate flowmeters usually have a working range within (0.3 and 1)$\cdot m_{max}$, i.e., 0.08 – 0.24 kg/s (The Flow and Level Handbook 2001).

In general, the constant C_{fl} is a function of Reynolds number (see Figure D5). However, this effect is minor within the investigated range of Reynolds numbers (**Re** = 57,000 – 1,130,000).

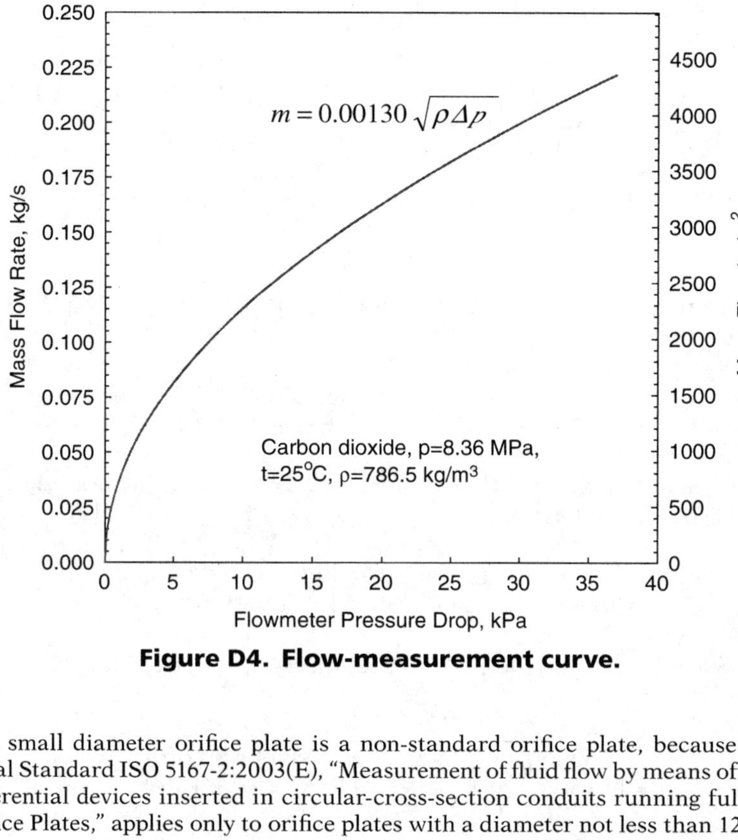

Figure D4. Flow-measurement curve.

[34] This small diameter orifice plate is a non-standard orifice plate, because International Standard ISO 5167-2:2003(E), "Measurement of fluid flow by means of pressure differential devices inserted in circular-cross-section conduits running full Part 2: Orifice Plates," applies only to orifice plates with a diameter not less than 12.5 mm.

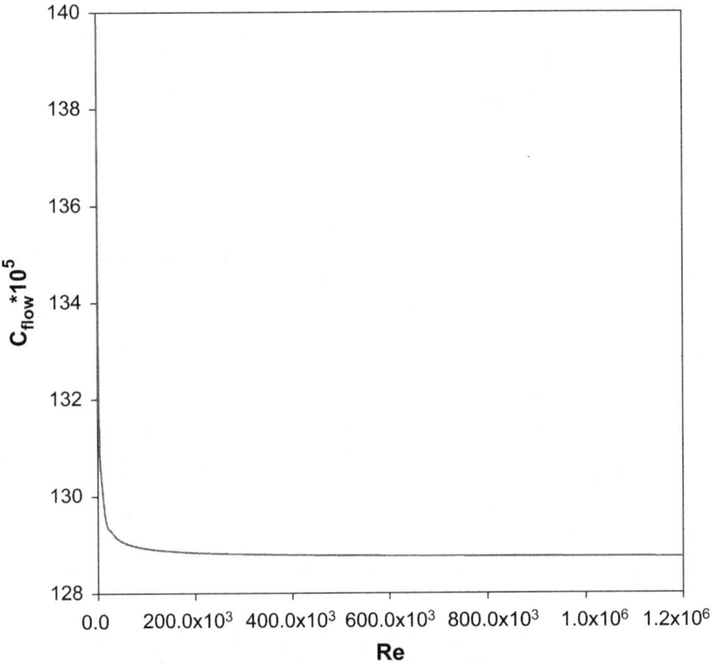

Figure D5. Effect of Reynolds number on flow constant.

We attempted to calibrate the flowmeter FM-1 with water using the direct weighting method (Hardy et al. 1999) within the supercritical CO_2 investigated Reynolds numbers range. Due to significantly different values of water dynamic viscosity compared to those of supercritical carbon dioxide and restrictions applied to the maximum water flow and its temperature, the flowmeter was calibrated (see Figure D6) within a lower range of Reynolds numbers (**Re** = 2,700 – 27, 000) compared to those of supercritical carbon dioxide (**Re** = 57,000 – 1,130,000).

However, the calibration results showed that Equation (D9) is reasonably accurate (a mean error is -0.15% and an RMS error is 0.5%) for flows that are not less than 0.045 kg/s. This finding is consistent with heat-balance error data obtained in supercritical CO_2. However, the heat-balance error data for $m <$ 0.045 kg/s show the opposite trend, i.e., steeper slope than that shown in Figure D6b. Mass-flow rates lower than 0.045 kg/s were calculated using:

$$m = \frac{POW}{H_{out} - H_{in}} \tag{D10}$$

In general, flow-rate measurement uncertainty within the range of m = 0.045 – 0.24 kg/s is given by:

$$\frac{\Delta m}{m} = \sqrt{\left(\frac{\Delta C_1}{C_1}\right)^2 + \left(\frac{0.5\,\Delta \rho}{\rho}\right)^2 + \left(\frac{0.5\,\Delta(\Delta p)}{\Delta p}\right)^2}. \tag{D11}$$

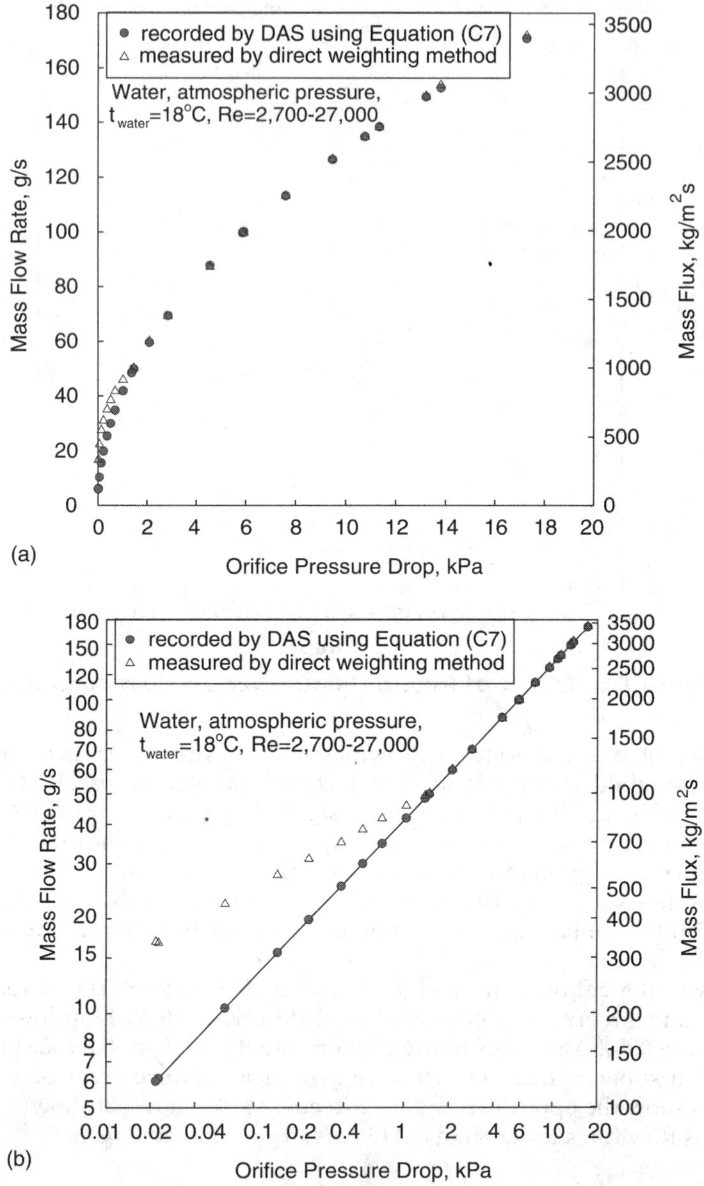

Figure D6. Calibration results for FM-1 orifice-plate flowmeter: (*a*) linear scale, and (*b*) logarithmic scale.

The estimated uncertainty in the constant C_1 is ±0.08% as a result of the minor effect of Reynolds number on the constant within the investigated range (White 1994).

Temperature, pressure (see Figure 10.5) and NIST software (2002) were used for the CO_2 density calculation. At pressures up to 30 MPa and temperatures up to 249.9°C (523 K), the estimated uncertainty in density (NIST 2002) varies up to 0.05%. Also, additional uncertainty in density arises from variations in density within the measured temperature uncertainty of ±1.1°C. This additional uncertainty is about ±1.1% at p = 8.36 MPa and t = 19°C, and ±5.0% at p = 8.8 MPa and t = 35°C. Therefore, the total uncertainty in density is:

$$\frac{\Delta\rho}{\rho}=\sqrt{\left(\frac{0.05}{100}\right)^2+\left(\frac{1.1}{100}\right)^2} = 0.011 \text{ at } p = 8.36 \text{ MPa and } t = 19^\circ\text{C} \qquad \text{(D12)}$$

and

$$\frac{\Delta\rho}{\rho}=\sqrt{\left(\frac{0.05}{100}\right)^2+\left(\frac{5.0}{100}\right)^2} = 0.05 \text{ at } p = 8.8 \text{ MPa and } t = 35^\circ\text{C}. \qquad \text{(D13)}$$

However, the vast majority of the experimental data were obtained at pressure of 8.36 MPa. Therefore, the uncertainty value of 0.011 was used below.

Pressure-drop measurement uncertainties for PDT-FT-1/1 are according to Section D.3.

Hence,

- for m_{min} = 46 g/s Δm = ±5.7 g/s (or ±12.5%) at t = 19°C and p = 8.36 MPa, and
- for m_{max} = 155 g/s Δm = ±2.4 g/s (or ±1.6%) at t = 19°C and p = 8.36 MPa.

D.5 MASS FLUX

Mass flux, G, is based on mass-flow rate measurements. The uncertainty, ΔG, includes an error in the estimation of the cross-sectional flow area, A_{fl} = 5.1 10^{-5} m². The test section is a tube of 8.058 mm ID and 10 mm OD, made of Inconel 600, with tolerances of ±0.02 mm. The uncertainties are as follows:

For ID $\quad \Delta D$ = ±0.02 mm (or ±0.25%),
For OD $\quad \Delta D_{ext}$ = ±0.02 mm (or ±0.20%), and
For A_{flow} $\quad \Delta A_{fl} = \pi D \Delta D / 2$ = ±0.253 mm² (or ±0.50%).

The uncertainty, ΔG, is obtained from the following equation:

$$\frac{\Delta G}{G}=\sqrt{\left(\frac{\Delta m}{A_{fl}G}\right)^2+\left(\frac{m\,\Delta A_{fl}}{A_{fl}^2G}\right)^2}=\sqrt{\left(\frac{\Delta m}{m}\right)^2+\left(\frac{\Delta A_{fl}}{A_{fl}}\right)^2}. \qquad \text{(D14)}$$

For the range of interest, the uncertainties, ΔG, are:

for G_{min} = 902 kg/m²s (m_{min} = 46 g/s) ΔG = ± 112.8 kg/m²s (or ± 12.5%),

and

for G_{max} = 3039 kg/m²s (m_{max} = 155 g/s) ΔG = ± 49.8 kg/m²s (or ± 1.6%).

D.6 ELECTRICAL RESISTIVITY

Electrical resistivity is a calculated value (for details, see Equation (C2)) that is based on measured values of electrical resistance, heated length and tube diameters.

The accuracy of the micro-ohmmeter used in test-section electrical resistance measurements is ±0.04% of the reading (its readings are in milliohms). The uncertainties in ID and OD are $\Delta D = \Delta D_{ext}$ = ±0.02 mm, and in L it is ΔL = ±0.5 mm.

For a given electrical resistivity, the uncertainty $\Delta \rho_{el}$ is given by:

$$\frac{\Delta \rho_{el}}{\rho_{el}} = \sqrt{\left(\frac{0.04}{100}\right)^2 + \left(\frac{\Delta D_{ext}}{D_{ext}}\right)^2 + \left(\frac{\Delta D}{D}\right)^2 + \left(\frac{\Delta L}{L}\right)^2} \tag{D15}$$

The uncertainty in $\Delta\rho_{el}$ (ρ_{el} = 104.3·10^{-8} Ohm·m) is:

- for L = 2461 mm $\Delta\rho_{el}$ = ±0.212·10^{-8} Ohm·m (or ±0.20%).

D.7 TOTAL TEST-SECTION POWER

The total test-section power is obtained by measuring the current through a 2000 A / 100 mV current shunt and the voltage across the test section. These signals are fed into a power-measuring unit (PMU), where the test-section voltage is scaled down to a 1-V level. Both the voltage and current signals are fed into isolation amplifiers and then into instrumentation amplifiers with outputs of 0 – 10 V. The amplifier outputs are fed to the computer analog inputs and represent a full-scale voltage of 175 V and a full-scale current of 2000 A. These signals are multiplied in the computer program to represent a 0 – 350 kW power level.

Calibration of the power measurement unit was performed *in situ*. Test-section voltage and current inputs were removed from the PMU. Simulated inputs were used to check the calibration of the unit. A comparison between the computer readings and the calibrated simulated inputs was used to create a curve fit for the DAS to correct for the differences. The voltage input from 0 – 110 V DC was simulated with a DC power supply and verified with a multimeter. The current shunt input was simulated with a calibrator for inputs from 10 to 100 mV, which represents 200 – 2000-A range:

Accuracy of current shunt	± 0.25% of reading
Error due to current shunt resistance change	± 0.02%
A/D accuracy	± 0.025% of f.s., 10.00 V

The uncertainty, $\Delta\, POW_{TS}$, in power measurements (the power is a product of U and I) is given by:

$$\frac{\Delta\, POW_{TS}}{POW_{TS}} = \sqrt{\left(\frac{0.25}{100}\right)^2 + \left(\frac{0.02}{100}\right)^2 + \left(\frac{0.1}{100}\right)^2 + \left(\frac{0.09}{100}\right)^2 + \left(\frac{\Delta\, I_1}{I}\right)^2 + \left(\frac{\Delta\, U_1}{U}\right)}. \quad \text{(D16)}$$

The first term is the accuracy of the current shunt, the second term is the effect of a temperature change on the current shunt, the third term is the error in the test-section voltage drop from the PMU output of $\Delta\, U$ = +0.1% (0.10 V) up to 100 V, the fourth term is the error in the test-section current from the PMU output with a maximum offset of $\Delta\, I$ = +0.09% (0.75 A) at 800 A, and the fifth and sixth terms are the ±0.025% uncertainties introduced by the AC/DC conversion process for reading the current ($\Delta\, I_1$ = ±0.5 A) and ($\Delta\, U_1$ = ±0.04 V) for reading the voltage, respectively.

For the power range, POW_{TS}, from 3.0 to 35.0 kW, and for L = 2.208 m, the corresponding values of voltage drop and current are:

- $POW_{TS\,min}$ = 3000 W U = 16.0 V, I = 188 A, and
- $POW_{TS\,max}$ = 35,000 W U = 54.6 V, I = 641 A.

The uncertainty in $\Delta\, POW_{TS}$ is as follows:.

- For $POW_{TS\,min}$ = 3000 W ΔPOW_{TS} = ±13.9 W (or ±0.46%), and
- For $POW_{TS\,max}$ = 35,000 W ΔPOW_{TS} = ±106.4 W (or ±0.30%).

D.8 AVERAGE HEAT FLUX

The uncertainty in heat flux, $\Delta\, q_{ave}$, involves the uncertainties in the total test-section power (see Section D.7) and in the heated area measurements, ΔA_h, where $\Delta\, A_h = \pi\, D\, L$. The uncertainty in ID is $\Delta\, D$ = ±0.02 mm, and in L it is ΔL = ±0.5 mm. Thus, A_h can be calculated from:

$$\frac{\Delta\, A_h}{A_h} = \sqrt{\left(\frac{\Delta\, D}{D}\right)^2 + \left(\frac{\Delta\, L}{L}\right)^2}. \quad \text{(D17)}$$

The uncertainty in A_h (A_h = 55,895.4 mm²) is:

- for L = 2208 mm and D = 8.058 mm $\Delta\, A_h$ = ±78.3 mm² (or ±0.14%).

Then, the uncertainty in q_{ave} can be computed from:

$$\frac{\Delta\, q_{ave}}{q_{ave}} = \sqrt{\left(\frac{\Delta\, POW_{TS}}{POW_{TS}}\right)^2 + \left(\frac{\Delta\, A_h}{A_h}\right)^2}, \quad \text{(D18)}$$

which, for the given power values, results in

- $q_{ave\,min}$ = 53.7 kW (POW_{TS} = 3.0 kW) Δq = ± 0.28 kW/m² (or ±0.53%), and
- $q_{ave\,max}$ = 626.2 kW (POW_{TS} = 35.0 kW) Δq = ±2.46 kW/m² (or ±0.39%).

However, Equation (D18) does not account for the uncertainties related to the heat loss, which are subtracted from the applied heat flux (for details, see Section 10.3.8), because the heat loss was negligible, i.e., less than 0.5%.

D.9 UNCERTAINTIES IN HEAT-TRANSFER COEFFICIENT

Local HTC is as follows:

$$HTC = \frac{q}{t_w^{\text{int}} - t_b}. \tag{D19}$$

Uncertainty in the temperature difference is:

$$\frac{\Delta\,(t_w^{\text{int}} - t_b)}{t_w^{\text{int}} - t_b} = \pm\frac{\sqrt{\left(\Delta\,t_w^{\text{int}}\right)^2 + \left(\Delta\,t_b\right)^2}}{t_w^{\text{int}} - t_b}, \tag{D20}$$

where uncertainty in t_w^{int} is taken as uncertainty in t_w^{ext} and uncertainty in t_b is taken as uncertainty in t_{out}.

And uncertainty $\Delta\,HTC$ is:

$$\frac{\Delta\,HTC}{HTC} = \sqrt{\left(\frac{\Delta\,q}{q}\right)^2 + \left(\frac{\Delta\,(t_w^{\text{int}} - t_b)}{t_w^{\text{int}} - t_b}\right)^2}. \tag{D21}$$

D.10 UNCERTAINTIES IN THERMOPHYSICAL PROPERTIES NEAR PSEUDOCRITICAL POINT

Uncertainties in thermophysical properties (NIST 2002) near the pseudocritical point within the uncertainty range of the measured value of bulk-fluid temperature ($\Delta\,t$ = ±0.4°C) are as follows (for example, at p = 8.38 MPa (t_{pc} = 36.7°C)):

$\Delta\rho$ = ±7%; $\Delta\,H$ = ±2.5%; Δc_p = 4.5%; $\Delta\,k$ = ±2%, and $\Delta\,\mu$ = ±7%.

D.11 HEAT-LOSS TESTS

Heat loss is an important component of the total heat-balance analysis. Heat loss from the test section, HL_{TS}, to the surrounding area was measured at various wall temperatures, with electrical power applied to the test section (the loop was previously evacuated to minimise heat removal through the coolant). This test provided (*i*) an indication of the difference between the measured external wall temperatures and ambient temperature, and (*ii*) data (voltage and current applied to the test section) to calculate the heat loss from the test section. To perform the heat loss power test, a small power supply was used.

The temperature difference between the external wall temperatures and ambient temperature at zero power was found to be ±0.2°C (i.e., within the accuracy range for the thermocouples); with an increase in power to the test section, the difference ($\Delta t = t_w^{ave} - t_{amb}$) increases. This temperature difference permits the evaluation of the heat loss from the test section to the surrounding area as follows:

$$HL_{TS} = POW_{TS} = f(\Delta t); \tag{D22}$$

or, as calculated,

$$HL_{TS} = POW_{TS} = U \cdot I; \tag{D23}$$

where U is the voltage drop over the test section, and I is the current through the test-section wall. This heat-loss test, compared to the usual zero-power test, eliminates uncertainties that are related to the estimation of the thermophysical properties of CO_2. This test also eliminates flow-measurement uncertainties and uncertainties that are incurred when measuring very small temperature differences (0.5°C – 1°C) between the inlet and outlet bulk-fluid temperatures.

The heat-loss power test was performed with the insulated reference test section (heated length of 2.208 m). The heat loss assessed from these tests, as a function of the wall-ambient temperature difference, ($t_w^{ave} - t_{amb}$), is shown in Figure D7, and can be approximated by the following equation:

$$HL_{TS} = 0.47\,(t_w^{ave} - t_{amb})\ [\mathrm{W}]. \tag{D24}$$

There were some non-uniformities in the temperature distribution along the heated length. These non-uniformities were caused by the power clamps and structural support elements for the test section, which acted as heat sinks. Therefore, a conservative approach (maximum possible heat loss and therefore, minimum HTC value) was taken, i.e., only two external wall thermocouples (TECO1 and TECO24), which are located in the same cross-sections as TECO2 and TECO23, respectively, but 180° apart, were not taken into account (see Figure 10.6).

For local heated lengths, the following formula would apply:

$$HL_{TS}\Big|_{L_\ell} = HL_{TS}\Big|_{L=2.208\,m} \frac{L_\ell}{2.208}\ [\mathrm{kW}]\ , \tag{D25}$$

where L_ℓ is in meters.

In general, heat loss was negligible, i.e., less than 0.5%.

D.12 HEAT-BALANCE EVALUATION NEAR PSEUDOCRITICAL REGION

For each run, an error in the heat balance was calculated using the following expression:

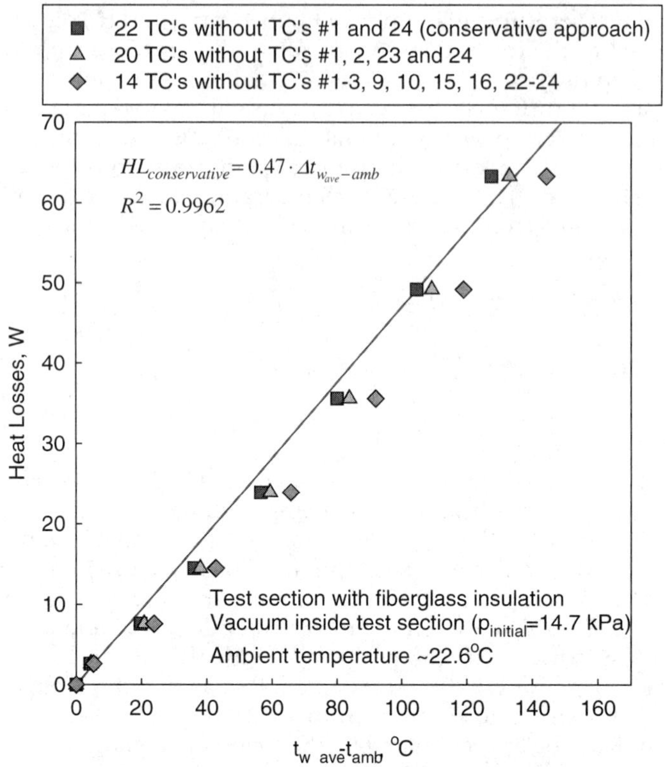

Figure D7. Heat loss from test section: Direct electrical heating of test section, heated length of 2.208 m, and loop vacuumed.

$$\Delta_{HB} = \frac{POW - HL - m\,(H_{out} - H_{in})}{POW} \cdot 100\% \qquad \text{(D26)}$$

In general, an analysis of errors in the heat-balance data shows that, at mass-flux values equal to or higher than 900 kg/m²s, at medium and high values of power ($POW \geq 5$ kW) and at the inlet and outlet bulk-fluid temperatures below or beyond the pseudocritical region (i.e., t_{in} and $t_{out} < t_{pc}$ - 2°C or t_{in} and $t_{out} > t_{pc}$ + 2°C), these errors are within ±4%.

Table D4. Maximum uncertainties in ΔH calculations near pseudocritical point (p_{out}= 8.36 MPa, t_{pc}=36.7°C, t_{in}=21°C, m=0.1 kg/s, and G=2000 kg/m²s).

t_b	H_b	Uncertainty at Δt_b= + 0.4°C	Uncertainty at Δt_b= – 0.4°C	$\Delta H_b = H_{out} - H_{in}$	Max uncertainty in ΔH_b
°C	kJ/kg	kJ/kg	kJ/kg	kJ/kg	%
21	248.94	1.18	–1.19	–	
35	313.72	4.29	–5.04	64.78	14.4
37	349.26	8.51	–7.82	100.32	16.3
41	395.75	2.56	–2.4	144.41	3.4

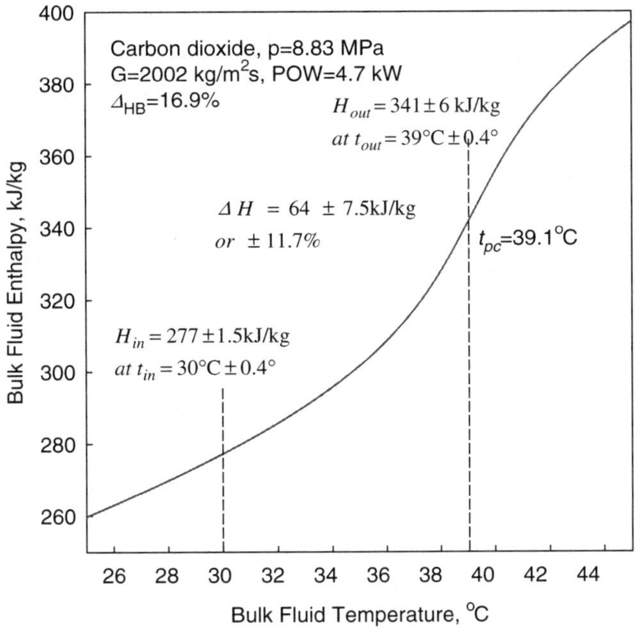

Figure D8. Heat-balance evaluation near pseudocritical region.

Increased values of heat-balance error (i.e., more than ±64%) at lower values of power (POW < 5 kW) and at inlet or outlet bulk-fluid temperatures within the pseudocritical region (i.e., t_{pc} - 2°C < t_{in} < t_{pc} + 2°C or < t_{pc} – 2°C < t_{out} < t_{pc} + 2°C) can be explained with the following (see Table D4 and Figure D8).

At lower values of power, the increase in bulk-fluid enthalpy is relatively small. However, uncertainties in bulk-fluid enthalpy within the pseudocritical region are larger for the same uncertainty range in bulk-fluid temperature, compared to the enthalpy values' uncertainties that correspond to temperatures far from the pseudocritical region.

Also, an additional error in the heat balance appears at mass-flux values below 900 kg/m²s (see Figure D6), where the flow-measuring curve is steep. Therefore, lower values of mass flux should be measured with a smaller diameter orifice flowmeter[35] or other type flowmeters.

Figure D8 shows an example of the heat-balance evaluation near the pseudocritical region. This graph shows that, at lower power values (POW < 5 kW) and at the outlet bulk-fluid temperature within the pseudocritical region, variations in bulk-fluid enthalpy difference can be up to 11.5% within the nominal uncertainty range for bulk-fluid temperatures (i.e., ±0.4°C).

[35] However, orifice-plate flowmeters with a diameter of the orifice less than 12.5 mm is considered a non-standard type.

Appendix E

SOME EXPERIMENTAL FEATURES OF VARIOUS SUPERCRITICAL-PRESSURE INSTALLATIONS

E.1 HEAT TRANSFER

E.1.1 Water

General

In addition to the heat transfer/pressure drop supercritical-water loops discussed in Chapter 10, there are several more supercritical-water test rigs in the world.

The Benson test rig (Kastner et al. 2000) was put into operation in Germany in 1975. This test rig has the following parameters: a system pressure of up to 33 MPa; a temperature of up to 600°C; a mass-flow rate of up to 28 kg/s; and a heating capacity of up to 2 MW. The test range was based on requirements for Benson boilers and gas heat-recovery steam generators.

The INEEL heat-transfer flow loop (McCreery et al. 2003) for development of SCWRs has the following parameters:

- Pressure 25 MPa;
- Test-section inlet bulk-fluid temperature 280°C – 488°C;
- Maximum test-section outlet bulk-fluid temperature 550°C;
- Maximum test-section heat flux 1.5 MW/m^2;
- Fuel-rod diameter 10.7 mm;
- Fuel-rod heated length 1 m;
- Maximum power per rod 50.4 kW;
- Maximum mass-flow rate per rod 0.060 kg/s-rod;
- Number of heated rods 3, 5 and 7; and
- Number of spacer grids (typical) 2.

The RDIPE supercritical-water loop (RDIPE 2004) for investigation of heat transfer and pressure drop.

The University of Wisconsin supercritical-water heat-transfer facility (Anderson et al. 2005) has the following range of parameters: a system pressure of up to 25 MPa; heater rod wall temperatures of up to 650°C; velocities of 1 – 10 m/s past a single 10.7-mm OD fuel pin, and a heating capacity of up to 1.5 MW.

There are also supercritical water loops of various capacities in China (Xi'an Jiatong University: loop pressure – 13 – 30 MPa; pump pressure up to 42 MPa and mass-flow rate up to 4500 kg/h; power up to 180 kW and preheaters power up to 660 kW (Chen 2004)), in Ukraine (Odessa National Politechnical University), in Canada (University of British Columbia), etc.

Quality of Water

Dickenson and Welch (1958) reported that the quality of the feedwater was the total solids of less than 1.0 ppm, dissolved oxygen of less than 0.005 ppm and *pH* of 9.5 during their experiments.

Smolin and Polyakov (1965) used supercritical water with the following parameters: $pH = 6 - 7$, hardness not more than 0.003 mg eqv/ℓ, C1 < 0.06 mg/ℓ, oxygen = 6 – 10 mg/ℓ, dry residue = 0.5 – 20 mg/ℓ and electrical conductance of $(0.6 - 10)\cdot 10^{-4}$ 1/Ohm · m.

Ackerman (1970) used deionized feedwater to maintain a specific electrical conductance of water between 5 and 20 1/Ohm·m. In his experiments, a chemical feed pump added ammonium hydroxide to maintain a *pH* of 9.5 at 25°C, and a deaerator reduced the oxygen to less than 5 ppb.

Test Sections

McAdams et al. (1950) performed experiments with water flowing through an annulus. In their experiments, a stainless steel inner heated tube ($D = 6.4$ mm, $L = 312$ mm, and $\delta_w = 0.84$ mm) was pressurized inside with nitrogen to allow minimal wall thickness.

Miropol'skiy and Shitsman (1957) reported that local HTC values within the entrance region of the channels might be higher than those beyond the entrance region. The length of the entrance region depends on how the fluid is supplied into the inlet of the test section. Also, it depends on the **Re** value. The length of the entrance region decreases as **Re** increasing. Therefore, it is important to have a stabilization section just upstream of the heated section to decrease the additional effects on HTC within the entrance region. Usually, the length of the stabilization section is about $(12 - 15)\cdot D$.

Goldmann (1961) used Hastelloy C tubes in his experiments.

Alferov et al. (1969) investigated heat transfer in vertical stainless-steel tubes ($D = 14$ and 20 mm, $L = 100\cdot D$ and $185\cdot D$) with supercritical water. In their setup, a heated part of the test section was preceded by an unheated section for hydraulic stabilization with a length of about $100\cdot D$.

Mode of Heating

Chakrygin and Lokshin (1957) used radiant heating in their experiments.

Ackerman (1970), Miropol'skii et al. (1970), Goldmann (1961) used a direct AC heating of the test-section wall.

Glushchenko and Gandzyuk (1972), Ornatskiy et al. (1972, 1971), Belyakov et al. (1971), McAdams et al. (1950) used a direct DC heating in their experiments.

Measuring Technique

Flow Rates Ackerman (1970) used a pressure drop across calibrated orifices to measure flow rates.

Glushchenko and Gandzyuk (1972) used a turbine-type flowmeter. The accuracy of flow measurements was ±1.5%.

Bulk-Fluid Temperatures In general, two bulk-fluid temperatures were measured at the test-section inlet and outlet; and local bulk-fluid temperatures along the heated length of a channel were calculated based on the inlet enthalpy, flow rate and wall heat flux.

Ackerman (1970) used thermocouples to measure bulk-fluid temperatures downstream of the mixing chambers.

Surface temperatures Chakrygin and Lokshin (1957) used six thermocouples per cross-section installed in intervals of 70 mm in their experiments with horizontal tubes.

Alferov et al. (1969) investigated heat transfer in vertical stainless-steel tubes (D = 14 and 20 mm, $L = 100 \cdot D$ and $185 \cdot D$) with supercritical water and used pairs of chromel-alumel thermocouples installed on a tube in 36 cross sections 180° apart.

Ackerman (1970) used chromel-alumel thermocouples resistance-welded to the wall to measure outside tube wall temperatures.

Miropol'skii et al. (1970) used three and four thermocouples per one cross-section to measure surface temperatures of plain-tube coils.

Belyakov et al. (1971) used thermocouples installed on the outer wall surface every 100 – 200 mm apart in their experiments with supercritical water flowing in vertical and horizontal tubes with D = 20 mm and L = 6.5 – 7.5 m (4 m for horizontal tube).

Ornatskiy et al. (1972, 1971) measured surface temperatures of the DC directly heated test sections through thin layers of mica (δ_w = 0.02 mm).

Glushchenko and Gandzyuk (1972) conducted experiments with an annulus and used a thermal probe consisting of a steel rod with eight brass pistons (thermocouple junctions were placed in each piston) press-fitted on to it for measuring internal wall temperatures of a heated inner tube. The measuring surface of the pistons was coated with a heat-resistant, electrically insulating paint.

Voltage Ackerman (1970) used voltage taps welded to the tube wall at 152.4 mm intervals to measure the incremental voltage drop along the 1.83 and 2.74 m heated lengths. These voltage drops were used to calculate local heat fluxes along the test section.

Data Reduction

Zhukovskiy et al. (1971) conducted experiments with a horizontal stainless-steel tube (D = 20 mm, D_{ext} = 28 mm, and L = 4 m). They calculated local heat fluxes accounting for the electrical current distribution inside the tube due to variations in electrical resistivity with temperature along and across the heated length. They considered a tube wall cross-section as parallel electrical resistances with different resistivity, but the same voltage drop. This method of data reduction decreases the effect of non-uniform cross-section temperature profile and thus, non-uniform heat generation in a tube wall on the calculated value of the local heat flux.

E.1.2 Carbon Dioxide

Quality of CO_2

Ikryannikov and Protopopov (1959) used dried carbon dioxide with 99.5% of purity in their experiments. The same purity carbon dioxide was used by Petukhov et al. (1961).

Kato et al. (1968) used highly pure (99.98%) and commercial (99.5% purity) CO_2 in their experiments. They did not find any measurable changes in heat transfer between these two grades of carbon dioxide.

Tanaka et al. (1971) used carbon dioxide of 99.9% purity.

Ankudinov and Kurganov (1981) used carbon dioxide with air impurity of ≤ 0.5 mol. %.

Test Sections

Bringer and Smith (1957) used a thin-wall Inconel tube (D = 6.35 mm, δ_w = 0.89 mm, and L = 610 mm). They pointed out that the use of Inconel made it possible to use a thin-wall tube due to the high tensile strength. Also, the Inconel temperature coefficient of electrical resistivity (about $6.9 \cdot 10^{-4}$ Ohm·ft/°F ($3.79 \cdot 10^{-4}$ Ohm·m/K) was sufficiently low not to have an appreciable effect on its electrical resistance. Therefore, a heat-flux profile was uniform over the heated length in spite of the non-uniform temperature profile.

Ikryannikov and Protopopov (1959) used a thin-wall copper tube (D = 6.7 mm, δ_w = 0.15 mm, and L = 670 mm) as the test section installed inside a pressurized case.

Petukhov et al. (1961) used an experimental test section consisting of a thin-wall copper tube (D = 6.7 mm, δ_w = 0.15 mm, and L = 670 mm). The use of a thin-wall copper tube brought the correction on the temperature drop across the tube wall to a negligible value. However, to relieve the stress on the tube wall from the inside pressure, the copper tube was installed inside a steel jacket; and the gap between the tube and the jacket was connected with a tube inside. Also, the test section was equipped with a stabilization section just upstream with a length of about 60 D.

Melik-Pashaev et al. (1968) used a stainless-steel tube (D = 4.5 mm, δ_w = 0.3 mm, and L = 135 mm) with a hydraulic stabilization length of $15 \cdot D$.

Bourke et al. (1970) used a thin-wall stainless-steel tube (D = 22.8 mm, δ_w = 1.27 mm, and L = 4.56 m). The test section was directly AC heated, with a straight, unheated entrance section of about 10 D long.

Bourke and Pulling (1971a,b) investigated heat transfer to carbon dioxide flowing inside a vertical stainless-steel tube (D = 22.5 mm and L = 4.57 m). They changed the heated length by moving the lower power clamp towards the upper power clamp. Also, direct AC heating was used.

Tanaka et al. (1971) used vertical tubes of 6-mm ID and with surface roughness of 0.2 μm and 14 μm.

Ikryannikov (1973) conducted experiments with supercritical carbon dioxide flowing through a stainless-steel tube (D = 29 mm, and δ_w = 1.5 mm) with an entrance (unheated) part of 0.72 m and a heated part of 2.25 m.

Adebiyi and Hall (1976) used a 25.4-mm OD horizontal stainless-steel tube with a 1.63-mm wall thickness in their experiments. The heated length was 110 diameters (2.44 m); and the upstream unheated length was 55 diameters (1.22 m). The test section was directly heated with AC.

Ankudinov and Kurganov (1981), used in their setup, an unheated section just upstream of the heated length of at least $20 \cdot D$.

Kurganov et al. (1986) used a polished inside stainless-steel tube (D=22.7 mm, δ_w = 1.3 mm, and L_{tot} = 5.215 m) with unheated sections of $50 \cdot D$ from both ends and a heated section of $130 \cdot D$.

Experimental Equipment and Mode of Heating

Krasnoshchekov and Protopopov (1966), Petukhov et al. (1961), Ikryannikov and Protopopov (1959) used a completely sealed electro-magnetically driven head circulation pump.

Ikryannikov (1973) and Ikryannikov et al. (1972) used an unpressurized centrifugal pump for CO_2 circulation.

Kurganov et al. (1986), Miropol'skiy and Baigulov (1974), Krasnoshchekov et al. (1971), Melik-Pashaev et al. (1968), Krasnoshchekov and Protopopov (1966), and Petukhov et al. (1961) used a direct AC heating of the test-section wall.

Measuring Technique

Flow Rates Bourke et al. (1970) used an orifice plate to measure the flow rate with an accuracy of 2%.

Adebiyi and Hall (1976) used an orifice plate to measure flow rate in a horizontal tube.

Bulk-Fluid Temperatures In general, two bulk-fluid temperatures were measured at the test-section inlet and outlet. Local bulk-fluid temperatures inside the heated part of the channel were calculated based on the inlet enthalpy, flow rate and wall heat flux.

Bringer and Smith (1957) used thermocouples installed inside mixing chambers to measure bulk-fluid temperatures at the inlet and outlet of the test section.

Ikryannikov and Protopopov (1959) used thermocouples installed just downstream of the mixing chambers to measure bulk temperatures at the inlet and outlet of the test section.

Bourke et al. (1970) used two thermocouples installed inside mixing chambers to measure the inlet and outlet bulk-fluid temperatures with an accuracy of 0.5°C. During the experiments, the flow temperatures were steady to within 0.5°C.

Krasnoshchekov et al. (1971) calculated a local bulk-fluid temperature along the channel by linear interpolation of the enthalpy between its values at the tube inlet and outlet.

Ikryannikov (1973) used a temperature probe installed inside the tube outlet to measure fluid temperature fields in a cross-section at the distance of $10 \cdot D$ from the outlet.

Adebiyi and Hall (1976) measured a bulk-fluid temperature at the inlet of the unheated part of the horizontal test section using a set of five thermocouples immersed in the fluid. These thermocouples were installed just downstream of the mixing chambers with a series of perforated copper discs. However, they found that this arrangement was not enough to have uniform temperature profile across the tube cross-section and, therefore, a suitable length of wire gauze was added to the mixer.

Surface Temperatures In the experiments of Bringer and Smith (1957), a pitch between copper-constantan thermocouples was about 50.8 mm.

Bourke et al. (1970) used 50 thermocouples along a 4.56 m heated length and around the tube diameter. The uncertainty in the inner wall temperature varied from less than 0.5°C at 30°C to 10°C at 300°C.

In the experiments of Ikryannikov (1973), the wall temperatures were measured in 21 cross-sections along the heated length of the tube. Three cross sections were equipped with four thermocouples; two cross-sections had six thermocouples and the rest only one.

Adebiyi and Hall (1976) measured surface temperatures along a horizontal tube (OD 25.4 mm, wall thickness 1.63 mm and heated length 2.44 m) using 196 chromel-alumel thermocouples welded on the outer tube surface. Sets of four thermocouples were installed every 76.2 mm along the tube at the cross-sectional locations of 0°, 90°, 180°, and 270° (where 0° is the tube top), and at every 152.4-mm intervals along the tube with the circumferential locations at 45°, 135°, 225°, and 315°.

Kurganov et al. (1986) measured wall temperatures with thermocouples installed every 113.5 mm.

Pressure In the experiments of Bourke et al. (1970) with forced convection of carbon dioxide through a tube, the pressure remained stable within the range of 3.5 kPa.

Kurganov et al. (1986) measured static pressure along the heated length in the cross-sections at 0, 25, 50, 65, 80, 105, and 130 D (D = 22.7 mm).

Heat Transfer Coefficient Krasnoshchekov and Protopopov (1966) reported that the maximum possible error in determining the HTC was between 5% and 10% at Δt = 5°C – 500°C and heat losses were not more than 2.4%.

E.1.3 Helium

Test Sections

Pron'ko et al. (1976) used nickel tubes (D = 1.04 mm and 0.7 mm, and δ_w = 0.05 mm) in their heat-transfer experiments.

Bogachev et al. (1983b) used a stainless-steel tube (D = 1.8 mm, L = 400 mm, and δ_w = 0.1 mm) with 78 mm unheated part just upstream of the heated length.

Experimental Equipment and Mode of Heating

Cairns and Brassington (1976) used a centrifugal pump to circulate supercritical helium in the flow loop.

Bogachev et al. (1983a,b) used a direct DC heating in their experiments.

Measuring Technique

Surface Temperatures Bogachev et al. (1983b) used Germanium RTDs installed at 15 cross-sections along the heated length (pitch 25 mm), which were pressed through a thin electrical insulating film (δ_w = 10 μm) on the tube surface.

REFERENCES

Abadzic, E. and Goldstein, R.J., 1970. Film boiling and free convection heat transfer to carbon dioxide near the critical state, *International Journal of Heat & Mass Transfer,* 13 (7), pp. 1163–1175.

Abdullaeva, G.K., Isaev, G.I., Mamedov, F.Kh., and Arabova, I.T., 1991. A study of heat transfer from a vertical and a horizontal surface to organic coolants with free convection under supercritical pressure conditions, *Thermal Engineering* (Теплоэнергетика, стр. 70–72) 38 (9), pp. 516–518.

Abdulagatov, I.M. and Alkhasov, A.B., 1998. Transformation of geothermal energy into electricity by means of a supercritical cycle in the secondary circuit, *Thermal Engineering* (Теплоэнергетика, стр. 53–56), 45 (4), pp. 320–324.

Ackerman, J.W., 1970. Pseudoboiling heat transfer to supercritical pressure water in smooth and ribbed tubes, *Journal of Heat Transfer,* Transactions of the ASME, 92 (3), pp. 490–498, (Paper No. 69-WA/HT-2, pp. 1–8).

Adebiyi, G.A. and Hall, W.B., 1976. Experimental investigation of heat transfer to supercritical pressure carbon dioxide in a horizontal tube, *International Journal of Heat & Mass Transfer,* 19 (7), pp. 715–720.

Adelt, M. and Mikielewicz, J., 1981. Heat transfer in a channel at supercritical pressure, *International Journal of Heat & Mass Transfer,* 24 (10), pp. 1667–1674.

Afonin, V.K. and Smirnov, O.K., 1985. The effect of thermal unsteadiness on heat transfer in the supercritical region, *Thermal Engineering* (Теплоэнергетика, стр. 67–69), 32 (3), pp. 163–166.

Akhmedov, F.D., Grigor'ev, V.A., and Dudkevich, A.S., 1974. The boiling of nitrogen at pressures from atmospheric to critical, *Thermal Engineering* (Теплоэнергетика, стр. 84–85), 21 (1), pp. 120–121.

Akol'zin, P.A., 1972. *Chemical-Water Regimes of Steam-Turbine Units at Supercritical Parameters,* (In Russian), Energiya Publishing House, Moscow, Russia, 174 pages.

Aksan, N., Schulenberg, T., Squarer, D., et al., 2003. Potential safety features and safety analysis aspects for High Performance Light Water Reactor (HPLWR), Proceedings of the Joint International Conference Global Environment and Nuclear Energy Systems/Advanced Nuclear Power Plants (GENES4/ANP2003), Kyoto, Japan, September 15–19, Paper No. 1223, 9 pages.

Alad'ev, I.T., Ivlev, A.A., and Turilina, E.S., 1980. Nomograms for predicting heat transfer with a heated turbulent flow of supercritical helium I in narrow channels, *Thermal Engineering* (Теплоэнергетика, стр. 74–76), 27 (11), pp. 651–654.

Alad'yev, I.T., Povarnin, P.I., and Malkina, L.I., 1963a. An investigation of the cooling properties of methyl alcohol at pressures (98–392)·10^5 N/m^2, *Journal of Engineering-Physics* (Инженерно-Физический Журнал (ИФЖ), стр. 83–87), 6 (10), pp. 91–96.

Alad'yev, I.T., Povarnin, P.I., Merkel', E.Yu., and Malkina, L.I., 1963b. Investigation of the cooling properties of ethyl alcohol at p ≤ 800 ata, (In Russian), *Thermal Engineering* (Теплоэнергетика, стр. 70–72), (8), pp. 70–72 (or Alad'yev, I.T., Povarnin, P.I., Malkina, L.I., and Merkel', E.Yu., 1967. Investigation of the cooling properties of ethyl alcohol at absolute pressure of up to 800 atm, Proceedings of the Second All-Soviet Union Conference on Heat and Mass Transfer, Edited by C. Gazley, Jr., Hartnett, J.P., and Eckert, E.R.G., Vol. I, RAND Report R-451-PR, The Rand Corporation, University of Ann Arbor, Michigan, Microfilms Library, pp. 57–60).

Alad'yev, I.T., Povarnin, P.I., Malkina, L.I., and Merkel', E.Yu., 1967. Investigation of the cooling properties of ethyl alcohol at absolute pressure of up to 800 atm, Proceedings of the 2nd All-Soviet Union Conference on Heat and Mass Transfer, Minsk, Belarus', May, 1964, Published as Rand Report R-451-PR, Edited by C. Gazley, Jr., J.P. Hartnett and E.R.C. Ecker, Vol. 1, pp. 57–60.

Alad'ev, I.T., Vas'yanov, V.D., Kafengauz, N.L., et al., 1976. Pseudoboiling mechanism for n-heptane, *Journal of Engineering-Physics* (Инженерно-Физический Журнал (ИФЖ), стр. 389–395), 31 (3), pp. 389–395.

Aleksandrov, A.A., 1980. International tables and equations for the thermal conductivity of water and steam, *Thermal Engineering* (Теплоэнергетика, стр. 70–74), 27 (4), pp. 235–240.

Aleksandrov, A.A., Ivanov, A.I., and Matveev, A.B., 1975. The dynamic viscosity of water and steam within a wide region of temperatures and pressures, *Thermal Engineering* (Теплоэнергетика, стр. 59–64), 22 (4), pp. 77–83.

Alekseev, P.N., Grishnanin, E.I., Zverkov, Yu.A., et al., 1989. Steam-water power reactor concept, Soviet-Japanese Seminar on Theoretical, Computational and Experimental Study of Physical Problems in Designing of Fast Reactors, July, pp. 90–97.

Alekseev, G.V. and Smirnov, A.M., 1976. Heat transfer at turbulent flow of fluid in channels at supercritical pressures, (In Russian), Preprint IPPE (Institute of Physics and Power Engineering) (Препринт ФЭИ), No. E–11 ОБ-20, Obninsk, Russia, 83 pages.

Alekseev, G.V., Silin, V.A., Smirnov, A.M., and Subbotin, V.I., 1976. Study of the thermal conditions on the wall of a pipe during the removal of heat by water at a supercritical pressure, *High Temperatures* (Теплофизика Высоких Температур, стр. 769–774), 14 (4), pp. 683–687.

Aleshchenkov, P.I., Zvereva, G.A., Kireev, G.A., et al., 1971. Start-up and operation of channel-type uranium-graphite reactor with tubular fuel elements and nuclear steam superheating, *Atomic Energy* (Атомная Энергия, стр. 137–144), 30 (2), pp. 163–170.

Alferov, N.S., Balunov, B.F., and Rybin, R.A., 1975. Calculating heat transfer with mixed convection, *Thermal Engineering* (Теплоэнергетика, стр. 71–75), 22 (6), pp. 96–100.

Al'ferov, N.S., Balunov, B.F., and Rybin, R.A., 1973. Reduction in heat transfer in the region of supercritical state variables of a liquid, *Heat Transfer–Soviet Research*, 5 (5), pp. 49–52.

Alferov, N.S., Balunov, B.F., Zhukovskiy, A.V., et al., 1973. Features of heat transfer due to combined free and forced convection with turbulent flow, *Heat Transfer–Soviet Research*, 5 (4), pp. 57–59.

Alferov, N.S., Rybin, R.A., and Balunov, B.F., 1969. Heat transfer with turbulent water flow in a vertical tube under conditions of appreciable influence of free convection, *Thermal Engineering* (Теплоэнергетика, стр. 66–70), 16 (12), pp. 90–95.

Allen, T.R., Tan, L., Chen, Y., et al., 2005a. Corrosion and radiation response of advanced ferritic-martensitic steels for Generation IV application, Proceedings of the International Conference GLOBAL-2005 "Nuclear Energy Systems for Future Generation and Global Sustainability," Tsukuba, Japan, October 9–13, Paper No. IL001, 6 pages.

Allen, T.R., Tan, L., Chen, Y., et al., 2005b. Corrosion of ferritic-martensitic alloys in supercritical water for GenIV application, Proceedings of the International Conference GLOBAL-2005 "Nuclear Energy Systems for Future Generation and Global Sustainability," Tsukuba, Japan, October 9–13, Paper No. 419, 6 pages.

Altunin, V.V., 1975. *Thermophysical Properties of Carbon Dioxide*, (In Russian), Izdatel'stvo Standartov Publishing House, Moscow, Russia, p. 551.

Altunin, V.V., Kuznetsov, D.O., and Bondarenko, V.F., 1973. The thermodynamic properties of commercial grade gaseous carbon dioxide, *Thermal Engineering* (Теплоэнергетика, стр. 61–63), 20 (4), pp. 88–90.

Altunin, V.A., Yagofarov, O.Kh., Zarifullin, M.E., and Zamaltdinov, Sh.Ya.-S., 1998. Characteristic properties of heat transfer to liquid hydrocarbon coolants under natural convection conditions at sub and supercritical pressures, *Izvestiya VUZov* (Известия ВУЗов), *Aviation Technique*, 41 (1), pp. 59–67.

Anderson, M.H., Licht, J.R., and Corradini, M.L., 2005. Progress on the University of Wisconsin supercritical water heat transfer facility, Proceedings of the International Topical Meeting on Nuclear Reactor Thermal Hydraulics (NURETH-11), Avignon, France, October 2–6, Paper No. 265, 18 pages.

Ankudinov, V.B. and Kurganov, V.A., 1981. Intensification of deteriorated heat transfer in heated tubes at supercritical pressures, *High Temperatures* (Теплофизика Высоких Температур, стр. 1208–1212), 19 (6), pp. 870–874.

Antikain, P.A., 1977. *Corrosion of Metal in Steam Generators*, (In Russian), Energiya Publishing House, Moscow, Russia.

Armand, A.A., Tarasova, N.V., and Kon'kov, A.S., 1959. Investigation of heat transfer from wall to steam near the critical point, (In Russian), In book: *Heat Transfer at High Heat Fluxes and Other Special Conditions*, Gosenergoizdat Publishing House, Moscow–Leningrad, Russia, pp. 41–50.

Arsent'ev, P.P. and Koledov, L.A., 1976. *Metal Melts and Their Properties*, (In Russian), Metallurgiya Publishing House, Moscow, Russia.

ASME International Steam Tables for Industrial Use, 2000. Prepared by W.T. Parry, J.C. Bellows, J.S. Galagher, A.H. Harvey, ASME Press, New York, NY, USA.

Bae, J.H., Yoo, J.Y., and Choi, H., 2003. Direct numerical simulation of heat transfer to CO_2 at supercritical pressure in a vertical tube, Proceedings of the International Topical Meeting on Nuclear Reactor Thermal Hydraulics (NURETH-10), Seoul, Korea, October 5–9, 20 pages.

Bae, J.H., Yoo, J.Y., Choi, H., and McEligot, D.M., 2005. Influence of fluid-property variation on turbulent convective heat transfer in vertical annular channel flows, Proceedings of the International Topical Meeting on Nuclear Reactor Thermal Hydraulics (NURETH-11), Avignon, France, October 2–6, Paper No. 018, 16 pages.

Bae, Yo.-Ye., Joo, H.-K., Jeong, Yo.H., et al., 2004. Research of a supercritical pressure water cooled reactor in Korea, Proceedings of the International Congress on Advances in Nuclear Power Plants (ICAPP'04), Pittsburgh, PA, USA, June 13–17, Paper No. 4247, pp. 480–491.

Baranaev, Yu.D., Kirillov, P.L., Poplavskii, V.M., and Sharapov, V.N., 2004. Supercritical-pressure water nuclear reactors, *Atomic Energy* (Атомная Энергия, стр. 374–380), 96 (5), pp. 345–351.

Barulin, Yu.D., Vikhrev, Yu.V., Dyadyakin, B.V., et al., 1971. Heat transfer during turbulent flow in vertical and horizontal tubes containing water with supercritical state parameters, *Journal of Engineering-Physics* (Инженерно-Физический Журнал (ИФЖ), стр. 929–930), 20 (5), p. 665.

Baskov, V.L., Kuraeva, I.V., and Protopopov, V.S., 1974. Heat transfer in the turbulent flow of carbon dioxide of supercritical pressure in a vertical tube under cooling conditions, (In Russian), *Transactions of MEI* (Moscow Power Engineering Institute (МЭИ)), Issue 179, pp. 140–149.

Baskov, V.L., Kuraeva, I.V., and Protopopov, V.S., 1977. Heat transfer with the turbulent flow of a liquid at supercritical pressure in tubes under cooling conditions, *High Temperatures* (Теплофизика Высоких Температур, стр. 96–102), 15 (1), pp. 81–86.

Bastron, A., Hofmeister, J., Meyer, L., and Schulenberg, T., 2005. Enhancement of heat transfer in HPLWR fuel assemblies, Proceedings of the International Conference GLOBAL-2005 "Nuclear Energy Systems for Future Generation and Global Sustainability," Tsukuba, Japan, October 9–13, Paper No. 36, 6 pages.

Baturov, B.B., Zvereva, G.A., Mityaev, Yu.I., and Mikhan, V.I., 1978. Nuclear superheating of steam, results and prospects at the present stage, *Atomic Energy* (Атомная Энергия, стр. 126–131), 44 (2), pp. 131–137.

Bazargan, M., Fraser, D., and Chatoorgoon, V., 2005. Effect of buoyancy on heat transfer in supercritical water flow in a horizontal round tube, *Journal of Heat Transfer*, 127, pp. 897–902.

Bellmore, C.P. and Reid, R.L., 1983. Numerical prediction of wall temperatures for near-critical para-hydrogen in turbulent upflow inside vertical tubes, *Journal of Heat Transfer*, Transactions of the ASME, 105 (3), pp. 536–541.

Belyakov, I.I., 1976. An investigation of the temperature conditions of tubes in supercritical boilers when there are internal iron oxide deposits, *Thermal Engineering* (Теплоэнергетика, стр. 64–66), 23 (4), pp. 42–44.

Belyakov, I.I., 1995. Supercritical boilers: The future for the development of thermal power stations, *Thermal Engineering* (Теплоэнергетика, стр. 9–12), 42 (8), pp. 618–621.

Belyakov, I.I., Krasyakova, L.Yu., Zhukovskii, A.V., and Fefelova, N.D., 1971. Heat transfer in vertical risers and horizontal tubes at supercritical pressure, *Thermal Engineering* (Теплоэнергетика, стр. 39–43), 18 (11), pp. 55–59.

Beschastnov, S.P., Kirillov, P.L., and Aladushev, V.P., 1976. Free-convection conditions near wires with supercritical pressures of CO_2, *High Temperatures* (Теплофизика Высоких Температур, стр. 775–780), 14 (4), pp. 688–692.

Beschastnov, S.P., Kirillov, P.L., and Petrov, V.P., 1973. Heat transfer by free convection to Freon-12 near the critical point, *Heat Transfer–Soviet Research*, 5 (4), pp. 72–75.

Beschastnov, S.P., Kirillov, P.L., and Saikin, A.M., 1973. Heat transfer by turbulent free convection from a flat plate to carbon dioxide at subcritical pressures, *High Temperatures* (Теплофизика Высоких Температур, стр. 346–351), 11 (2), pp. 299–304.

Beschastnov, S.P., and Petrov, V.P., 1973. Heat transfer by free convection from a horizontal cylinders to CO_2 under near-critical conditions, *High Temperatures* (Теплофизика Высоких Температур, стр. 588–592), 11 (3), pp. 524–528.

Bezrodny, M.K., Pioro, I.L., and Kostyuk, T.O., 2005. *Transfer Processes in Two-Phase Thermosyphon Systems: Theory and Practice*, (In Russian), 2nd edition, Augmented and Revised, Fact Publ. House, Kiev, Ukraine, 704 pages.

Bishop, A.A., Efferding, L.E., and Tong, L.S., 1962. A review of heat transfer and fluid flow of water in the supercritical region and during "once-through" operation, Report WCAP-2040, Westinghouse Electric Corporation, Atomic Power Division, Pittsburgh, PA, USA, December, 106 pages.

Bishop, A.A., Sandberg, R.O., and Tong, L.S., 1964. Forced convection heat transfer to water at near-critical temperatures and super-critical pressures, Report WCAP-2056, Westinghouse Electric Corporation, Atomic Power Division, Pittsburgh, PA, USA, December, 85 pages.

Bishop, A.A., Sandberg, R.O., and Tong, L.S., 1965. Forced convection heat transfer to water at near-critical temperature and supercritical pressures, *AIChE*, J. *Chemical Engineering Symposium Series*, No. 2, London, Institute of Chemical Engineers.

Bittermann, D., Squarer, D., Schulenberg, T., and Oka Y., 2003a. Economic prospects of the HPLWR, Proceedings of the Joint International Conference Global Environment and Nuclear Energy Systems/Advanced Nuclear Power Plants (GENES4/ANP2003), Kyoto, Japan, September 15–19, Paper No. 1003, 7 pages.

Bittermann, D., Squarer, D., Schulenberg, T., et al., 2003b. Potential plant characteristics of a High Performance Light Water Reactor (HPLWR), Proceedings of the International Congress on Advances in Nuclear Power Plants (ICAPP'03), Córdoba, Spain, May 4–7, Paper No. 3046, 9 pages.

Bloem, W.B., 1986. Transient heat transfer to a forced flow of supercritical helium at 4.2 K, *Cryogenics*, 26, pp. 300–308.

Boehm, C., Starflinger, J., Schulenberg, Th., and Oeynhausen, H., 2005. Supercritical steam cycle for lead cooled nuclear systems, Proceedings of the International Conference GLOBAL-2005 "Nuclear Energy Systems for Future Generation and Global Sustainability," Tsukuba, Japan, October 9–13, Paper No. 035, 6 pages.

Bogachev, V.A. and Eroshenko, V.M., 1986. The generality of equations for mixed-convective heat transfer to liquids at supercritical pressure in vertical pipes, *Journal of Engineering Physics* (Инженерно-Физический Журнал (ИФЖ), стр. 946–951), 50 (6), pp. 666–670.

Bogachev, V.A., Eroshenko, V.M., and Kuznetsov, E.V., 1986. Experimental study of thermally induced oscillations and heat transfer in an ascending flow of supercritical helium in a vertical tube, *Journal of Engineering-Physics* (Инженерно-Физический Журнал (ИФЖ), стр. 719–723), 51 (5), pp. 1265–1268.

Bogachev, V.A., Eroshenko, V.M., and Kuznetsov, E.V., 1988. Experimental study of thermalhydraulic stability and heat transfer in the descending flow of supercritical helium in a vertical tube, *Journal of Engineering-Physics* (Инженерно-Физический Журнал (ИФЖ), стр. 181–185), 55 (2), pp. 831–834.

Bogachev, V.A., Yeroshenko, V.M., Snyitina, O.F., and Yaskin, L.A., 1985. Measurements of heat transfer to supercritical helium in vertical tubes under forced and mixed convection conditions, *Cryogenics*, 25, pp. 198–201.

Bogachev, V.A., Yeroshenko, V.M., Snytina, O.F., and Yaskin, L.A., 1986. Mixed–convection heat transfer in downtake flow of supercritical helium in a heated tube at transition Reynolds numbers, *Heat Transfer–Soviet Research*, 18 (6), pp. 91–96.

Bogachev, V.A., Eroshenko, V.M., and Yaskin, L.A., 1983a. Heat transfer associated with an ascending flow of supercritical helium in a heated tube with Re<2300 at the entry, *High Temperatures* (Теплофизика Высоких Температур, стр. 101–106), 21 (1), pp. 89–94.

Bogachev, V.A., Eroshenko, V.M., and Yaskin, L.A., 1983b. Relative increase in heat transfer in viscous-inertial regimes of flow of helium at supercritical pressure in a heated tube, *Journal of Engineering-Physics* (Инженерно-Физический Журнал (ИФЖ), стр. 544–548), 44 (4), pp. 363–366.

Bogoslovskaia, G.P., Karpenko, A.A., Kirillov, P.L., and Sorokin, A.P., 2005. MIF-SCD computer code for thermal hydraulic calculation of supercritical water cooled reactor core, Proceedings of the International Topical Meeting on Nuclear Reactor Thermal Hydraulics (NURETH-11), Avignon, France, October 2–6, Paper No. 204, 14 pages.

Bojinov, M., Heikinheimo, L., Saario, T., and Tuurna, S., 2005. Characterization of corrosion films on steels after long-term exposure to simulated supercritical water conditions, Proceedings of the International Congress on Advances in Nuclear Power Plants (ICAPP'05), Seoul, Korea, May 15–19, Paper No. 5349, 9 pages.

Bonilla, C.F. and Sigel, L.A., 1961. High-intensity natural-convection heat transfer near the critical point, *Chemical Engineering Progress, Symposium Series,* 57 (32), pp. 87–95.

Bourke, P.J. and Denton, W.H., 1967. An unusual phenomenon of heat transfer near the critical point, Memorandum AERE–M 1946, Chemical Engineering and Process Technology Division, U.K.A.E.A. Research Group, Atomic Energy Research Establishment, Harwell, Berkshire, August, 9 pages.

Bourke, P.J. and Pulling, D.J., 1971a. An experimental explanation of deterioration in heat transfer to supercritical carbon dioxide, Report AERE–R.6765, Chemical Engineering Division, U.K.A.E.A. Research Group, Atomic Energy Research Establishment, Harwell, March, 12 pages.

Bourke, P.J. and Pulling, D.J., 1971b. Experimental explanation of deterioration in heat transfer to supercritical carbon dioxide, ASME Paper No. 71–HT–24, 7 pages.

Bourke, P.J., Pulling, D.J., Gill, L.E., and Denton, W.H., 1968. The measurement of turbulent velocity fluctuations and turbulent temperature fluctuations in the supercritical region by hot wire anemometer and a "cold" wire resistance thermometer, *Institute of Mechanical Engineers,* Vol. 182, Part 31, Paper No. 3, pp. 58–67.

Bourke, P.J., Pulling, D.J., Gill, L.E., and Denton, W.H., 1970. Forced convective heat transfer to turbulent CO_2 in the supercritical region, *International Journal Heat & Mass Transfer,* 13 (8), pp. 1339–1348.

Brassington, D.J. and Cairns, D.N.H., 1977. Measurements of forced convective heat transfer to supercritical helium, *International Journal of Heat & Mass Transfer,* 20 (3), pp. 207–214.

Breus, V.I. and Belyakov, I.I., 1990. Heat transfer in helical coils at supercritical pressure, *Thermal Engineering* (Теплоэнергетика, стр. 49–51), 21 (1), pp. 189–191.

Bringer, R.P. and Smith, J.M., 1957. Heat transfer in the critical region, *AIChE Journal*, 3 (1), pp. 49–55.

Budnevich, S.S. and Uskenbaev, S.V., 1972. Heat transfer in liquefied gases in the supercritical range, *Journal of Engineering-Physics* (Инженерно-Физический Журнал (ИФЖ), стр. 446–452), 23 (3), pp. 1117–1121.

Buongiorno, J., 2004. The supercritical water cooled reactor: Ongoing research and development in the U.S., Proceedings of the International Congress on Advances in Nuclear Power Plants (ICAPP'04), Pittsburgh, PA, USA, June 13–17, Paper No. 4229, pp. 451–458.

Buongiorno, J., 2003. An alternative SCWR design based on vertical power channels and hexagonal fuel assemblies, Proceedings of the ANS/ENS International Winter Meeting and Nuclear Technology Expo, Embedded Topical Meeting GLOBAL 2003 "Advanced Nuclear Energy and Fuel Cycle Systems," New Orleans, LA, USA, November 16–20, pp. 1155–1162.

Buongiorno, J., Corwin, W., MacDonald, Ph.E., et al., 2003. Supercritical Water Reactor (SCWR): Survey of Materials Experience and R&D Needs to Assess Viability, Report INEEL/EXT-03-00693 Rev. 1, Idaho National Engineering and Environmental Laboratory, USA, September, 63 pages.

Buongiorno, J. and MacDonald, Ph.E., 2003a. Supercritical water reactor (SCWR). Progress report for the FY-03 Generation-IV R&D activities for the development of the SCWR in the U.S., Report INEEL/EXT-03-01210, Idaho National Engineering and Environmental Laboratory, USA, September, 38 pages.

Buongiorno, J. and MacDonald, Ph.E., 2003b. Study of solid moderators for the thermal-spectrum supercritical water reactor (neutronics), Proceedings of the 11th International Conference on Nuclear Engineering (ICONE-11), Shinjuku, Tokyo, Japan, April 20–23, Paper No. 36571, 7 pages.

Buongiorno, J. and MacDonald, Ph.E., 2003c. Study of solid moderators for the thermal-spectrum supercritical water reactor (thermo-mechanics and cost), Proceedings of the International Congress on Advances in Nuclear Power Plants (ICAPP'03), Córdoba, Spain, May 4–7, Paper No. 3329, 7 pages.

Buongiorno, J., Sterbentz, J.W., and MacDonald, Ph.E., 2006. Study of solid moderators for the thermal-spectrum supercritical water-cooled reactor, *Nuclear Technology*, 153, pp. 282–303.

Bushby, S.J., Dimmick, G.R., Duffey, R.B., et al., 1999. Conceptual design for a supercritical-water-cooled CANDU, Global'99: Nuclear Technology: Bridging the Millennia, Proceedings of the International Conference on Future Nuclear Systems, Jackson Hole, WY, USA, August 29–September 3, Paper No. 103.

Bushby, S.J., Dimmick, G.R., Duffey, R.B., et al., 2000a. Conceptual designs for advanced, high-temperature CANDU reactors, Proceedings of the 1st International Symposium on Supercritical Water-Cooled Reactor Design and Technology (SCR-2000), Tokyo, Japan, November 6–8, Paper No. 103.

Bushby, S.J., Dimmick, G.R., Duffey, R.B., et al., 2000b. Conceptual designs for advanced, high-temperature CANDU reactors, Proceedings of the 8th International Conference on Nuclear Engineering (ICONE-8), Baltimore, MD, USA, April 2–6, Paper No. 8470, 12 pages.

Cairns, D.N.H. and Brassington, D.J., 1976. A pumped supercritical helium flow loop for heat transfer studies, *Cryogenics*, 16 (8), pp. 465–468.

Chakrygin, V.G., 1964. Calculating pressure losses in heated tubes at supercritical pressures, *Thermal Engineering* (Теплоэнергетика, стр. 78–80), 11 (10), pp. 95–100.

Chakrygin, V.G., 1965. Calculations of pressure drop in heated tubes at supercritical pressures (p=230–350 ata), (In Russian), *Transactions of TsKTI "Boiler-Turbine Engineering"*(Труды ЦКТИ "Котло-Турбостроение", стр. 82–89), Issue 59, Leningrad, Russia, pp. 82–89.

Chakrygin, V.G., 1967. The effect of non-uniform heating on pressure losses at supercritical pressures, *Thermal Engineering* (Теплоэнергетика, стр. 58–62), 14 (1), pp. 77–83.

Chakrygin, V.G., Agafonov, M.B., and Letyagin, I.P., 1974. Experimental determination of the limits of aperiodic instability at very high and supercritical pressures, *Thermal Engineering* (Теплоэнергетика, стр. 12–16), 21 (1), pp. 17–22.

Chakrygin, V.G. and Lokshin, V.A., 1957. Temperature regime of horizontal steam generating tubes at supercritical pressures (p=230–250 ata), (In Russian), *Thermal Engineering* (Теплоэнергетика, стр. 27–30), 4 (10), pp. 27–30.

Chalfant, A.I., 1954. Heat transfer and fluid friction experiments for the supercritical water reactor, Report PWAC-109, June 1.

Chatoorgoon, V., 2001. Stability of supercritical fluid flow in a single-channels natural-convection loop, *International Journal of Heat & Mass Transfer*, 44, pp. 1963–1972.

Chatoorgoon, V. and Upadhye, P., 2005. Analytical studies of supercritical flow instability in natural convection loops, Proceedings of the International Topical Meeting on Nuclear Reactor Thermal Hydraulics (NURETH-11), Avignon, France, October 2–6, Paper No. 165, 19 pages.

ChemicalLogic Corporation, 1999. Thermodynamic and transport properties of carbon dioxide. Software (CO_2 Tables, Version 1.0).

Chen T., 2004. *Two-Phase Flow and Heat Transfer Study*, (In Chinese), Xi'an Jiaotong University Press, China, ISBN 7-5605-1892-3.

Cheng, X., Lauren, E., and Yang, Y.H., 2005. CFD analysis of heat transfer in supercritical water in different flow channels, Proceedings of the International Conference GLOBAL-2005 "Nuclear Energy Systems for Future Generation and Global Sustainability," Tsukuba, Japan, October 9–13, Paper No. 369, 6 pages.

Cheng, X. and Schulenberg, T., 2001. Heat transfer at supercritical pressures – literature review and application to a HPLWR, Report Forschungszentrum Karlsruhe, Technik und Umwelt, Wissenschaftliche Berichte, FZKA 6609, Institute für Kern- und Energietechnik, Mai 2001, 45 pages.

Cheng, X., Schulenberg, T., Bittermann, D., and Rau, P., 2003. Design analysis of core assemblies for supercritical pressure conditions, *Nuclear Engineering and Design*, 223, pp. 279–294.

Cheng, X., Schulenberg, T., Koshizuka, S., et al., 2002. Thermal-hydraulic analysis of supercritical pressure light water reactors, Proceedings of the International Congress on Advances in Nuclear Power Plants (ICAPP'02), Hollywood, FL, USA, June 9–13, Paper No. 1015, 11 pages.

Cho, B.H., Kim, Yo.I., and Bae, Yo.Ye., 2005. Numerical study on the heat transfer to CO_2 flowing upward in a heated vertical tube at supercritical pressure,

Proceedings of the International Conference GLOBAL-2005 "Nuclear Energy Systems for Future Generation and Global Sustainability," Tsukuba, Japan, October 9–13, Paper No. 215, 6 pages.

Coleman, H.W. and Steele, W.G., 1999. *Experimentation and Uncertainty Analysis for Engineers,* 2nd edition, J. Wiley & Sons, Inc., New York, USA, 275 pages.

Cornelius, A.J. and Parker, J.D., 1965. Heat transfer instabilities near the thermodynamic critical point, Proceedings of the 1965 Heat Transfer and Fluid Mechanics Institute, Stanford University Press, pp. 317–329.

Dashkiyev, Yu.G. and Rozhalin, V.P., 1975. Thermo-hydraulic stability of a system of steam-generating channels with super-critical pressure, *Heat Transfer–Soviet Research,* 7 (5), pp. 102–110.

Dashevskiy, Yu.M., Mal'kovskiy, V.I., Miropol'skiy, Z.L., et al., 1986. Effect of thermogravitation on heat transfer in up or down pipe flow of supercritical fluids, *Heat Transfer–Soviet Research,* 18 (6), pp. 109–114.

Dashevskii, Yu.M., Miropol'skii, Z.L., and Khasanov-Agaev, L.R., 1987. Heat transfer with flow of medium at supercritical parameters of state in channels with different spatial orientation, *Thermal Engineering* (Теплоэнергетика, стр. 68–70), 34 (8), pp. 454–457.

Davis, C.B., Buongirno, J, and MacDonald, Ph.E., 2003. A parametric study of the thermal-hydraulic response of supercritical light water reactors during loss-of-feedwater and turbine-trip events, Proceedings of the Joint International Conference Global Environment and Nuclear Energy Systems/Advanced Nuclear Power Plants (GENES4/ANP2003), Kyoto, Japan, September 15–19, Paper No. 1009, 9 pages.

Deev, V.I., Kondratenko, A.K., Petrovichev, V.I., et al., 1978. Natural convection heat transfer from a vertical plate to supercritical helium, Proceedings of the 6th International Heat Transfer Conference, Toronto, Ontario, Canada, August 7–11, Vol. 2, Paper No. NC-4.

Deissler, R.G., 1954. Heat transfer and fluid friction for fully developed water with variable fluid properties, *Transactions of the ASME,* 76 (1), January, pp. 73–85.

Deissler, R.G. and Taylor, M.F., 1953. Analysis of heat transfer and fluid friction for fully developed water with variable fluid properties in a smooth tube, Research Memorandum, National Advisory Committee for Aeronautics, April 9, 29 pages.

Dement'ev, B.A., 1990. *Nuclear Power Reactors,* (In Russian), Energoatomizdat Publishing House, Moscow, Russia, 351 pages.

Dickinson, N.L. and Welch, C.P., 1958. Heat transfer to supercritical water, *Transactions of the ASME,* 80, pp. 746–752.

Dimmick, G.R., Spinks, N.J., and Duffey, R., 1998. An advanced CANDU reactor with supercritical water coolant: Conceptual design features, Proceedings of the 6th International Conference on Nuclear Engineering (ICONE-6), San Diego, CA, USA, May 10–15, Paper No. 6501.

Dittus, F.W. and Boelter, L.M.K., 1930. University of California, Berkeley, Publications on Engineering, Vol. 2, p. 443.

Dobashi, K., Kimura, A., Oka, Y., and Koshizuka, S., 1998a. Conceptual design of a high temperature power reactor cooled and moderated by supercritical light water, *Annals of Nuclear Energy,* 25 (8), pp. 487–505.

Dobashi, K., Oka, Y., and Koshizuka, S., 1998b. Conceptual design of a high temperature power reactor cooled and moderated by supercritical water,

Proceedings of the 6th International Conference on Nuclear Engineering (ICONE-6), San Diego, CA, USA, May 10–15, Paper No. 6232.

Dobashi, K., Oka, Y., and Koshizuka, S., 1997. Core and plant design of the power reactor cooled and moderated by supercritical light water with single tube water rods, *Annals of Nuclear Energy*, 24 (16), pp. 1281–1300.

Department of Energy (DOE) USA, 2005. BONUS, Puerto Rico, Decommissioned reactor, Fact Sheet, 2 pages.

Dolgoy, M.L., Kirichenko, Yu.A., Sklovsky, Yu.B., et al., 1983. Heat transfer to near-critical helium in horizontal channels, *Cryogenics*, 23, pp. 125–126.

Dolezhal', N.A., Aleshchenkov, P.I., Bulankov, Yu.V., and Knyazeva, G.D., 1971. Construction of uranium-graphite channel-type reactors with tubular fuel elements and nuclear-superheated steam, *Atomic Energy* (Атомная Энергия, стр. 149–155), 30 (2), pp. 177–182.

Dollezhal', N.A. and Emel'yanov, I.Va., 1980. *Channel Nuclear Power Reactor*, (In Russian), Energoatomizdat Publishing House, Moscow, Russia.

Dollezhal, N.A., Krasin, A.K., Aleshchenkov, P.I., et al., 1958. Uranium-graphite reactor with superheated high pressure steam, Proceedings of the 2nd International Conference on the Peaceful Uses of Atomic Energy, United Nations, Vol. 8, Session G-7, P/2139, pp. 398–414.

Dollezhal', N.A., Malyshev, V.M., Shirokov, S.V., et al., 1974. Some results of operation of the I.V. Kurchatov nuclear power station at Belyi Yar, *Atomic Energy* (Атомная Энергия, стр. 432–438), 36 (6), pp. 556–564.

Domin, G, 1963. Wärmeübergang in kritischen und überkritischen Bereichen von Wasser in Rohren, *Brennstoff-Warme-Fraft* (BWK), 15 (11), pp. 527–532.

Doroshchuk, V.E., Lel'chuk, V.L., and Modnikova, V.V., 1959. Heat transfer to water at high pressure, (In Russian), In book: *Heat Transfer at High Heat Fluxes and Other Special Conditions*, Gosenergoizdat Publishing House, Moscow–Leningrad, Russia, pp. 30–40.

Dreitser, G.A., 1993. High-effective tubular heat exchangers, In book: Aerospace Heat Exchanger Technology, Editors: R.K. Shah and A. Hashemi, Elsevier Science Publishers, Proceedings of the 1st International Conference on Aerospace Heat Exchanger Technology, Palo Alto, CA, USA, February, 15–17, pp. 581–610.

Dreitser, G.A. and Lobanov, I.E., 2002. Modelling of heat transfer and resistance of propulsion fuel of supercritical pressure at conditions of heat transfer intensification, (In Russian), Proceedings of the 3rd Russian National Heat Transfer Conference, Moscow, Russia, October, 21–25, MEI Publishing House, Moscow, Vol. 1, pp. 53–58.

Dreitser, G.A. and Lobanov, I.E., 2004. Simulation of enhanced heat transfer and hydraulic resistance under turbulent flow of a supercritical-pressure jet-propulsion propellant in channels, *Thermal Engineering* (Теплоэнергетика, стр. 63–68), 51 (1), pp. 66–71.

Dreitser, G.A., Myakochin, A.S., Janovski, L.S., and Podvorin, I.V., 1993. Investigation of hydrocarbon fuels as coolants of spacecraft high-temperature structures, In book: Aerospace Heat Exchanger Technology, Editors: R.K. Shah and A. Hashemi, Elsevier Science Publishers, (Proceedings of the First International Conference on Aerospace Heat Exchanger Technology, Palo Alto, CA, USA, February, 15–17), pp. 31–43.

Dubrovina, E.N. and Skripov, V.P., 1967. Convective heat transfer to carbon dioxide in the supercritical region, Proceedings of the 2nd All-Soviet Union Conference on Heat and Mass Transfer, Minsk, Belarus', May, 1964, Published as Rand Report R-451-PR, Edited by C. Gazley, Jr., J.P. Hartnett and E.R.C. Ecker, Vol. 1, pp. 36–45.

Dubrovina, É.N., Skripov, V.P., and Shuravenko, N.A., 1969. Pseudoboiling of carbon dioxide in the presence of free convection, *High Temperatures* (Теплофизика Высоких Температур, стр. 730–735), 7 (4), pp. 671–675.

Duffey, R.B., Khartabil, H.F., Pioro, I.L., and Hopwood, J.M., 2003. The future of nuclear: SCWR Generation IV high performance channels, Proceedings of the 11th International Conference on Nuclear Engineering (ICONE-11), Shinjuku, Tokyo, Japan, April 20–23, Paper No. 36222, 8 pages.

Duffey, R.B. and Pioro, I.L., 2005a. Supercritical water-cooled nuclear reactors: Review and status, In Nuclear Materials and Reactors from Encyclopedia of Life Support Systems (EOLSS), developed under the Auspices of the UNESCO, EOLSS Publishers, Oxford, UK, http://www.eolss.net, August 9.

Duffey, R.B. and Pioro, I.L., 2005b. Experimental heat transfer of supercritical carbon dioxide flowing inside channels (survey), *Nuclear Engineering and Design*, 235 (8), pp. 913–924.

Duffey, R.B. and Pioro, I.L., 2004. Experimental heat transfer and pressure drop of supercritical carbon dioxide flowing inside channels (survey), Proceedings of the International Conference on "Pressure Tube Reactors: Problems and Solutions," Moscow-Kurchatov, Russia, October 19–22, Session 1_18–31, Paper No. 26, 11 pages.

Duffey, R.B., Pioro, I.L., and Khartabil, H.F., 2005. Supercritical water-cooled pressure channel nuclear reactors: Review and status, Proceedings of the International Conference GLOBAL-2005 "Nuclear Energy Systems for Future Generation and Global Sustainability," Tsukuba, Japan, October 9–13, Paper No. 20, 12 pages.

Duffey, R.B. and Pioro, I.L., 2006. Advanced high-temperature concepts for pressure-tube reactors, including co-generation and sustainability, Proceedings of the 3rd International Topical Meeting on High Temperature Reactor Technology – HTR 2006, October 1–4, Johannesburg, South Africa, Paper F167, 12 pages.

Duffey, R.B., Pioro, I.L., Gabaraev, B.A., and Kuznetsov, Yu.N., 2006. SCW pressure-channel nuclear reactors: Some design features, Proceedings of the 14th International Conference on Nuclear Engineering (ICONE-14), July 17–20, 2006, Miami, FL, USA, Paper #89609.

Dumaz, P. and Antoni, O., 2003. The extension of the CATHARE2 computer code above the critical point, applications to a supercritical light water reactor, Proceedings of the International Topical Meeting on Nuclear Reactor Thermal Hydraulics (NURETH-10), Seoul, Korea, October 5–9, 11 pages.

Dyadyakin, B.V. and Popov, A.S., 1977. Heat transfer and thermal resistance of tight seven-rod bundle, cooled with water flow at supercritical pressures, (In Russian), *Transactions of VTI* (Труды ВТИ), No. 11, pp. 244–253.

Fedorov, E.P., Yanovskiy, L.S., Krasnyuk, V.I., Dreitser, G.A., et al., 1986. Intensification of heat transfer in heated channels at flow of hydrocarbons of supercritical pressure, (In Russian), In book: *Modern Problems of Hydrodynamics and Heat*

Transfer in Elements of Power Equipment and Cryogenic Technique, All-Union by Correspondence Machine Building Institute, Moscow, Russia, pp. 10–17.

Fewster, J., 1976. Mixed Forced and Free Convective Heat Transfer to Supercritical Pressure Fluids Flowing in Vertical Pipes, Ph.D. Thesis, University of Manchester, UK.

Fewster, J. and Jackson, J.D., 2004. Experiments on supercritical pressure convective heat transfer having relevance to SCWR, Proceedings of the International Congress on Advances in Nuclear Power Plants (ICAPP'04), Pittsburgh, PA, USA, June 13–17, Paper No. 4342, pp. 537–551.

Filippov, G.A., Avdeev, A.A., Bogoyavlensky, R.G., et al., 2003. Case micro-fuel-element nuclear reactor with supercritical pressure of light water coolant for single-loop power units of NPP, International Scientific-Engineering Conference "Nuclear Power Engineering and Fuel Cycles," Moscow-Dimitrovgrad, Russia, December 1–5, 7 pages.

Filonenko, G.K., 1954. Hydraulic resistance of pipelines, (In Russian), *Thermal Engineering*, No. 4, pp. 40–44.

Fischer, B., Smolinski, M., and Buongiorno, J., 2004. Nitrogen-16 generation and transport and associated shielding requirements in a supercritical-water-cooled reactor, *Nuclear Technology*, 147, pp. 269– 283.

Fong, R., 2002. Private communications. Chalk River Laboratories, AECL, Chalk River, Ontario, Canada.

Fritsch, C.A. and Grosh, R.J., 1963. Free convective heat transfer to supercritical water experimental measurements, *Journal of Heat Transfer*, Transactions of the ASME, 85 (4), pp. 289–294.

Gabaraev, B.A., Ganev, I.Kh., Davydov, V.K., et al., 2003a. Vessel and channel fast reactors cooled by boiling water or water with supercritical parameters, *Atomic Energy* (Атомная Энергия, стр. 243–251), 95 (4), pp. 655–662.

Gabaraev, B.A., Ganev, I.Kh., Kuznetsov, Yu.N., et al., 2003b. The three-target channel-type uranium water fast reactor with direct flow of supercritical water to solve the problem of weapon-plutonium and power generation at high efficiency, Proceedings of the 11th International Conference on Nuclear Engineering (ICONE-11), Tokyo, Japan, April 20–23, Paper #36021.

Gabaraev, B.A., Vikulov, V.K., Yermoshin, F.Ye., et al., 2004. Pressure tube once-through reactor with supercritical coolant pressure, (In Russian), Proceedings of the International Scientific-Technical Conference "Channel Reactors: Problems and Solutions," Moscow, Russia, October 19–22, Paper #42.

Gabaraev, B.A., Vikulov, V.K., Ermoshin, F.E., et al., 2005. Direct-flow channel reactor with supercritical coolant pressure, *Atomic Energy* (Атомная Энергия, стр. 243–253), 98 (4), pp. 233–241.

Galitseyskiy, B.M., Ryzhov, Yu.A., and Yakush, E.V., 1977. *Thermal and Hydrodynamic Processes in Pulsating Flows*, (In Russian), Moscow, Russia, 255 pages.

Gerliga, V.A. and Skalozubov, V.I., 1992. *Nucleate Boiling Flows in Power Equipment of Nuclear Power Plants*, (In Russian), Energoatomizdat, Moscow, Russia, 430 pages.

Ghajar, A.J. and Asadi, A., 1986. Improved forced convective heat-transfer correlations for liquids in the near-critical region, *AIAA Journal*, 24 (12), pp. 2030–2037.

Ghajar, A.J. and Parker, J.D., 1981. Reference temperatures for supercritical laminar free convection on a vertical flat plate, *Journal of Heat Transfer*, Transactions of the ASME, 103, pp. 613–616.

Ghorbani-Tari, S. and Ghajar, A.J., 1985. Improved free convective heat-transfer correlations in the near-critical region, *AIAA Journal*, 23 (10), pp. 1647–1649.

Giarratano, P.J., Arp, V.D., and Smith, R.V., 1971. Forced convection heat transfer to supercritical helium, *Cryogenics*, 11, October, pp. 385–393.

Giarratano, P.J. and Jones, M.C., 1975. Deterioration of heat transfer to supercritical helium at 2·5 atmospheres, *International Journal of Heat & Mass Transfer*, 18 (5), pp. 649–653.

Giarratano, P.J. and Steward, W.G., 1983. Transient forced convection heat transfer to helium during a step in heat flux, *Journal of Heat Transfer*, 105, May, pp. 350–357.

Glebov, V.P., Eskin, N.B., Trubachev, V.M., et al., 1983. *Deposits in Tubes of Supercritical Pressure Boilers*, (In Russian), Energoatomizdat Publishing House, Moscow, Russia, 239 pages.

Glebov, V.P., Estin, N.B., Zusman, V.M., et al., 1978. Heat conduction of iron oxide deposits formed in tubes of radiant heating surfaces in supercritical steam generators, *Thermal Engineering* (Теплоэнергетика, стр. 55–59), 25 (4), pp. 36–40.

Glushchenko, L.F. and Gandzyuk, O.F., 1972. Temperature conditions at the wall of an annular channel with internal heating at supercritical pressures, *High Temperatures* (Теплофизика Высоких Температур, стр. 820–825), 10 (4), pp. 734–738.

Glushchenko, L.F., Kalachev, S.I., and Gandzyuk, O.F., 1972. Determining the conditions of existence of deteriorated heat transfer at supercritical pressures of the medium, *Thermal Engineering* (Теплоэнергетика, стр. 69–72), 19 (2), pp. 107–111.

Gnielinski, V., 1976. New equation for heat and mass transfer in turbulent pipe and channel flow, *International Chemical Engineering*, 16 (2), pp. 359–368.

Goldmann, K., 1954. Heat transfer to supercritical water and other fluids with temperature-dependent properties, *Chemical Engineering Progress, Symposium Series*, Nuclear Engineering, Part III, Vol. 50, No. 11, pp. 105–113.

Goldmann, K., 1961. Heat transfer to supercritical water at 5000 psi flowing at high mass-flow rates through round tubes, In book: *International Developments in Heat Transfer: Papers presented at the 1961 International Heat Transfer Conference*, ASME, University of Colorado, Boulder, CO, USA, January 8–12, Part III, Paper No. 66, pp. 561–568.

Goldstein, R.J. and Aung, W., 1967. Heat transfer by free convection from a horizontal wire to carbon dioxide in the critical region, Paper No. 67-WA/HT-2, *Journal of Heat Transfer*, Transactions of the ASME, pp. 1–5.

Gorban', L.M., Pomet'ko, R.S., and Khryaschev, O.A., 1990. Modeling of water heat transfer with Freon of supercritical pressure, (In Russian), ФЭИ-1766, Institute of Physics and Power Engineering (ФЭИ), Obninsk, Russia, 19 pages.

Gortyshov, Yu.F., Dresvyannikov, F.N., Idiatullin, N.S., et al., 1985. *Theory and Technique of Thermophysical Experiment, Textbook for Technical Universities*, (In Russian), Editor V.K. Shchukin, Energoatomizdat Publishing House, Moscow, Russia, 360 pages.

Grabezhnaya, V.A. and Kirillov, P.L., 2004. Heat transfer in pipes and rod bundles during flow of supercritical-pressure water, *Atomic Energy* (Атомная Энергия, стр. 387–393), 96 (5), pp. 358–364.

Graham, R.W., 1969. Penetration model explanation for turbulent forced-convection heat transfer observed in near-critical fluids, Report NASA TND-5522, (Accession number N69-40437), Lewis Research Center, NASA, Cleveland, OH, USA, October, 31 pages.

Green, J.R. and Hauptmann, E.G., 1971. Forced convection heat transfer from a cylinder in carbon dioxide near the thermodynamic critical point, (Paper No. 70-HT. SpT-36), *Journal of Heat Transfer*, Transactions of the ASME, August, pp. 290–296.

Griem, H., 1996. A new procedure for the prediction of forced convection heat transfer at near- and supercritical pressure, *Heat and Mass Transfer* (Warme- und Stoffubertragung), Springer-Verlag Publishing House, 31 (5), pp. 301–305.

Griffith, J.D. and Sabersky, R.H., 1960. Convection in a fluid at supercritical pressures, *American Rocket Society Journal*, 30 (3), pp. 289–291.

Grigor'ev, V.S., Polyakov, A.F., and Rosnovskii, S.V., 1977. Heat transfer in pipes under conditions of supercritical pressures of the heat carrier and thermal load varying along the length, *High Temperatures* (Теплофизика Высоких Температур, стр. 1241–1247), 15 (6), pp. 1062–1068.

Grigor'yants, A.N., Baturov, B.B., Malyshev, V.M., et al., 1979. Tests on zirconium superheating channels in the first unit at the Kurchatov Beloyarsk nuclear power station, *Atomic Energy* (Атомная Энергия, стр. 55–56), 46 (1), pp. 58–60.

Guide to the Expression of Uncertainty in Measurement, 1995. Corrected and Reprinted, International Bureau of Weights and Measures and other International Organizations, 101 pages.

Gunson, W.E. and Kellogg, H.B., 1966. Engineering application technique for supercritical-pressure heat-transfer correlations, ASME, Paper No. 66-WA/HT-11.

Guo, Y., Bullock, D.E., Pioro, I.L., and Martin, J., 2006. Measurements of sheath temperature profiles in LVRF bundles under post-dryout heat transfer conditions in Freon, Proceedings of the 14th International Conference on Nuclear Engineering (ICONE-14), July 17-20, Miami, Florida, USA, Paper #89621, 9 pages.

Hahne, E., Feurstein, G., and Grigull, U., 1974. Free convective heat transfer in the supercritical region, Proceedings of the 5th International Heat Transfer Conference, Tokyo, Japan, September 3–7, Vol. III, Paper No. NC1.2, pp. 5–9.

Hall, W.B., 1971. Heat transfer near the critical point, In book : *Advances in Heat Transfer*, Editors Th.F. Irvine, Jr. and J.P. Hartnett, Vol. 7, Academic Press, New York, NY, USA, pp. 1–86.

Hall, W.B. and Jackson, J.D., 1978. Heat transfer near the critical point, Keynote lecture, Proceedings of the 6th International Heat Transfer Conference, Toronto, Ontario, Canada, August 7–11, Vol. 6, pp. 377–392.

Hall, W.B. and Jackson, J.D., 1969. Laminarization of a turbulent pipe flow by buoyancy forces, ASME Paper No. 69-HT-55, 8 pages.

Hall, W.B., Jackson, J.D., and Khan, S.A., 1966. An investigation of forced convection heat transfer to supercritical pressure carbon dioxide, Proceedings of the 3rd International Heat Transfer Conference, Chicago, IL, USA, August 7–12, Vol. 1, pp. 257–266.

Hall, W.B., Jackson, J.D., and Watson, A., 1968. A review of forced convection heat transfer to fluids at supercritical pressures, Proceedings of the Symposium on Heat Transfer and Fluid Dynamics of Near Critical Fluids, Institute of Mechanical Engineers, Vol. 182, Part 31, Paper No. 3, pp. 10–22.

Hanna, B.N., 1998. CATHENA: A thermalhydraulic code for CANDU analysis, *Nuclear Engineering and Design,* 180, pp. 113–131.

Harden, D.G. and Boggs, J.H., 1964. Transient flow characteristics of a natural-circulation loop operated in the critical region, Proceedings of the Heat Transfer and Fluid Mechanics Institute, Stanford University Press, pp. 33–50.

Hardy, J.E., Hylton, J.O., McKnight, T.E., et al., 1999. *Flow Measurement Methods and Applications*. J. Wiley & Sons, Inc., New York, NY, USA, pp. 207–208.

Harrison, G.S. and Watson, A., 1976a. Similarity and the formation of correlations for forced convection to supercritical pressure fluids, *Cryogenics,* 16 (3), pp. 147–151.

Harrison, G.S. and Watson, A., 1976b. An experimental investigation of forced convection to supercritical pressure water in heated small bore tubes, Proceedings of the Institute of Mechanical Engineers, 190 (40/76), pp. 429–435.

Harvey, A.H., 2001. Private communications.

Hasegawa, Sh. and Yoshioka, K., 1966, An analysis for free convective heat transfer to supercritical fluids, Proceedings of the 3rd International Heat Transfer Conference, Chicago, IL, USA, August 7–12, Vol. 2, pp. 214–222.

He, S., Jiang, P.-X., Xu, Y.-J., et al., 2005. A computational study of convection heat transfer to CO_2 at supercritical pressures in a vertical mini tube, *International Journal of Thermal Sciences,* 44, pp. 521–530.

He, S., Kim, W.S., Jiang, P.-X., and Jackson, J.D., 2004. Simulation of mixed of convection heat transfer to carbon dioxide at supercritical pressure, Proceedings of the Institution of Mechanical Engineers, Part C, *Journal of Mechanical Engineering Science*, 218, pp. 1281–1296.

Heitmüller, R.J., Fischer, H., Sigg, J., et al., 1999. Lignite-fired Niederauβem K aims for efficiency of 45 percent and more, *Modern Power Systems,* May, pp. 59–66.

Hejzlar, P., Dostal, V., and Driscoll, M.J., 2005. Assessment of gas cooled fast reactor with indirect supercritical CO_2 cycle, Proceedings of the International Congress on Advances in Nuclear Power Plants (ICAPP'05), Seoul, Korea, May 15–19, Paper No. 5090, 11 pages.

Hendricks, R.C., Graham, R.W., Hsu, Y.Y., and Friedman, R., 1966. Experimental heat-transfer results for cryogenic hydrogen flowing in tubes at subcritical and supercritical pressures to 800 pounds per square inch absolute, Report NASA TN D-3095, (Accession number N66-20934), Lewis Research Center, NASA, Cleveland, OH, USA, March, 158 pages.

Hendricks, R.C., Graham, R.W., Hsu, Y.Y., and Medeiros, A.A., 1962. Correlation of hydrogen heat transfer in boiling and supercritical pressure state, *American Rocket Society Journal* (ARS), 32 (2), pp. 244–252.

Hendricks, R.C., Simoneau, R.J., and Smith, R.V., 1970a. Survey of heat transfer to near-critical fluids, Report No. NASA TN-D-5886, Lewis Research Center (Report No. E-5084), NASA, Cleveland, OH, USA, 116 pages.

Hendricks, R.C., Simoneau, R.J., and Smith, R.V., 1970b. Survey of heat transfer to near-critical fluids, *Advances in Cryogenic Engineering,* Cryogenic Engineering Conference, Plenum Press, USA, 15, pp. 197–237.

Hess, H.L. and Kunz, H.R., 1965. A study of forced convection heat transfer to supercritical hydrogen, *Journal of Heat Transfer,* Transactions of the ASME, 87 (1), February, pp. 41–48.

Heusener, G., Muller, U., Schulenberg, T., and Squarer, D., 2000. A European development program for a High Performance Light Water Reactor (HPLWR),

Proceedings of the 1st International Symposium on Supercritical Water-Cooled Reactor Design and Technology (SCR-2000), Tokyo, Japan, November 6–8, Paper No. 102.

Hewitt, G.F. and Collier, J.G., 2000. *Introduction to Nuclear Power,* 2nd edition, Taylor & Francis, New York, NY, USA, pp. 40, 67, 68.

Hewitt, G.F., Shires, G.L., and Bott, T.R., 1994. *Process Heat Transfer,* CRC Press, Boca Raton, Florida / Begell House, New York, NY, USA, pp. 981–1017.

Hilal, M.A., and Boom, R.W., 1980. An experimental investigation of free convection heat transfer in supercritical helium, *International Journal of Heat & Mass Transfer,* 23, pp. 697–705.

Hines, W.S. and Wolf, H., 1962. Pressure oscillations associated with heat transfer to hydrocarbon fluids at supercritical pressures and temperatures, *American Rocket Society Journal* (ARS), 32 (3), pp. 361–366.

Hofmeister, J., Schulenberg, Th. and Starflinger, Jo., 2005. Optimization of a fuel assembly for a HPLWR, Proceedings of the International Congress on Advances in Nuclear Power Plants (ICAPP'05), Seoul, Korea, May 15–19, Paper No. 5077, 8 pages.

Holman, J.P., 1994. *Experimental Methods for Engineers,* 6th edition, McGraw-Hill, Inc., New York, NY, USA, 616 pages.

Holman, J.P. and Boggs, J.H., 1960. Heat transfer to Freon 12 near the critical state in natural-circulation loop, *Journal of Heat Transfer,* Transactions of the ASME, 82 (3), pp. 221–226.

Hong, S.D., Chun, S.Y., Kim, S.Y., and Baek, W.P., 2004. Heat transfer characteristics of an internally-heated annulus cooled with R-134a near the critical pressure, *Journal of the Korean Nuclear Society,* 36 (5), pp. 403–414.

Hong, S.D., Chun, S.Y., Yoon, Y.J., and Baek, W.P., 2003. Heat transfer characteristics of R-134a fluid during the pressure transient from supercritical pressure to subcritical pressure, Proceedings of the International Topical Meeting on Nuclear Reactor Thermal Hydraulics (NURETH-10), Seoul, Korea, October 5–9, 12 pages.

Howell, J.R. and Lee, S.H., 1999. Convective heat transfer in the entrance region of a vertical tube for water near the thermodynamic critical point, *International Journal of Heat and Mass Transfer,* 42, pp. 1177–1187.

Hsu, Yi.-Yu. and Graham, R.W., 1986. *Transport Processes in Boiling and Two-Phase Systems. Including Near-Critical Fluids,* American Nuclear Society, La Grange Park, IL, USA, pp. 477–509.

Hsu, Y.-Y. and Smith, J.M., 1961. The effect of density variation on heat transfer in the critical region, *Journal of Heat Transfer,* Transactions of the ASME, 83 (2), pp. 176–182.

Hwang, S.S., Lee, B.H., Kim, J.-G., and Jang, J., 2005. Corrosion behaviour of F/M/ steels and high Ni alloys in supercritical water, Proceedings of the International Conference GLOBAL-2005 "Nuclear Energy Systems for Future Generation and Global Sustainability," Tsukuba, Japan, October 9–13, Paper No. 043, 4 pages.

International Atomic Energy Agency (IAEA), 1999. Hydrogen as an energy carrier and its production by nuclear power, Vienna, Austria, May.

Ikryannikov, N.P., 1973. An experimental investigation of heat transfer in the single-phase near-critical region with viscous-inertial-gravitational flow in tubes, (In Russian), Ph.D. Thesis, Moscow Power Institute (МЭИ), Russia, 27 pages.

Ikryannikov, N.P., Petukhov, B.S. and Protopopov, V.S., 1972. An experimental investigation of heat transfer in the single-phase near-critical region with combined forced and free convection, *High Temperatures* (Теплофизика Высоких Температур, стр. 96–100), 10 (1), pp. 80–83.

Ikryannikov, N.P., Petukhov, B.S. and Protopopov, V.S., 1973. Calculation of heat transfer in the single-phase, near-critical region in the case of a viscosity-inertia-gravity flow, *High Temperatures* (Теплофизика Высоких Температур, стр. 1068–1075), 11 (5), pp. 949–955.

Incropera, F.P. and DeWitt, D.P., 2002. *Fundamentals of Heat and Mass Transfer*, 5th edition, John Wiley & Sons, New York, NY, USA, pp. 500–502.

International Encyclopedia of Heat & Mass Transfer, 1998. Edited by G.F. Hewitt, G.L. Shires and Y.V. Polezhaev, CRC Press, Boca Raton, FL, USA, pp. 1112–1117 (Title "Supercritical heat transfer").

Isayev, G.I., 1983. Investigation of forced–convection heat transfer to supercritical n-heptane, *Heat Transfer–Soviet Research*, 15 (1), pp. 119–123.

Isaev, G.I., Arabova, I.T. and Mamedov, F.Kh., 1995. Heat transfer during transition flows of hydrocarbons at supercritical pressures in channels, *Heat Transfer Research*, 26 (3–8), pp. 412–414.

Isaev, G.I. and Kalbaliev, F.I., 1979. Heat transfer of aromatic hydrocarbons in a downward flow at supercritical pressure, *Journal of Engineering-Physics* (Инженерно-Физический Журнал (ИФЖ), стр. 742–755), 36 (4), pp. 498–499.

Ishigai, S., Kadgi, M. and Nakamoto, M., 1976. Heat transfer and friction for water flow in tubes at supercritical pressures, (In Russian), Teplomassoobmen (Heat-Mass-Transfer)-V, Proceedings of the Vth All-Union Conference on Heat Mass Transfer, Minsk, Belarus', May, Vol. 1, Part 1, pp. 261–269.

Ishiwatari, Y., Oka, Y., and Koshizuka, S., 2004. Safety design principle of supercritical water cooled reactors, Proceedings of the International Congress on Advances in Nuclear Power Plants (ICAPP'04), Pittsburgh, PA, USA, June 13–17, Paper No. 4319, pp. 492–501.

Ishiwatari, Yu., Oka, Yo., and Koshizuka, S., 2003a. Control of a high temperature supercritical pressure light water cooled and moderated reactor with water rods, *Journal of Nuclear Science and Technology*, 40 (5), pp. 298–306.

Ishiwatari, Yu., Oka, Yo., and Koshizuka, S., 2003b. Plant control of a high temperature reactor cooled and moderated by supercritical light water, Proceedings of the Joint International Conference Global Environment and Nuclear Energy Systems / Advanced Nuclear Power Plants (GENES4/ANP2003), Kyoto, Japan, September 15–19, Paper No. 1158, 9 pages.

Ishiwatari, Yu., Oka, Yo., and Koshizuka, S., 2003c. Safety analysis of high temperature reactor cooled and moderated by supercritical light water, Proceedings of the Joint International Conference Global Environment and Nuclear Energy Systems / Advanced Nuclear Power Plants (GENES4/ANP2003), Kyoto, Japan, September 15–19, Paper No. 1159, 10 pages.

Ishiwatari, Yu., Oka, Yo., and Koshizuka, S., 2003d. LOCA analysis of high temperature reactor cooled and moderated by supercritical light water, Proceedings of the Joint International Conference Global Environment and Nuclear Energy Systems / Advanced Nuclear Power Plants (GENES4/ANP2003), Kyoto, Japan, September 15–19, Paper No. 1160, 8 pages.

Ishiwatari, Y., Oka, Y., and Koshizuka, S., 2003e. ATWS analysis of supercritical pressure light water cooled reactor, Proceedings of the ANS/ENS International

Winter Meeting and Nuclear Technology Expo, Embedded Topical Meeting GLOBAL 2003 "Advanced Nuclear Energy and Fuel Cycle Systems", New Orleans, LA, USA, November 16–20, pp. 2335–2341.

Ishiwatari, Yu., Oka, Yo., and Koshizuka, S., 2002. Safety analysis of a high temperature supercritical pressure light water cooled and moderated reactor, Proceedings of the International Congress on Advances in Nuclear Power Plants (ICAPP'02), Hollywood, FL, USA, June 9–13, Paper No. 1045, 9 pages.

Ishiwatari, Yu., Oka, Yo., and Koshizuka, S., 2001. Breeding ratio analysis of a fast reactor cooled by supercritical light water, *Journal of Nuclear Science and Technology*, 38 (9), pp. 703–710.

Ishizuka, T., Kato, Ya., Muto, Ya., et al., 2005. Thermal-hydraulic characteristics of a printed circuit heat exchanger in a supercritical CO_2 loop, Proceedings of the International Conference GLOBAL-2005 "Nuclear Energy Systems for Future Generation and Global Sustainability," Tsukuba, Japan, October 9–13, Paper No. 218, 15 pages.

Ito, T., Takata, Y., Yamaguchi, M., et al., 1986. Forced-convection heat transfer to supercritical helium flowing in a vertical straight circular tube, *AIChE*, Symposium Series, 82, pp. 86–91.

Ivlev, A.A., 1979. Calculation of heat transfer to supercritical helium by the Melik-Pashaev technique, *Heat Transfer–Soviet Research*, 11 (4), pp. 113–120.

Ivlev, A.A., Alad'ev, I.T., Voskresenskiy, K.D., and Turilina, E.S., 1980. Ultimate regimes of improved and deteriorated heat transfer at turbulent flow of fluid of supercritical pressure (SCP), (In Russian), Materials of the Interindustrial Conference "Heat Transfer and Hydrodynamics in the Parts of Nuclear Power Installations at Single-Phase Flow of Working Fluid," Institute of Physics and Power Engineering (ФЭИ), Obninsk, Russia, pp. 113–121.

Jackson, J.D., 2001. Some striking features of heat transfer with fluids at pressures and temperatures near the critical point, Proceedings of the International Conference on Energy Conservation and Application (ICECA 2001), Wuhan, China, 14 pages.

Jackson, J.D., 2002. Consideration of the heat transfer properties of supercritical pressure water in connection with the cooling of advanced nuclear reactors, Proceedings of the 13th Pacific Basin Nuclear Conference, Shenzhen City, China, October 21–25.

Jackson, J.D. and Fewster, J., 1975. Forced convection data for supercritical pressure fluids, HTFS 21540.

Jackson, J.D. and Hall, W.B., 1979a. Forced convection heat transfer to fluids at supercritical pressure, In book: *Turbulent Forced Convection in Channels and Bundles*, Editors S. Kakaç and D.B. Spalding, Hemisphere Publishing Corp., New York, NY, USA, Vol. 2, pp. 563–612.

Jackson, J.D. and Hall, W.B., 1979b. Influences of buoyancy on heat transfer to fluids in vertical tubes under turbulent conditions, In book: *Turbulent Forced Convection in Channels and Bundles*, Editors S. Kakaç and D.B. Spalding, Hemisphere Publishing Corp., New York, NY, USA, Vol. 2, pp. 613–640.

Jackson, J.D., Lutterodt, E., and Weinberg, R., 2003. Experimental studies of buoyancy-influenced convective heat transfer in heated vertical tubes at pressures just above and just below the thermodynamic critical value, Proceedings of the Joint International Conference Global Environment and Nuclear Energy Systems / Advanced Nuclear Power Plants (GENES4/ANP2003), Kyoto, Japan, September 15–19, Paper No. 1177, 14 pages.

Jang, J., Han, Ch.H., Lee, B.H., et al., 2005. Corrosion behaviour of 9Cr F/M steels in supercritical water, Proceedings of the International Congress on Advances in Nuclear Power Plants (ICAPP'05), Seoul, Korea, May 15–19, Paper No. 5136, 8 pages.

Jevremovich, T., Oka, Yo., and Koshizuka, S.-I., 1996. UO_2 core design of a direct-cycle fast converter reactor cooled by supercritical water, *Nuclear Energy*, 114, pp. 273–284.

Jevremovich, T., Oka, Y., and Koshizuka, S., 1994. Core design of a direct-cycle, supercritical-water-cooled fast breeder reactor, *Nuclear Technology*, 108, pp. 24–32.

Jevremovich, T., Oka, Y., and Koshizuka, S., 1993a. Design of an indirect-cycle fast breeder reactor cooled by supercritical steam, *Nuclear Engineering and Design*, 144, pp. 337–344.

Jevremovich, T., Oka, Y., and Koshizuka, S., 1993b. Conceptual design of an indirect-cycle supercritical-steam-cooled fast breeder reactor with negative coolant void reactivity characteristics, *Annals of Nuclear Energy*, 20 (5), pp. 305–313.

Jiang, P.-X., Ren, Z.-P., and Wang, B.-X., 1995. Convective heat and mass transfer in water at super-critical pressures under heating or cooling conditions in vertical tubes, *Journal of Thermal Science*, 4 (1), pp. 15–25.

Jing, P.-X., Xu, Y.-J., Lv, J., et al., 2004. Experimental investigation of convection heat transfer of CO_2 at super-critical pressures in vertical mini-tubes and in porous media, *Applied Thermal Engineering*, 24, pp. 1255–1270.

Johannes, C., 1972. Studies of forced convection heat transfer to helium I, *Cryogenics*, 17, pp. 352–360.

Joo, H.-K., Bae, K.-M., Lee, H.-C., et al., 2005. A conceptual design with a rectangular fuel assembly for a thermal SCWR system, Proceedings of the International Congress on Advances in Nuclear Power Plants (ICAPP'05), Seoul, Korea, May 15–19, Paper No. 5223, 8 pages.

Kafengauz, N.L., 1967. Heat transfer peculiarities to turbulent fluid flow at supercritical pressure, (In Russian), Reports of the USSR Academy of Sciences (Доклады Академии Наук СССР), *Technical Physics*, 173 (3), pp. 557–559.

Kafengauz, N.L., 1969. On physical nature of heat transfer at supercritical pressure with pseudo-boiling, *Heat Transfer–Soviet Research*, 1 (12), pp. 88–93.

Kafengaus, N.L., 1975. The mechanism of pseudoboiling, *Heat Transfer-Soviet Research*, 7 (4), pp. 94–100.

Kafengauz, N.L., 1983. Heat transfer to turbulent stream in pipes under supercritical pressures, *Journal of Engineering-Physics* (Инженерно-Физический Журнал (ИФЖ), стр. 14–19), 44 (1), pp. 9–12.

Kafengauz, N.L., 1986. About some peculiarities in fluid behaviour at supercritical pressure in conditions of intensive heat transfer, *Applied Thermal Sciences*, (Промышленная Теплотехника, стр. 6–10), 8 (5), pp. 26–28.

Kafengauz, N.L. and Borovitskii, A.B., 1985. Influence of solid deposits on the inception of self-exited thermoacoustic oscillations in heat transfer to turbulent fluid flow in tubes, *Journal of Engineering-Physics* (Инженерно-Физический Журнал (ИФЖ), стр. 571–573), 49 (4), pp. 1155–1157.

Kafengauz, N.L. and Fedorov, M.I., 1966. Excitation of high-frequency pressure oscillations during heat exchange with diisopropylcyclohexane, *Journal of Engineering-Physics* (Инженерно-Физический Журнал (ИФЖ), стр. 99–104), 11 (1), pp. 63–67.

Kafengauz, N.L. and Fedorov, M.I., 1968. Pseudoboiling and heat transfer in a turbulent flow, *Journal of Engineering-Physics* (Инженерно-Физический Журнал (ИФЖ), стр. 923–924), 14 (5), pp. 489–490.

Kafengauz, N.L. and Fedorov, M.I., 1968. Interrelation of temperature of cooled surface and frequency of natural pressure oscillations in turbulent heat transfer, *Journal of Engineering-Physics* (Инженерно-Физический Журнал (ИФЖ), стр. 455–458), 15 (3), pp. 825–828.

Kafengauz, N.L. and Fedorov, M.I., 1970. Effect of fuel element dimensions on heat emission under supercritical pressure, *Atomic Energy* (Атомная Энергия, стр. 293–294), 29 (4), pp. 1022–1023.

Kakaç, S., Shah, R.K., and Aung, W., 1987. *Handbook of Single-Phase Convective Heat Transfer*, J. Wiley & Sons, New York, NY, USA.

Kakarala, Ch.R. and Thomas, L.C., 1974. A theoretical analysis of turbulent convective heat transfer for supercritical fluids, Proceedings of the 5th International Heat Transfer Conference, Tokyo, Japan, September 3–7, Vol. II, Paper No. FC1.10, pp. 45–49.

Kalbaliev, F.I., 1978. Heat transfer in laminar flow at supercritical pressure, *Journal of Engineering-Physics* (Инженерно-Физический Журнал (ИФЖ), стр. 12–14), 34 (1), pp. 6–8.

Kalbaliev, F.I. and Babayev, F.K., 1986. Heat transfer in the thermal inlet region in laminar flow of toluene at supercritical velocities, *Heat Transfer–Soviet Research*, 18 (1), pp. 138–143.

Kalbaliev, F.I., Babayev, F.K., and Mamedova, D.P., 1983. Experimental investigation of the HTC of aromatic hydrocarbons at supercritical pressure and small values Reynolds number, (In Russian), In book: *Thermophysical Properties of High Temperature Boiling Fluids*, Baku, Azerbaijan, pp. 11–15.

Kalbaliev, F.I., Mamedova, S.G., Sultanov, R.A., and Tavakkuli, D., 2002. Heat transfer in coil tubes at supercritical pressures of hydrocarbons, (In Russian), Proceedings of the 3rd Russian National Heat Transfer Conference, Moscow, Russia, October, 21–25, MEI Publishing House, Moscow, Vol. 2, pp. 155–158.

Kalinin, E.K., Dreitser, G.A., Kopp, I.Z., and Myakochin, A.S., 1998. *Effective Heat Transfer Surfaces*, (In Russian), Energoatomizdat Publishing House, Moscow, Russia, 408 pages.

Kamenetskii, B.Ya., 1973. Heat-transfer regimes for mixed convection in vertical tubes, *High Temperatures* (Теплофизика Высоких Температур, стр. 352–356), 11 (2), pp. 305–309.

Kamenetskii, B.Ya., 1974. Special features of heat transfer with mixed convection in horizontal tubes, *Thermal Engineering* (Теплоэнергетика, стр. 61–63), 21 (6), pp. 84–87.

Kamenetskii, B.Ya., 1975. Heat-transfer characteristics of a nonuniformly circumferentially heated pipe, *High Temperatures* (Теплофизика Высоких Температур, стр. 671–674), 13 (3), pp. 613–616.

Kamenetskii, B.Ya., 1980. The effectiveness of turbulence promoters in tubes with nonuniformity heated perimeters under conditions of impaired heat transfer, *Thermal Engineering* (Теплоэнергетика, стр. 57–58), 27 (4), pp. 222–223.

Kamenetsky, B. and Shitsman, M., 1970. Experimental investigation of turbulent heat transfer to supercritical water in a tube with circumferentially varying heat flux, Proceedings of the 4th International Heat Transfer Conference, Paris-Versailles, France, Elsevier Publishing Company, Vol. VI, Paper No. B 8.10.

Kamei, K., Yamaji, A., Ishiwatari, Yu., et al., 2005. Fuel and core design of super LWR with stainless steel cladding, Proceedings of the International Congress on Advances in Nuclear Power Plants (ICAPP'05), Seoul, Korea, May 15–19, Paper No. 5527.

Kaneda, Ju., Kasahara, Sh., Kuniya, Ji., et al., 2005. Corrosion film of the candidate materials for the fuel claddings of the supercritical-water cooled power reactor, Proceedings of the International Congress on Advances in Nuclear Power Plants (ICAPP'05), Seoul, Korea, May 15–19, Paper No. 5594, 10 pages.

Kaplan, Sh.G., 1971. Heat-transfer mechanism in turbulent flow of fluid at supercritical pressures, *Journal of Engineering-Physics* (Инженерно-Физический Журнал (ИФЖ), стр. 431–437), 21 (3), pp. 1111–1115.

Kaplan, Sh.G., Grinevich, K.P., Fastova, K.N., and Tolchinskaya, R.E., 1974. Heat transfer in a pumped polymethylphenylsiloxane liquid near the critical pressure, *Journal of Engineering-Physics* (Инженерно-Физический Журнал (ИФЖ), стр. 776–784), 27 (5), pp. 1310–1316.

Kaplan, Sh.G. and Tolchinskaya, R.E., 1974a. Energy dissipation in heat transfer with a turbulent liquid at near-critical pressures, *High Temperatures* (Теплофизика Высоких Температур, стр. 1209–1214), 12 (6), pp. 1065–1068.

Kaplan, Sh.G. and Tolchinskaya, R.E., 1974b. One possible model for the transfer process in the near-critical state of a liquid, *Journal of Engineering-Physics* (Инженерно-Физический Журнал (ИФЖ), стр. 409–415), 27 (3), pp. 1064–1068.

Kaplan, Sh.G. and Tolchinskaya, R.E., 1971. Nature of heat exchange and hydraulic resistance under conditions of forced movement of a liquid at supercritical pressure, *Journal of Engineering-Physics* (Инженерно-Физический Журнал (ИФЖ), стр. 219–224), 21 (2), pp. 965–969.

Kaplan, Sh.G. and Tolchinskaya, R.E., 1969. Development of high-frequency pressure oscillations during heat transfer with forced liquid motion, *Journal of Engineering-Physics* (Инженерно-Физический Журнал (ИФЖ), стр. 486–490), 17 (3), pp. 1121–1123.

Kasao, D. and Ito, T., 1989. Review of existing experimental findings on forced convection heat transfer to supercritical, *Cryogenics,* Vol. 29, June, pp. 630–636.

Kasatkin, A.P., Labuntsov, D.A., and Soziev, R.I., 1984. An experimental investigation of heat transfer with turbulent flow of helium at supercritical parameters of state, *Thermal Engineering* (Теплоэнергетика, стр. 68–70), 31 (10), pp. 578–580.

Kastner, W., Köhler, W., and Schmidt, H., 2000. High-pressure test facility – 25 years of operation, *VGB Power Tech,* 80 (6), pp. 25–31.

Kataoka, K. and Oka, Yo., 1991. Neutronic feasibility of supercritical steam cooled fast breeder reactor, *Journal of Nuclear Science and Technology,* 28 (6), pp. 585–587.

Kataoka, K., Shiga, S., Moriya, K., et al., 2003. Progress of development project of supercritical-water cooled power reactor, 2003, Proceedings of the International Congress on Advances in Nuclear Power Plants (ICAPP'03), Córdoba, Spain, May 4–7, Paper No. 3258, 9 pages.

Kataoka, K., Shiga, S., Moriya, K., et al., 2002. Development project of supercritical-water cooled power reactor, Proceedings of the International Congress on

Advances in Nuclear Power Plants (ICAPP'02), Hollywood, FL, USA, June 9–13, Paper No. 10134, 9 pages.

Kato, Ya., Muto, Ya., Ishizuka, T., and Mito, M., 2005. Design of recuperator for the supercritical CO_2 gas turbine fast reactor, Proceedings of the International Congress on Advances in Nuclear Power Plants (ICAPP'05), Seoul, Korea, May 15–19, Paper No. 5196, 10 pages.

Kato, H., Nishiwaki, N., and Hirata, M., 1968. Studies on the heat transfer of fluids at a supercritical pressure, *Bulletin of JSME,* 11 (46), pp. 654–663.

Kaye, G.W.C. and Laby, T.H., 1973. *Tables of Physical and Chemical Constants and Some Mathematical Functions,* 14th edition, Longman, London, UK, pp. 183–186.

Kelbaliev, R.F., Rzaev, M.A., Bayramov, N.M., and Ashurova, U.I., 2002. Heat transfer at supercritical pressures of aromatic hydrocarbons, (In Russian), Proceedings of the 3rd Russian National Heat Transfer Conference, Moscow, Russia, October, 21–25, MEI Publishing House, Moscow, Vol. 2, pp. 167–169.

Kelbaliev, R.F., Sultanov, R.A., and Tavakkuli, D., 2003. Specifics of heat transfer in coil tubes at supercritical pressures of substance, (In Russian), Transactions of the XIV School-Seminar of Young Scientists and Specialists under the Leadership of Academician RAS A.I. Leont'ev "Problems of Gas Dynamics and Heat-Mass-Transfer in Power Installations," Rybinsk, Russia, May 26–30, Vol. 1, MEI Publishing House, Moscow, pp. 402–405.

Khabenskii, V.B., Kvetnyi, M.A., Tyntarev, É.M., and Kalinin, R.I., 1974. Temperature fields in a pipe wall in the case of radiative heating and a supercritical working-medium pressure, *Journal of Engineering-Physics* (Инженерно-Физический Журнал (ИФЖ), стр. 303–309), 27 (2), pp. 993–997.

Khabenskii, V.B., Gorosh, Ya.V., and Kalinin, P.I., 1971. Heat transfer to water of supercritical pressure in vertical tubes, (In Russian), Transactions of the IVth All-Union Conference on Heat Transfer and Hydraulics at Movement of Two-Phase Flow inside Elements of Power Engineering Machines and Apparatuses, Leningrad, Russia, pp. 41–56.

Khartabil, H.F., 1998. A flashing-driven moderator cooling system for CANDU reactors: Experimental and computational results, Presented at the IAEA Technical Committee Meeting on Experimental Tests and Qualification of Analytical Methods to Address Thermohydraulic Phenomena in Advanced Water Cooled Reactors, PSI, Switzerland.

Khartabil, H.F., 2002. CANDU SCWCR, Information Exchange Meeting on SCWCRs, Washington, D.C., USA, November 19.

Khartabil, H.F., Duffey, R.B., Spinks, N., and Diamond, W., 2005. The pressure-tube concept of Generation IV Supercritical Water-Cooled Reactor (SCWR): Overview and status, Proceedings of ICAPP'05, Seoul, Korea, May 15–19, Paper #5564.

Kichigin, A.M. and Pioro, I.L., 1992. Analysis of methods for detecting dryout and burnout fluxes, *Heat Transfer Research,* 24 (7), pp. 957–964.

Kim, S.H., Kim, Y.I., Bae, Y.Y., and Cho, B.H., 2004. Numerical simulation of the vertical upward flow of water in a heated tube at supercritical pressure, Proceedings of the International Congress on Advances in Nuclear Power Plants (ICAPP'04), Pittsburgh, PA, USA, June 13–17, Paper No. 4047, pp. 1527–1534.

Kim, H.Ye., Kim, H., Song, J.H., et al., 2005a. Heat transfer test in a tube using CO_2 at supercritical pressures, Proceedings of the International Conference GLOBAL-2005 "Nuclear Energy Systems for Future Generation and Global Sustainability, Tsukuba," Japan, October 9–13, Paper No. 103, 6 pages.

Kim, J.K., Jeon, H.K., Yoo, J.Y., and Lee J.S., 2005b. Experimental study on heat transfer characteristics of turbulent supercritical flow in vertical circular/ non-circular tubes, Proceedings of the International Topical Meeting on Nuclear Reactor Thermal Hydrauics (NURETH-11), Avignon, France, October 2–6, 10 pages.

Kirillov, P.L., 2000. Heat and mass transfer at supercritical parameters (The short review of researches in Russia. Theory and experiments), Proceedings of the 1st International Symposium on Supercritical Water-Cooled Reactor Design and Technology (SCR-2000), Tokyo, Japan, November 6–8, Paper No. 105.

Kirillov, P.L., 2001a. Transition to supercritical parameters — way to improving NPP with water-cooled reactors, *Thermal Engineering* (Теплоэнергетика, стр. 6–10), 48 (12), pp. 973–978.

Kirillov, P.L., 2001b. Supercritical parameters — future for reactors with water coolant and NPP, (In Russian), *Atomic Energy Abroad*, (6), pp. 6–8.

Kirillov, P.L., 2003. About structure and some water properties at supercritical pressures (review), (In Russian), ФЭИ-0296, Institute of Physics and Power Engineering (ФЭИ), Obninsk, Russia, 23 pages.

Kirillov, P.L., Kolosov, A.A., Petrova, E.N., et al., 1986. Temperature distribution in turbulent flow of water at supercritical pressures (circular tube), (In Russian), ФЭИ-1766, Institute of Physics and Power Engineering (ФЭИ), Obninsk, Russia, 10 pages.

Kirillov, P.L., Pomet'ko, R.S., Smirnov, et al., 2005. Experimental study on heat transfer to supercritical water flowing in vertical tubes, Proceedings of the International Conference GLOBAL-2005 "Nuclear Energy Systems for Future Generation and Global Sustainability," Tsukuba, Japan, October 9–13, Paper No. 518, 8 pages.

Kirillov, P.L., Yur'ev, Yu.S., and Bobkov, V.P., 1990. *Handbook of Thermal-Hydraulics Calculations*, (In Russian), Energoatomizdat Publishing House, Moscow, Russia, "3.2. Flow hydraulic resistance of the working fluids with significantly changing properties," pp. 66–67, "8.4. Working fluids at near-critical state," pp. 130–132.

Kitoh, K., Koshizuka, S., and Oka, Yo., 2001. Refinement of transient criteria and safety analysis for a high-temperature reactor cooled by supercritical water, *Nuclear Technology*, 135, pp. 252–264.

Kitoh, K., Koshizuka, S., and Oka, Yo., 1999. Refinement of transient criteria and safety analysis for a high temperature reactor cooled by supercritical water, Proceedings of the 7th International Conference on Nuclear Engineering (ICONE-7), Tokyo, Japan, April 19–23, Paper No. 7234.

Kitoh, K., Koshizuka, S., and Oka, Yo., 1998. Pressure- and flow-induced accident and transient analyses of a direct-cycle, supercritical-pressure, light-water-cooled fast reactor, *Nuclear Technology*, 123, pp. 233–244.

Kitou, K., Chaki, M., Ishii, Yo., et al., 2005. Three–dimensional heat transmission simulation of supercritical pressure fluid, Proceedings of the International

Congress on Advances in Nuclear Power Plants (ICAPP'05), Seoul, Korea, May 15–19, Paper No. 5428, 10 pages.

Kitou, K. and Ishii, Yo., 2005. Conceptual design of indirect-cycle, natural circulation cooling supercritical water reactor for hydrogen production, Proceedings of the International Conference GLOBAL-2005 "Nuclear Energy Systems for Future Generation and Global Sustainability," Tsukuba, Japan, October 9–13, Paper No. 463, 5 pages.

Klimov, N.N., Kuraeva, I.V., and Protopopov, V.S., 1985. Heat transfer with natural convection of carbon dioxide at supercritical pressure, *Thermal Engineering* (Теплоэнергетика, стр. 67–70), 12, pp. 698–702.

Klochkov, V.N., 1975. Evaluating the corrosivity of the medium in the condensate-feed circuit of high pressure and supercritical pressure power generating units, *Thermal Engineering* (Теплоэнергетика, стр. 55–58), 22 (4), pp. 72–76.

Knapp, K.K. and Sabersky, R.H., 1966. Free convection heat transfer to carbon dioxide near the critical point, *International Journal of Heat & Mass Transfer,* 9 (1), pp. 41–51.

Ko, H.S., Sakurai, K., Okamoto, K., and Madarame, H., 2000. Analysis of forced convection heat transfer to supercritical carbon dioxide, Proceedings of the 1st International Symposium on Supercritical Water-Cooled Reactor Design and Technology (SCR-2000), Tokyo, Japan, November 6–8, Paper No. 304.

Komita, H., Morooka, S., Yoshida, S., and Mori, H., 2003. Study of the heat transfer to the supercritical pressure fluid for supercritical water cooled power reactor development, Proceedings of the International Topical Meeting on Nuclear Reactor Thermal Hydraulics (NURETH-10), Seoul, Korea, October 5–9, 11 pages.

Kondrat'ev, N.S., 1969. Heat transfer and hydraulic resistance with supercritical water flowing in tubes, *Thermal Engineering* (Теплоэнергетика, стр. 49–51), 16 (8), pp. 73–77.

Kondrat'ev, N.S., 1971. About regimes of the deteriorated heat transfer at flow of supercritical pressure water in tubes, (In Russian), Transactions of the IVth All-Union Conference on Heat Transfer and Hydraulics at Movement of Two-Phase Flow inside Elements of Power Engineering Machines and Apparatuses, Leningrad, Russia, pp. 71–74.

Konobeev, Yu.V. and Birzhevoi, G.A., 2004. Prospects for using high-nickel alloys in power reactors with supercritical-pressure water, *Atomic Energy* (Атомная Энергия, стр. 394–403), 96 (5), pp. 365–373.

Koppel, L.B. and Smith, J.M., 1961. Turbulent heat transfer in the critical region, In book: *International Developments in Heat Transfer*: Papers presented at the 1961 International Heat Transfer Conference, ASME, August 28 – September 1, University of Colorado, Boulder, CO, USA, January 8–12, Paper No. 196, pp. 585–590.

Koppel, L.B. and Smith, J.M., 1962. Laminar flow heat transfer for variable physical properties, *Journal of Heat Transfer,* Transactions of the ASME, 84 (2), pp. 157–163.

Kornbichler, H., 1964. Superheat reactor development in the Federal Republic of Germany, International UN Conference on the Peaceful Uses of Atomic Energy, Session 1.5, Paper No. 535, pp. 266–275.

Koshizuka, S. and Oka, Yo., 2000. Computational analysis of deterioration phenomena and thermal-hydraulic design of SCR, Proceedings of the 1st

International Symposium on Supercritical Water-Cooled Reactor Design and Technology (SCR-2000), Tokyo, Japan, November 6–8, Paper No. 302.

Koshizuka, S. and Oka, Y., 1998. Supercritical-pressure, light-water-cooled reactors for economical nuclear power plants, *Progress in Nuclear Energy*, 32 (3/4), pp. 547–554.

Koshizuka, S., Oka, Yo., and Suhwan, J., 2003. Stability analysis of a high temperature reactor cooled by supercritical light water, Proceedings of the Joint International Conference Global Environment and Nuclear Energy Systems/ Advanced Nuclear Power Plants (GENES4/ANP2003), Kyoto, Japan, September 15–19, Paper No. 1166, 8 pages.

Koshizuka, S., Takano, N., and Oka, Y., 1995. Numerical analysis of deterioration phenomena in heat transfer to supercritical water, *International Journal of Heat & Mass Transfer*, 38 (16), pp. 3077–3084.

Kovalevskiy, V.B. and Miropol'skiy, V.B., 1978. Heat transfer and drag in the flow of water supercritical pressure in curved channels, *Heat Transfer–Soviet Research*, 10 (4), pp. 69–78.

Krasnoshchekov, E.A., Kuraeva, I.V., and Protopopov, V.S., 1969. Local heat transfer of carbon dioxide at supercritical pressure under cooling conditions, *High Temperatures* (Теплофизика Высоких Температур, стр. 922–930), 7 (5), pp. 856–861.

Krasnoshchekov, E.A. and Protopopov, V.S., 1959. Heat transfer at supercritical region in flow of carbon dioxide and water in tubes, (In Russian), *Thermal Engineering* (Теплоэнергетика, стр. 26–30), No. 12, pp. 26–30.

Krasnoshchekov, E.A. and Protopopov, V.S., 1960. About heat transfer in flow of carbon dioxide and water at supercritical region of state parameters, (In Russian), *Thermal Engineering* (Теплоэнергетика, стр. 94), No. 10, p. 94.

Krasnoshchekov, E.A. and Protopopov, V.S., 1966. Experimental study of heat exchange in carbon dioxide in the supercritical range at high temperature drops, *High Temperatures* (Теплофизика Высоких Температур, стр. 389–398), 4 (3), pp. 375–382.

Krasnoshchekov, E.A. and Protopopov, V.S., 1971. A generalized relationship for calculation of heat transfer to carbon dioxide at supercritical pressure (π = 1.02 – 5.25), *High Temperatures* (Теплофизика Высоких Температур, стр. 1314), 9 (6), pp. 1215.

Krasnoshchekov, E.A., Protopopov, V.S., Parkhovnik, I.A., and Silin, V.A., 1971. Some results of an experimental investigation of heat transfer to carbon dioxide at supercritical pressure and temperature heads of up to 850°C, *High Temperatures* (Теплофизика Высоких Температур, стр. 1081–1084), 9 (5), pp. 992–995.

Krasnoshchekov, E.A., Protopopov, V.S., Van, F., and Kuraeva, I.V., 1967. Experimental investigation of heat transfer for carbon dioxide in the supercritical region, Proceedings of the 2nd All-Soviet Union Conference on Heat and Mass Transfer, Minsk, Belarus', May, 1964, Published as Rand Report R-451-PR, Edited by C. Gazley, Jr., J.P. Hartnett and E.R.C. Ecker, Vol. 1, pp. 26–35.

Krasyakova, L.Yu., Belyakov, I.I., and Fefelova, N.D., 1973. Hydraulic resistance with isothermal and nonisothermal flow of medium at supercritical pressure, *Thermal Engineering* (Теплоэнергетика, стр. 31–35), 20 (4), pp. 45–50.

Krasyakova, L.Yu., Belyakov, I.I., and Fefelova, N.D., 1977. Heat transfer with a downward flow of water at supercritical pressure, *Thermal Engineering* (Теплоэнергетика, стр. 8–13), 24 (1), pp. 9–14.

Krasyakova, L.Yu. and Glusker, B.N., 1965. Investigated flow stability in parallel serpentine tubes with upward-downward flow of medium at subcritical and supercritical pressure, (In Russian), *Transactions of TsKTI Boiler-Turbine Engineering* (Труды ЦКТИ Котло-Турбостроение), Issue 59, Leningrad, Russia, pp. 198–217.

Krasyakova, L.Yu., Raykin, Ya.M., Belyakov, I.I., et al., 1967. Investigation of temperature regime of heated tubes at supercritical pressure, (In Russian), *Soviet Energy Technology* (Энергомашиностроение), No. 1, pp. 1–4.

Krasyakova, L.Yu., Raykin, Ya.M., Belyakov, I.I., et al., 1968. Temperature regime of vertical and horizontal heated tubes at supercritical pressure, (In Russian), *Transactions of TsKTI Boiler-Turbine Engineering* (Труды ЦКТИ Котло-Турбостроение, стр. 105–118), Issue 90, Leningrad, Russia, pp. 105–118.

Kreglewski, A., 1984. *Equilibrium Properties of Fluids and Fluid Mixtures*, Texas A&M University Press, TX, USA, pp. 223–230.

Kritzer, P., 2001. Corrosion in high-temperature and supercritical water and aqueous solutions: influence of solution and material parameters, Proceedings of the 1st International Symposium on Supercritical Water-Cooled Reactor Design and Technology (SCR-2000), Tokyo, Japan, November 6–8, Paper No. 204.

Kruzhilin, G.N., 1974. Heat transfer from wall to steam flow at supercritical pressures, Proceedings of the 5th International Heat Transfer Conference, Tokyo, Japan, September 3–7, Vol. II, Paper No. FC4.11, pp. 173–176.

Kuraeva, I.V., Klimov, N.N., Protopopov, V.S., and Ferubko, N.E., 1985. Free convection heat transfer of supercritical pressure fluids at temperatures close to the pseudocritical one, *Heat Transfer–Soviet Research*, 17 (2), pp. 39–44.

Kuraeva, I.V. and Protopopov, V.S., 1974. Mean friction coefficients for turbulent flow of a liquid at a supercritical pressure in horizontal circular tubes, *High Temperatures* (Теплофизика Высоких Температур, стр. 218–220), 12 (1), pp. 194–196.

Kurganov, V.A., 1982. Calculation of heat transfer in smooth pipes with turbulent flow of gaseous heat-carriers with constant and variable physical properties, *High Temperatures* (Теплофизика Высоких Температур, стр. 705–711), 20 (4), pp. 587–593.

Kurganov, V.A., 1991. Predicting normal and deteriorated heat transfer with mixed convection of heat carrier at supercritical pressure in vertical tubes, *Thermal Engineering* (Теплоэнергетика, стр. 63–68), 38 (1), pp. 40–45.

Kurganov, V.A., 1998a. Heat transfer and pressure drop in tubes under supercritical pressure of the coolant. Part 1: Specifics of the thermophysical properties, hydrodynamics, and heat transfer of the liquid. Regimes of normal heat transfer, *Thermal Engineering* (Теплоэнергетика, стр. 2–10), 45 (3), pp. 177–185.

Kurganov, V.A., 1998b. Heat transfer and pressure drop in tubes under supercritical pressure. Part 2. Heat transfer and friction at high heat fluxes. The influence of additional factors. Enhancement of deteriorated heat transfer, *Thermal Engineering* (Теплоэнергетика, стр. 35–44), 45 (4), pp. 301–310.

Kurganov, V.A. and Ankundinov, V.B., 1985. Calculation of normal and deteriorated heat transfer in tubes with turbulent flow of liquids in the near-critical and vapour region of state, *Thermal Engineering* (Теплоэнергетика, стр. 53–57), 32 (6), pp. 332–336.

Kurganov, V.A., Ankundinov, V.B., and Kaptil'nyi, A.G., 1986. Experimental study of velocity and temperature fields in an ascending flow of carbon dioxide at supercritical pressure in a heated vertical tube, *High Temperatures* (Теплофизика Высоких Температур, стр. 1104–1111), 24 (6), pp. 811–818.

Kurganov, V.A. and Kaptil'nyi, A.G., 1992. Velocity and enthalpy fields and eddy diffusivities in a heated supercritical fluid flow, *Experimental Thermal and Fluid Science*, 5, pp. 465–478.

Kurganov, V.A. and Kaptil'nyi, A.G., 1993. Flow structure and turbulent transport of a supercritical pressure fluid in a vertical heated tube under the conditions of mixed convection. Experimental data, *International Journal of Heat & Mass Transfer*, 36 (13), pp. 3383–3392.

Kurganov, V.A., Kaptil'nyi, A.G., and Ankundinov, V.B., 1989. Total flow resistance and fluid friction associated with ascending and descending supercritical fluid flow in heated pipes, *High Temperatures* (Теплофизика Высоких Температур, стр. 94–103), 27 (1), pp. 87–94.

Kurganov, V.A. and Zeigarnik, Yu.A., 2005. Results of studying of turbulent heat transfer deterioration in channels and their application in practical calculations of heat transfer and pressure drop of supercritical pressure coolants, Proceedings of the International Topical Meeting on Nuclear Reactor Thermal Hydraulics (NURETH-11), Avignon, France, October 2–6, Paper No. 235, 19 pages.

Kurihara, R., Watanabe, K., and Konishi, S., 2003. Literature survey of thermal-hydraulic studies on super-critical pressurized water, (In Japanese), Japanese Atomic Energy Research Institute, JAERI–Review, 2003–020, 43 pages.

Kuznetsov, Yu.N., 2004. Private communications, August.

Kuznetsov, Yu.N. and Gabaraev, B.A., 2004. Channel type reactors with supercritical water coolant. Russian experience, Proceedings of the International Congress on Advances in Nuclear Power Plants (ICAPP'04), Pittsburgh, PA, USA, June 13–17, Paper 4232, pp. 562–566.

Kutateladze, S.S. and Leont'ev, A.I., 1964. *Turbulent Boundary Layers in Compressible Gases*, Edward Arnold (Publishers) Ltd., London, UK, pp. 141–145.

Kuznetsov, Yu.N. and Gabaraev, B.A., 2004. Channel type reactors with supercritical water coolant. Russian experience, Proceedings of the International Congress on Advances in Nuclear Power Plants (ICAPP'04), Pittsburgh, PA, USA, June 13–17, Paper No. 4232, pp. 562–566.

Labuntsov, D.A. and Mirzoyan, P.I., 1983. Analysis of boundaries of stability of motion of helium at supercritical parameters in heated channel, *Thermal Engineering* (Теплоэнергетика, стр. 2–4), 30 (3), pp. 121–123.

Labuntsov, D.A. and Mirzoyan, P.A., 1986. Stability of flow of helium at supercritical pressure with non-uniform distribution of heat flux over the length of a channel, *Thermal Engineering* (Теплоэнергетика, стр. 53–56), 33 (4), pp. 208–211.

Larson, J.R. and Schoenhals, R.J., 1966. Turbulent free convection in near-critical water, *Journal of Heat Transfer*, Transactions of the ASME, 88 (4), pp. 407–414.

Latanision, R.M. and Mitton, D.B., 2001. Stress corrosion cracking in supercritical water systems, Proceedings of the 1st International Symposium on Supercritical Water-Cooled Reactor Design and Technology (SCR-2000), Tokyo, Japan, November 6–8, Paper No. 205.

Latgé, C., Rodriguez, G., and Simon, N., 2005. Supercritical CO_2 Brayton cycle for SFR: Na–CO_2 interaction and consequences on design and operation, Proceedings of the International Conference GLOBAL-2005 "Nuclear Energy Systems for Future Generation and Global Sustainability," Tsukuba, Japan, October 9–13, Paper No. 535DF, 6 pages.

Lee, A., 1997. Effect of shape and orientation on the performance of supercritical water oxidation reactors, ASME Proceedings of the 32nd National Heat Transfer Conference, Baltimore, MD, USA, August 8–12, HTD-Vol. 350 (Vol. 12), pp. 99–106.

Lee, R.A. and Haller, K.H., 1974. Supercritical water heat transfer developments and applications, Proceedings of the 5th International Heat Transfer Conference, Tokyo, Japan, September 3–7, Vol. IV, Paper No. B7.7, pp. 335–339.

Lee, S.H. and Howell, J.R., 1997. Convective heat transfer near the entrance region of a tube for water near the critical point, ASME Proceedings of the 32nd National Heat Transfer Conference, Baltimore, MD, USA, August 8–12, HTD-Vol. 350 (Vol. 12), pp. 121–129.

Lee, J.H., Koshizuka, S., and Oka, Y., 1998. Development of a LOCA analysis code for the supercritical-pressure light water cooled reactors, *Annals of Nuclear Energy*, 25 (16), pp. 1341–1361.

Lee, J.H., Oka, Y., and Koshizuka, S., 1999. Safety system consideration of a supercritical-water cooled fast reactor with simplified PSA, *Reliability Engineering and System Safety*, 64, pp. 327–338.

Leontiev, A.I., 1969. To the question of the impairment of turbulent convection heat transfer at supercritical pressures at forced flow of fluid in the vertical channels, Paper No. 69–HT–60, ASME, pp. 1–4.

Letyagin, I.P. and Chakrygin, V.G., 1976. Use of generalized dependences in calculations of the hydraulics of heated elements at supercritical pressures, *Thermal Engineering* (Теплоэнергетика, стр. 13–17), 23 (3), pp. 15–19.

Leung, L.K.H., Pioro, I.L., and Bullock, D.E., 2003. Post-dryout surface-temperature distributions in a vertical freon-cooled 37-element bundle, Proceedigns of the 10th International Topical Meeting on Nuclear Reactor Thermal Hydraulics (NURETH-10), Seoul, Korea, October 5–9, Paper C00201, 13 pages.

Levelt Sengers, J.M.H.L., 2000. Supercritical fluids: Their properties and applications, Chapter 1, In book: *Supercritical Fluids*, editors: E. Kiran et al., NATO Advanced Study Institute on Supercritical Fluids — Fundamentals and Application, NATO Science Series, Series E, Applied Sciences, Kluwer Academic Publishers, Netherlands, Vol. 366, pp. 1–29.

Li, L.J., Lin, C.X., and Ebadian, M.A., 1999. Turbulent heat transfer to near-critical water in a heated curved pipe under the conditions of mixed convection, *International Journal of Heat & Mass Transfer*, 42, pp. 3147–3158.

Liao, S.M. and Zhao, T.S., 2002a. An experimental investigation of convection heat transfer to supercritical carbon dioxide in miniature tubes, *International Journal of Heat and Mass Transfer*, 45, pp. 5025–5034.

Liao, S.M. and Zhao, T.S., 2002b. Heat transfer of supercritical carbon dioxide flowing in heated horizontal and vertical miniature tubes, Proceedings of the 12th International Heat Transfer Conference, Elsevier SAS, pp. 501–506.

Lipets, A.U., Kuznetsova, S.M., Dirina, L.V., et al., 1998. Increasing efficiency of boiler and energy unit for ultra-supercritical steam parameters, *Thermal Engineering* (Теплоэнергетика, стр. 31–37), 45 (6), pp. 473–480.

Lokshin, V.A., Semenovker, I.E., and Vikhrev, Yu.V., 1968. Calculating the temperature conditions of the radiant heating surfaces in supercritical boilers, *Thermal Engineering* (Теплоэнергетика, стр. 21–24), 15 (9), pp. 34–39.

Lomperski, S., Cho, D., Jain, R., and Corradini, M.L., 2004. Stability of a natural circulation loop with a fluid heated through the thermodynamic pseudocriticial point, Proceedings of the International Congress on Advances in Nuclear Power Plants (ICAPP'04), Pittsburgh, PA, USA, June 13–17, Paper No. 4268, pp. 1736–1741.

Lorentzen, G., 1994. Revival of carbon dioxide as a refrigerant, *International Journal of Refrigeration,* 17 (5), pp. 292–301.

Lorentzen, G. and Pettersen, J., 1993. A new, efficient and environmentally benign system for car air-conditioning, *International Journal of Refrigeration,* 16 (1), pp. 4–12.

Löwenberg, M., Starflinger, J., Lauren, E., and Schulenberg, Th., 2005. A look-up table for heat transfer of supercritical water, Proceedings of the International Conference GLOBAL-2005 "Nuclear Energy Systems for Future Generation and Global Sustainability," Tsukuba, Japan, October 9–13, Paper No. 037, 6 pages.

MacDonald, Ph., Buongiorno, J., Sterbentz, J.W., et al., 2005. Feasibility study of supercritical light water cooled reactors for electric power production, Final Report INEEL/EXT-04-02530, Idaho National Engineering and Environmental Laboratory, Bechtel, BWXT, Idaho, LLC, 174 pages.

Malandrone, M., Panella, B., Pedrelli, G., and Sobrero, G., 1987a. Fenomeni di deterioramento nello scambio termico a pressione ipercritica, *L'Energia Elettrica,* 64 (2), pp. 45–56.

Malandrone, M., Panella, B., and Sobrero, G., 1987b. Critri per la previsione del deterioramento dello scambio termico par acqua a pressione ipercritica, *L'Energia Elettrica,* 64 (2), pp. 57–72.

Malkina, L.I., Maksimova, G.P., Kafengauz, N.L., and Fedorov, M.I., 1972. Heat transfer to water with pseudoboiling, *Heat Transfer–Soviet Research,* 4 (5), pp. 23–26.

Malyshev, G.P. and Pron'ko, V.G., 1972. Characteristics of heat transfer during the turbulent flow of supercritical helium, *Journal of Engineering-Physics* (Инженерно-Физический Журнал (ИФЖ), стр. 814–818), 23 (5), pp. 1382–1385.

Mamedov, A.M., Kalbaliev, F.I., and Isaev, G.I., 1976. Heat transfer to ascending turbulent streams of aromatic hydrocarbons at near-critical pressures, *Journal of Engineering-Physics* (Инженерно-Физический Журнал (ИФЖ), стр. 281–287), 30 (2), pp. 183–187.

Mamedov, A.M., Kalbaliev, F.I., and Mirkadirov, Yu.M., 1977. Heat transfer to some aromatic hydrocarbons at supercritical pressure under conditions of free convection, *High Temperatures* (Теплофизика Высоких Температур, стр. 436–440), 15 (2), pp. 373–375.

Man'kina, N.N., 1977. *Physico-Chemical Processes in Steam-Water Cycle of Electrical Power Station,* (In Russian), Energiya Publishing House, Moscow, Russia.

Marchaterre, J.F. and Petrick, M., 1960. Review of the status of supercritical water reactor technology, Report ANL-6202, Argonne National Laboratory, August, 37 pages.

Margen, P.H., 1961. Current status of work in Sweden on a boiling and superheating heavy-water reactor, IAEA Symposium on Power Reactor Experiments, Vienna, Austria, SM – 21/6, pp. 97–133.

Margulova, T.Kh. and Martynova, O.I., 1987. *Water Regimes of Thermal and Nuclear Power Plants, 2nd edition*, (In Russian), Vysshaya Shkola Publishing House, Moscow, Russia, 319 pages.

Mark's Standard Handbook for Mechanical Engineers, 1996. Editors: Eu.A. Avallone and Th. Baumeister III, McGraw-Hill, New York, NY, USA, p. 16-2.

Marsault, Ph., Renault, C., Rimpault, G., et al., 2004. Pre-design studies of SCWR in fast neutron spectrum: Evaluation of operating conditions and analysis of the behaviour in accidental situations, Proceedings of the International Congress on Advances in Nuclear Power Plants (ICAPP'04), Pittsburgh, PA, USA, June 13–17, Paper No. 4078, pp. 552–561.

Martynova, O.I. and Rogatskin, B.S., 1969. Calcium sulphate deposits in supercritical boiler circuits, *Thermal Engineering* (Теплоэнергетика, стр. 66–69), 16 (8), pp. 99–102.

McAdams, W.H., 1942. *Heat Transmission,* 2nd edition, McGraw-Hill, New York, NY, USA, 459 pages.

McAdams, W.H., Kennel, W.E., and Addoms, J.N., 1950. Heat transfer to superheated steam at high pressures, *Transactions of the ASME,* 72 (4), pp. 421–428.

McCarthy, J.R., Seader, J.D., and Trebes, D.M., 1967. Heat transfer to supercritical nitrogen tetroxide at high heat fluxes and in axially curved flow passages, ASME, Paper No. 67-HT-59, pp. 1–12.

McGreery, G.E., Buongiorno, J., Condie, K.G., et al., 2003. The INEEL heat transfer flow loop for development of supercritical-pressure water reactors (SCWRs), Proceedings of the Joint International Conference Global Environment and Nuclear Energy Systems/Advanced Nuclear Power Plants (GENES4/ANP2003), Kyoto, Japan, September 15–19, Paper No. 1010, 7 pages.

Mechanical Engineers' Handbook, 1998. 2nd edition, Edited by M. Kutz, J. Wiley & Sons, Inc., New York, NY, USA.

Melik-Pashaev, N.I., 1966. Calculation of convective heat transfer at supercritical pressure, *High Temperatures* (Теплофизика Высоких Температур, стр. 853–864), 4 (6), pp. 789–798.

Melik-Pashaev, N.I., Kobel'kov, V.N., and Plyugin, M.D., 1968. Investigation of convective heat transfer at supercritical pressures, *High Temperatures* (Теплофизика Высоких Температур, стр. 272–276), 6 (2), pp. 263–266.

Mikheev, M.A., 1956. *Fundamentals of Heat Transfer,* (In Russian), Gosenergoizdat Publishing House, Moscow, Russia.

Miropol'skiy, Z.L. and Baigulov, V.I., 1974. Investigation in heat transfer, velocity and temperature profiles with carbon dioxide flow in a tube over the nearly critical region of parameters, Proceedings of the 5th International Heat Transfer Conference, Tokyo, Japan, September 3–7, Vol. II, Paper No. FC-4.12, pp. 177–181.

Miropolskiy, Z.L. and Pikus, V.Yu., 1967–1968. Heat transfer in supercritical flows through curvilinear channels, Proceedings of the Institute of Mechanical Engineers, 182, Part 31, pp. 6–9.

Miropol'skii, Z.L., Pikus, V.Yu., and Kovalevskii, V.B., 1970. Heat transfer with a flow of medium at supercritical pressure in curvilinear channels, *Thermal Engineering* (Теплоэнергетика, стр. 70–72), 17 (11), pp. 101–104.

Miropolskiy, Z.L., Picus, V.J., and Shitsman, M.E., 1966. Regimes of deteriorated heat transfer at forced flow of fluids in curvilinear channels, Proceedings of the 3rd International Heat Transfer Conference, Chicago, IL, USA, August 7–12, Vol. 2, pp. 95–101.

Miropol'skiy, Z.L. and Shitsman, M.E., 1957. Heat transfer to water and steam at variable specific heat (at near-critical region), (In Russian), *Journal of Technical Physics* (Журнал Технической Физики), XXVII (10), pp. 2359–2372.

Miropol'skiy, Z.L. and Shitsman, M.E., 1958a. Investigation of heat transfer to water and steam at pressures up to 280 ata, (In Russian), In book: *Investigation of Heat Transfer to Steam and Water Boiling in Tubes at High Pressures*, Editor N.A. Dollezhal', Atomizdat Publishing House, Moscow, Russia, pp. 54–70.

Miropol'skiy, Z.L. and Shitsman, M.E., 1958b. About calculation methods of heat transfer to water and steam at near-critical region, (In Russian), *Soviet Energy Technology* (Энергомашиностроение), No. 1, pp. 8–11.

Modro, S.M., 2005. The supercritical water cooled reactor research and development in the U.S., Proceedings of the International Congress on Advances in Nuclear Power Plants (ICAPP'05), Seoul, Korea, May 15–19, Paper No. 5694, 13 pages.

Moffat, R.J., 1988. Describing the uncertainties in experimental results, *Experimental Thermal and Fluid Science*, 1, pp. 3–17.

Mori, M., Maschek, W., and Rineiski, A., 2005. Heterogeneous cores for improved safety performance. A case study: The supercritical water fast reactor, Proceedings of the International Conference on Nuclear Engineering (ICONE-13), Beijing, China, May 16–20, Paper No. 50441, 8 pages.

Mori, H. and Ogata, H., 1991. Natural convection heat transfer of supercritical helium in a channel, (In Japanese), *Transactions of the JSME*, Series B, 57 (542), pp. 151–154.

Mori, H., Yoshida, S., Morooka, S., and Komita, H., 2005. Heat transfer study under supercritical pressure conditions for single rod test section, Proceedings of the International Congress on Advances in Nuclear Power Plants (ICAPP'05), Seoul, Korea, May 15–19, Paper No. 5303, 9 pages.

Mukohara, T., Koshizuka, S., and Oka, Yo., 2000a. Subchannel analysis of a fast reactor cooled by supercritical light water, *Progress in Nuclear Energy*, 37 (1–4), pp. 197–204.

Mukohara, T., Koshizuka, S., and Oka, Yo., 2000b. Subchannel analysis of supercritical water cooled reactors, Proceedings of the 1st International Symposium on Supercritical Water-Cooled Reactor Design and Technology (SCR-2000), Tokyo, Japan, November 6–8, Paper No. 301.

Mukohara, T., Koshizuka, S.-I., and Oka, Yo., 1999. Core design of a high-temperature fast reactor cooled by supercritical light water, *Annals of Nuclear Energy*, 26, pp. 1423–1436.

Nakatsuka, T., Oka, Yo., and Koshizuka, S., 2001. Startup thermal considerations for supercritical-pressure light water-cooled reactors, *Nuclear Technology*, 134, pp. 221–230.

Nakatsuka, T., Oka, Yo., and Koshizuka, S., 2000. Start-up of supercritical-pressure light water cooled reactors, Proceedings of the 8th International Conference on Nuclear Engineering (ICONE-8), Baltimore, MD, USA, April 2–6, Paper No. 8304.

Nakatsuka, T., Oka, Yo., and Koshizuka, S., 1998. Control of a fast reactor cooled by supercritical water, *Nuclear Technology*, 121, pp. 81–92.

Le Neindre, B., Tufeu, R., Bury, P., and Sengers, J.V., 1973. Thermal conductivity of carbon dioxide and steam in the supercritical region, *International Journal Physical Chemistry*: *Berichte der Bunsen-Gesellschaft*, 77 (4), pp. 262–275.

Nesterenko, V.B., Devoino, A.N., Kolykhan, L.I., et al., 1974. Investigation of heat transfer with turbulent flows of nitrogen tetroxide in a round heated tube, *Thermal Engineering* (Теплоэнергетика, стр. 72–75), 21 (11), pp. 105–110.

Neumann, R.J. and Hahne, E.W.P., 1980. Free convective heat transfer to supercritical carbon dioxide, *International Journal of Heat & Mass Transfer*, 23, pp. 1643–1652.

Nishikawa, K., Ito, T., and Yamashita, H., 1973. Free-convective heat transfer to a supercritical fluid, *Journal of Heat Transfer*, Transactions of the ASME, 95 (2), pp. 187–191.

NIST/ASME Steam Properties, 1997. NIST Standard Reference Database 10 (on diskette: Executable with Source), Version 2.11, A.H. Harvey, A.P. Peskin and S.A. Klein, U.S. Department of Commerce.

NIST/ASME Steam Properties, 1996. NIST Standard Reference Database 10 (on diskette: Executable with Source), Version 2.2, A.H. Harvey, A.P. Peskin and S.A. Klein, U.S. Department of Commerce.

NIST Reference Fluid Thermodynamic and Transport Properties—REFPROP, 2002. NIST Standard Reference Database 23 (on CD: Executable with Source plus Supplemental Fluids in ZIP File), Version 7.0, E.W. Lemmon, M.O. McLinden and M.L. Huber, National Institute of Standards and Technology, Boulder, CO, U.S. Department of Commerce, August.

Noer, M. and Kjaer, S., 1998. Development of ultra super critical PF power plants in Denmark, Proceedings of the 17th Congress of the World Energy Council, Houston, TX, USA, September 13–18, 1998, pp. 295–311.

Nowak, E.S. and Konanur, A.K., 1970. An analytical investigation of free convection heat transfer to supercritical water, *Journal of Heat Transfer*, Transactions of the ASME, 92 (3), pp. 345–350.

Nozdrenko, G.V., 1968. Estimation of the heat transfer to supercritical Freons, *Journal of Engineering-Physics* (Инженерно-Физический Журнал (ИФЖ), стр. 1091–1095), 14 (6), pp. 566–568.

Ogata, H. and Sato, S., 1972. Measurements of forced convection heat transfer to supercritical helium, Proceedings of the 4th International Engineering Conference, Eindhoven, Netherlands, pp. 291–294.

Oka, Yo., 2003. Research and development of the supercritical-pressure light water cooled reactors, Proceedings of the International Topical Meeting on Nuclear Reactor Thermal Hydraulics (NURETH-10), Seoul, Korea, October 5–9, 11 pages.

Oka, Yo., 2002. Review of high temperature water and steam cooled reactor concepts, Proceedings of the International Congress on Advances in Nuclear Power Plants (ICAPP'02), Hollywood, FL, USA, June 9–13, Paper No. 1006, 20 pages.

Oka, Yo., 2000. Review of high temperature water and steam cooled reactor concepts, Proceedings of the 1st International Symposium on Supercritical Water-Cooled Reactor Design and Technology (SCR-2000), Tokyo, Japan, November 6–8, Paper No. 104.

Oka, Yo., Jevremovich, T., and Koshizuka, S.-I., 1994a. A direct-cycle, supercritical-water-cooled fast breeder reactor, *Journal of Nuclear Science and Technology*, 31 (1), pp. 83–85.

Oka, Yo., Jevremovich, T., and Koshizuka, S.-I., 1994b. Indirect-cycle FBR cooled by supercritical steam — concept and design, *Transactions of the American Nuclear Society,* 68 (Part A), pp. 448–449.

Oka, Yo., Jevremovich, T., Koshizuka, S.-I., and Okano, Ya., 1993. A direct-cycle FBR, SCFBR-D, *Transactions of the American Nuclear Society,* 69, p. 357.

Oka, Y. and Kataoka, K., 1992. Conceptual design of a fast breeder reactor cooled by supercritical steam, *Annals of Nuclear Energy,* 19 (4), pp. 243–247.

Oka, Y., Kataoka, K., and Jevremovich, 1992a. Supercritical steam-cooled FBR with negative reactivity characteristics against voiding and flooding, *Transactions of the American Nuclear Society,* 64, p. 576.

Oka, Yo. and Koshizuka, S., 2002. Status and prospects of high temperature (supercritical-pressure) light water cooled reactor research and development, Proceedings of the 13th Pacific Basin Nuclear Conference, Shenzhen City, China, October 21–25.

Oka, Yo. and Koshizuka, S., 2001, Supercritical-pressure, once-through cycle light water cooled reactor concept, *Journal of Nuclear Science and Technology,* 38 (12), pp. 1081–1089.

Oka, Yo. and Koshizuka, S., 2000. Design concept of once-through cycle supercritical-pressure light–water–cooled reactors, Proceedings of the 1st International Symposium on Supercritical Water-Cooled Reactor Design and Technology (SCR-2000), Tokyo, Japan, November 6–8, Paper No. 101.

Oka, Y. and Koshizuka, S., 1998. Conceptual design study of advanced power reactors, *Progress in Nuclear Energy,* 32 (1/2), p. 163–177.

Oka, Yo. and Koshizuka, S., 1993. Concept and design of a supercritical-pressure, direct-cycle light-water-cooled reactor, *Nuclear Technology,* 103, September, pp. 295–302.

Oka, Yo., Koshizuka, S., Ishiwatari, Yu., and Yamaji, A., 2003a. Overview of design studies of high temperature reactor cooled by supercritical light water at the University of Tokyo, Proceedings of the Joint International Conference Global Environment and Nuclear Energy Systems/Advanced Nuclear Power Plants (GENES4/ANP2003), Kyoto, Japan, September 15–19, Paper No. 1168, 10 pages.

Oka, Yo., Koshizuka, S., Ishiwatari, Yu., and Yamaji, A., 2003b. Conceptual design of high temperature reactors cooled by supercritical light water, Proceedings of the International Congress on Advances in Nuclear Power Plants (ICAPP'03), Córdoba, Spain, May 4–7, Paper No. 3257, 10 pages.

Oka, Yo., Koshizuka, S., Ishiwatari, Yu., and Yamaji, A., 2003c. Conceptual design of high temperature reactors cooled by supercritical light water, Proceedings of the International Congress on Advances in Nuclear Power Plants (ICAPP'03), Cordoba, Spain, May 4–7, Paper No. 3257, 10 pages.

Oka, Yo., Koshizuka, S., Ishiwatari, Yu., et al., 2003d. High temperature LWR operating at supercritical pressure, Proceedings of the ANS/ENS International Winter Meeting and Nuclear Technology Expo, Embedded Topical Meeting GLOBAL 2003 "Advanced Nuclear Energy and Fuel Cycle Systems," New Orleans, LA, USA, November 16–20, pp. 1128–1135.

Oka, Yo., Koshizuka, S., Ishiwatari, Yu., and Yamaji, A., 2002. Elements of design of once-through cycle, supercritical-pressure light water cooled reactor, Proceedings of the International Congress on Advances in Nuclear Power Plants (ICAPP'02), Hollywood, FL, USA, June 9–13, Paper No. 1005, 12 pages.

Oka, Y., Koshizuka, S., Jevremovich, T., and Okano, Y., 1995a. Supercritical-pressure, light-water-cooled reactors for improving economy, safety, plutonium utilization and environment, *Progress in Nuclear Energy,* 29 (Supplement), pp. 431–438.

Oka, Y., Koshizuka, S., Jevremovich, T., and Okano, Y., 1995b. Systems design of direct-cycle supercritical-water-cooled fast reactors, *Nuclear Technology,* 109, pp. 1–10.

Oka, Y., Koshizuka, S., Okano, Y., et al., 1996. Design concepts of light water cooled rectors operating at supercritical pressure for technological innovation, Proceedings of the 10th Pacific Basin Nuclear Conference, Kobe, Japan, October 20–25, pp. 779–786.

Oka, Yo., Koshizuka, S., and Yamasaki, T., 1992b. Direct cycle light water reactor operating at supercritical pressure, *Journal of Nuclear Science and Technology,* 29 (6), pp. 585–588.

Oka, Yo., Nomura, K., Koshizuka, S.-I., and Mukohara, T., 2000. Supercritical-pressure light water cooled fast reactors, A competitive way of FR over LWR, Proceedings of the 13th International Conference on Nuclear Engineering (ICONE-8), Baltimore, MD, USA, April 2–6, Paper No. 8216.

Oka, Yo. and Yamada, K., 2004. Research and development of high temperature light water cooled reactor operating at supercritical-pressure in Japan, Proceedings of the International Congress on Advances in Nuclear Power Plants (ICAPP'04), Pittsburgh, PA, USA, June 13–17, Paper No. 4233, pp. 470–479.

Okano, Ya., Koshizuka, S., Kitoh, K., and Oka, Yo., 1996a. Flow-induced accident and transient analyses of a direct-cycle, light-water-cooled, fast breeder reactor operating at supercritical pressure, *Journal of Nuclear Science and Technology*, 33 (4), pp. 307–315.

Okano, Ya., Koshizuka, S., and Oka, Yo., 1996b. Core design of a direct-cycle light water reactor with double tube water rods, *Journal of Nuclear Science and Technology,* 33 (5), pp. 365–373.

Okano, Ya., Koshizuka, S.-I., and Oka, Yo., 1994. Design of water rod cores of a direct cycle supercritical-pressure light water reactor, *Annals of Nuclear Energy,* 21 (10), pp. 601–611.

Ookawa, M., Sakurai, S., and Yamada, K., 2005. Optimization method for design of the supercritical-water-cooled reactor fuel assembly, Proceedings of the International Congress on Advances in Nuclear Power Plants (ICAPP'05), Seoul, Korea, May 15–19, Paper No. 5216, 7 pages.

Ornatskiy, A.P., Dashkiev, Yu.G., and Perkov, V.G., 1980. *Supercritical Steam Generators,* (In Russian), Vyshcha Shkola Publishing House, Kiev, Ukraine, 287 pages.

Ornatskiy, A.P., Glushchenko, L.F., and Gandzyuk, O.F., 1972. An experimental study of heat transfer in externally-heated annuli at supercritical pressures, *Heat Transfer–Soviet Research*, 4 (6), pp. 25–29.

Ornatskiy, A.P., Glushchenko, L.F., and Kalachev, S.I., 1971. Heat transfer with rising and falling flows of water in tubes of small diameter at supercritical pressures, *Thermal Engineering* (Теплоэнергетика, стр. 91–93), 18 (5), pp. 137–141.

Ornatsky, A.P., Glushchenko, L.P., Siomin, E.T., et al., 1970. The research of temperature conditions of small diameter parallel tubes cooled by water under supercritical pressures, Proceedings of the 4th International Heat Transfer Conference, Paris-Versailles, France, Elsevier Publishing Company, Vol. VI, Paper No. B 8.11.

Petersen, H. and Kaiser, N.E., 1974. Heat transfer and pressure drop measurements in cooling channels with helium at a pressure of 20 bar and with temperature up to 1000°C, Proceedings of the 5th International Heat Transfer Conference, Tokyo, Japan, September 3–7, Vol. II, Paper No. FC4.8, pp. 160–164.

Pettersen, J., Rieberer, R., and Leister, A., 2000. Heat transfer and pressure drop characteristics of super-critical carbon dioxide in microchannel tubes under cooling, Proceedings of the 4th IIR-Gustav Lorentzen conference on Natural Working Fluids at Purdue, USA–2000/1, pp. 315–323.

Petrov, V.P., Beschastnov, S.P., and Belozerov, V.I., 1978. Natural-convection heat transfer from supercritical CO_2 to a cooled horizontal cylinder, *Heat Transfer–Soviet Research*, 10 (5), pp. 150–153.

Petrov, N.E. and Popov, V.N., 1985. Heat transfer and resistance of carbon dioxide being cooled in the supercritical region, *Thermal Engineering* (Теплоэнергетика, стр. 16–19), 32 (3), pp. 131–134.

Petrov, N.E. and Popov, V.N., 1988. Heat transfer and hydraulic resistance with turbulent flow in a tube of water at supercritical parameters of state, *Thermal Engineering* (Теплоэнергетика, стр. 45–48), 35 (10), pp. 577–580.

Petukhov, B.S., 1968. Heat transfer in a single-phase medium under supercritical conditions (survey), *High Temperatures* (Теплофизика Высоких Температур, стр. 732–745), 6 (4), pp. 696–709.

Petukhov, B.S., 1970. Heat transfer and friction in turbulent pipe flow with variable physical properties, In book: *Advances in Heat Transfer*, Editors J.P. Hartnett and Th.F. Irvine, Jr., Academic Press, New York, NY, USA, Vol. 6, pp. 504–564.

Petukhov, B.S., 1993. *Heat Transfer in Single-Phase Flow. Laminar Boundary Layer*, (In Russian), Editor A.F. Polyakov, MEI Publishing House, Moscow, Russia, "Natural convection near vertical plate under near-critical conditions", pp. 312–317.

Petukhov, B.S., Genin, L.G., and Kovalev, S.A., 1986. *Heat Transfer in Nuclear Power Installations*, 2nd edition, (In Russian), Energoatomizdat, Moscow, Russia, 472 pages.

Petukhov, B.S. and Kirillov, 1958. About heat transfer at turbulent fluid flow in tubes, (In Russian), *Thermal Engineering* (Теплоэнергетика, стр. 63–68), (4), pp. 63–68.

Petukhov, B.S., Krasnoschekov, E.A., and Protopopov, V.S., 1961. An investigation of heat transfer to fluids flowing in pipes under supercritical conditions, In book: *International Developments in Heat Transfer: Papers presented at the 1961 International Heat Transfer Conference*, ASME, University of Colorado, Boulder, CO, USA, January 8–12, Part III, Paper No. 67, pp. 569–578.

Petukhov, B.S., Kurganov, V.A., and Ankudinov, V.B., 1983. Heat transfer and flow resistance in the turbulent pipe flow of a fluid with near-critical state parameters, *High Temperatures* (Теплофизика Высоких Температур, стр. 92–100), 21 (1), pp. 81–89.

Petukhov, B.S., Kurganov, V.A., Ankudinov, V.B., and Grigor'ev, V.S., 1980. Experimental investigation of drag and heat transfer in a turbulent flow of fluid at supercritical pressure, *High Temperatures* (Теплофизика Высоких Температур, стр. 100–111), 18 (1), pp. 90–99.

Petukhov, B.S., Kurganov, V.A., Ankudinov, V.B., and Kaptil'nyi, A.G., 1985. Effect of dissolved gas on heat transfer during turbulent flow of carbon dioxide at supercritical pressure, *High Temperatures* (Теплофизика Высоких Температур, стр. 742–747), 23 (4), pp. 594–599.

Petukhov, B.S. and Medvetskaya, N.V., 1978. Turbulent flow and heat exchange in vertical pipes under conditions of strong influence of upward forces, *High Temperatures* (Теплофизика Высоких Температур, стр. 778–786), 16 (4), pp. 665–672.

Petukhov, B.S. and Medvetskaya, N.V., 1979. Calculation of turbulent flow and heat transfer in the heated tubes for single-phase working fluids at near-critical state, *High Temperatures* (Теплофизика Высоких Температур, стр. 343–350), 17 (2), pp. 287–294.

Petukhov, B.S. and Polyakov, A.F., 1974. Boundaries of regimes with "worsened" heat transfer for supercritical pressure of coolant, *High Temperatures* (Теплофизика Высоких Температур, стр. 221–224), 12 (1), pp. 197–200.

Petukhov, B.S. and Polyakov, A.F., 1988. *Heat Transfer in Turbulent Mixed Convection*, Chapter 7.1. Heat transfer at supercritical pressures, Hemisphere Publishing Corp., New York, NY, USA, pp. 183–201.

Petukhov, B.S., Polyakov, A.F., and Rosnovskii, S.V., 1976. New approach to calculation of heat transfer with supercritical pressures of the heat-transfer medium, *High Temperatures* (Теплофизика Высоких Температур, стр. 1326–1329), 14 (6), pp. 1189–1192.

Petukhov, B.S., Protopopov, V.S., and Silin, V.A., 1972. Experimental investigation of worsened heat-transfer conditions with the turbulent flow of carbon dioxide at supercritical pressure, *High Temperatures* (Теплофизика Высоких Температур, стр. 347–354), 10 (2), pp. 304–310.

Petukhov, B.S., Vilenskii, V.D., and Medvetskaya, N.V., 1977. Semiempirical models in heat-transfer calculations for turbulent flow in a tube of a single-phase heat carrier with near-critical parameters, *High Temperatures* (Теплофизика Высоких Температур, стр. 554–565), 15 (3), pp. 464–471.

Pioro, I.L., 1982. Investigation of closed two-phase thermosyphons used for cooling of melting converters, (In Russian), *Applied Problems of Heat Transfer and Hydrodynamics*, Naukova Dumka, Publishing House, Kiev, Ukraine, pp. 76–79.

Pioro, I.L., 1992. Maximum heat-transferring capacity of two-phase thermosiphons with separate vapor and condensate streams, *Heat Transfer Research*, 24 (4), pp. 535–542.

Pioro, I.L., 1999. Experimental evaluation of constants for the Rohsenow pool boiling correlation, *International Journal of Heat and Mass Transfer*, 42, pp. 2003–2013.

Pioro, I.L. and Cheng, S.C., 1998. Concise literature survey of the Russian and selected western publications devoted to the heat transfer and hydraulic resistance of a fluid at near critical and supercritical pressures, Report UO-MCG-TH-98003, prepared for AECL CRL, Mechanical Engineering Department, University of Ottawa, November, 148 pages.

Pioro, I.L., Cheng, S.C., Vasić, A., and Felisari, R., 2000. Some problems for bundle CHF prediction based on CHF measurements in simple flow geometries, *Nuclear Engineering and Design*, 201 (2–3), pp. 189–207.

Pioro, I. and Duffey, R., 2005. Experimental heat transfer to supercritical water flowing inside channels (survey), *Nuclear Engineering and Design*, 235 (22), pp. 2407–2430.

Pioro, I. and Duffey, R., 2003a. Experimental heat transfer to water flowing in channels at supercritical pressures (survey), Proceedings of the ANS/ENS International Winter Meeting and Nuclear Technology Expo, Embedded Topical

Meeting GLOBAL 2003 "Advanced Nuclear Energy and Fuel Cycle Systems," New Orleans, LA, USA, November 16–20, Paper No. 2342, 20 pages.

Pioro, I.L. and Duffey, R.B., 2003b. Literature survey of the heat transfer and hydraulic resistance of water, carbon dioxide, helium and other fluids at supercritical and near-critical pressures, Report AECL-12137/FFC-FCT-409, Chalk River Laboratories, AECL, April, ISSN 0067-0367, 182 pages.

Pioro, I., Duffey, R., and Dumouchel, T., 2004a. Hydraulic resistance of fluids flowing in channels at supercritical pressures (survey), *Nuclear Engineering and Design*, 231 (2), pp. 187–197.

Pioro, I.L., Groeneveld, D.C., Leung, L.K.H., et al., 2002a. Comparison of CHF measurements in horizontal and vertical tubes cooled with R-134a, *International Journal of Heat and Mass Transfer*, 45 (22), pp. 4435–4450.

Pioro, I.L., Groeneveld, D.C., Doerffer, S.S., et al., 2002b. Effects of flow obstacles on the critical heat flux in a vertical tube cooled with upward flow of R-134a, *International Journal of Heat and Mass Transfer*, 45 (22), pp. 4417–4433.

Pioro, I.L., Groeneveld, D., Cheng, S.C., et al., 2001. Comparison of CHF measurements in R-134a cooled tubes and the water CHF look-up table, *International Journal of Heat and Mass Transfer*, 44 (1), pp. 73–88.

Pioro, I.L., Khartabil, H.F., and Duffey, R.B., 2004b. Heat transfer to supercritical fluids flowing in channels — empirical correlations (survey), *Nuclear Engineering and Design*, 230 (1–3), pp. 69–91.

Pioro, I.L. and Kalashnikov, A.Yu., 1988. Maximum heat transfer in two-phase thermosyphons with external down-flow channel, *Applied Thermal Sciences*, 1 (2), pp. 62–67.

Pioro, I.L. and Khartabil, H.F., 2005. Experimental study on heat transfer to supercritical carbon dioxide flowing upward in a vertical tube, Proceedings of the 13th International Conference on Nuclear Engineering (ICONE-13), Beijing, China, May 16–20, Paper No. 50118, 9 pages.

Pioro, I.L., Khartabil, H.F., and Duffey, R.B., 2003. Heat transfer at supercritical pressures (survey), Proceedings of the 11th International Conference on Nuclear Engineering (ICONE-11), Shinjuku, Tokyo, Japan, April 20–23, Paper No. 36454, 13 pages.

Pioro, L.S. and Pioro, I.L., 1997. *Industrial Two-Phase Thermosyphons*, Begell House, New York, NY, USA, 288 pages.

Pis'mennyy, E.N., Razumovskiy, V.G., Maevskiy, E.M., et al., 2005. Experimental study on temperature regimes to supercritical water flowing in vertical tubes at low mass fluxes, Proceedings of the International Conference GLOBAL-2005 "Nuclear Energy Systems for Future Generation and Global Sustainability," Tsukuba, Japan, October 9–13, Paper No. 519, 9 pages.

Pitla, S.S., Groll, E.A., and Ramadhyani, S., 2002. New correlation to predict the HTC during in-tube cooling of turbulent supercritical CO_2, *International Journal of Refrigeration*, 25, pp. 887–895.

Pitla, S.S., Groll, E.A., and Ramadhyani, S., 2001a. Convective heat transfer from in-tube cooling of turbulent supercritical carbon dioxide: Part 2 — Experimental data and numerical predictions, *International Journal of HVAC&R Research*, 7 (4), pp. 367–382.

Pitla, S.S., Ramadhyani, S., and Groll, E.A., 2001b. Convective heat transfer from in-tube flow of turbulent supercritical carbon dioxide, Part 1— Numerical analysis, *International Journal of HVAC&R Research*, 7 (4), pp. 345–366.

Pitla, S.S., Robinson, D.M., Groll, E.A., and Ramadhyani, S., 1998. Heat transfer from supercritical carbon dioxide in tube flow: A critical review, *International Journal of HVAC&R Research*, 4 (3), pp. 281–301.

Pitla, S.S., Robinson, D.M., Zingerly, A., et al., 2000. Heat transfer and pressure drop characteristics during in-tube gas cooling of supercritical carbon dioxide, Report ASHRAE, 913-RP, August, 153 pages.

Polyakov, A.F., 1975. Mechanism and limits on the formation of conditions for impaired heat transfer at a supercritical coolant pressure, *High Temperatures* (Теплофизика Высоких Температур, стр. 1210–1219), 13 (6), pp. 1119–1126.

Polyakov, A.F., 1991. Heat transfer under supercritical pressures, In book—*Advances in Heat Transfer*, Edited by J.P. Hartnett, Th.F. Irvine, and Y.I. Cho, Academic Press, Inc., Vol. 21, pp. 1–53.

Popov, V.N., 1966. Theoretical calculation of the heat transfer and friction drag for carbon dioxide in the supercritical region during laminar flow in a circular tube, *High Temperatures* (Теплофизика Высоких Температур, стр. 689–697), 4 (5), pp. 646–652.

Popov, V.N., 1967. Theoretical calculation of heat transfer and friction resistance for supercritical carbon dioxide, Proceedings of the 2nd All-Soviet Union Conference on Heat and Mass Transfer, Minsk, Belarus', May, 1964, Published as Rand Report R-451-PR, Edited by C. Gazley, Jr., J.P. Hartnett and E.R.C. Ecker, Vol. 1, pp. 46–56.

Popov, V.N., 1977. Heat-transfer process and turbulent flow of compressible fluid in circular tube, *High Temperatures* (Теплофизика Высоких Температур, стр. 795–801), 15 (4), pp. 670–675.

Popov, V.N., 1983. The effects of free convection on turbulent transport for a liquid flowing in a vertical channel, *High Temperatures* (Теплофизика Высоких Температур, стр. 515–523), 21 (3), pp. 393–401.

Popov, V.N., Belyaev, V.M., and Valueva, E.P., 1977. Calculation of heat exchange and resistance of turbulent flow in a round pipe of liquid with various types of dependence of physical properties on the temperature, *High Temperatures* (Теплофизика Высоких Температур, стр. 1220–1229), 15 (6), pp. 1043–1052.

Popov, V.N., Belyaev, V.M., and Valueva, E.P., 1978. Heat transfer and hydraulic drag in a turbulent flow of helium in a circular tube at supercritical pressure, *High Temperatures* (Теплофизика Высоких Температур, стр. 1018–1027), 16 (5), pp. 864–871.

Popov, V.N., Belyayev, V.M., and Valuyeva, Ye.P., 1979. Calculation of heat transfer to turbulent pipe flow of supercritical helium, *Heat Transfer–Soviet Research*, 11 (3), pp. 108–113.

Popov, V.N., Belyayev, V.M., and Valuyeva, Ye.P., 1980. Calculation of HTCs and drag in turbulent pipe flow of supercritical helium with allowance for the effect of free convection, *Heat Transfer–Soviet Research*, 12 (2), pp. 123–128.

Popov, V.N. and Petrov, N.E., 1985. Calculation of heat transfer and flow resistance in the turbulent pipe flow of cooled carbon dioxide in the supercritical region, *High Temperatures* (Теплофизика Высоких Температур, стр. 309–316), 23 (2), pp. 255–262.

Popov, V.N. and Valueva, E.P., 1986. Heat transfer and turbulent flow of water at supercritical parameters of state in a vertical tube with a significant effect of free convection, *Thermal Engineering* (Теплоэнергетика, стр. 22–29), 33 (4), pp. 174–181.

Popov, V.N. and Valueva, E.P., 1988. Numerical modelling of mixed turbulent convection of helium at supercritical parameters of state in a vertical tube, *Thermal Engineering* (Теплоэнергетика, стр. 54–59), 35 (7), pp. 399–404.

Popov, V.N. and Yan'kov, G.G., 1979. Calculations of heat transfer in laminar natural convection of supercritical carbon dioxide and helium about a vertical wall, *Heat Transfer–Soviet Research*, 11 (3), pp. 114–120.

Popov, V.N. and Yankov, G.G., 1982. Heat transfer in laminar free convection near a vertical plate for fluids in the supercritical regions of state variables, *High Temperatures* (Теплофизика Высоких Температур, стр. 1110–1118), 20 (6), pp. 878–886.

Popov, V.N. and Yan'kov, G.G., 1985. Turbulent free-convection of supercritical helium, *Thermal Engineering* (Теплоэнергетика, стр. 30–35), 32 (3), pp. 138–142.

Powel, W.B., 1957. Heat transfer to fluids in the region of critical temperature, *Jet Propulsion*, 27 (7), pp. 776–783.

Proceedings of the 5th International Symposium on Supercritical Fluids, 2000, Atlanta, GA, USA, April 8–12.

Pron'ko, V.G. and Malyshev, G.P., 1972. Heat transfer in a turbulent helium flow at supercritical pressure in small-bore tubes, *High Temperatures* (Теплофизика Высоких Температур, стр. 1039–1042), 10 (5), pp. 932–935.

Pron'ko, V.G., Malyshev, G.P., and Migalinskaya, L.N., 1976. Modes of "normal" and "deteriorated" heat exchange in the single-phase near-critical region in the turbulent flow of helium in tubes, *Journal of Engineering-Physics* (Инженерно-Физический Журнал (ИФЖ), стр. 606–612), 30 (4), pp. 389–395.

Protopopov, V.S., 1977. Generalizing relations for the local heat-transfer coefficients in turbulent flows of water and carbon dioxide at supercritical pressure in a uniformly heated circular tube, *High Temperatures* (Теплофизика Высоких Температур, стр. 815–821), 15 (4), pp. 687–692.

Protopopov, V.S. and Igamberdyev, A.T., 1972. Results of an experimental investigation of local heat transfer at supercritical pressure in a rectangular channel heated from one side, *High Temperatures* (Теплофизика Высоких Температур, стр. 1242–1247), 10 (6), pp. 1116–1121.

Protopopov, V.S., Kuraeva, I.V., and Antonov, A.M., 1973. An approach to the determination of the conditions of occurrence of deteriorated heat transfer regimes at supercritical pressures, *High Temperatures* (Теплофизика Высоких Температур, стр. 593–597), 11 (3), pp. 529–532.

Protopopov, V.S. and Sharma, G.K., 1976. Experimental investigation of heat transfer from a vertical surface to carbon dioxide with natural convection in the supercritical region of the parameters of state, *High Temperatures* (Теплофизика Высоких Температур, стр. 781–878), 14 (4), pp. 693–698.

Protopopov, V.S. and Silin, V.A., 1973. Approximate method of calculating the start of local deterioration of heat transfer at supercritical pressure, *High Temperatures* (Теплофизика Высоких Температур, стр. 445–447), 11 (2), pp. 399–401.

Randall, D.G., 1956. Some heat transfer and fluid friction experiments with supercritical water, Report NDA 2051, November (also TID-7529, Part 3, November, 1957).

Razumovskiy, V.G., 1984. Heat transfer and hydraulic resistance of smooth channels at turbulent flow of water of supercritical pressure, (In Russian), Ph.D.

Thesis, Institute of Engineering Thermal Physics, Academy of Sciences, Kiev, Ukraine, 24 pages.

Razumovskiy, V.G., 2003. Private communications, State Technical University "KPI," Kiev, Ukraine.

Razumovskiy, V.G., Ornatskiy, A.P., and Maevskiy, E.M., 1984. Hydraulic resistance and heat transfer of smooth channels with turbulent flow of water of supercritical pressure, *Thermal Engineering* (Теплоэнергетика, стр. 69–72), 31 (2), pp. 109–113.

Razumovskiy, V.G., Ornatskiy, A.P., and Maevskiy, Ye.M., 1990. Local heat transfer and hydraulic behaviour in turbulent channel flow of water at supercritical pressure, *Heat Transfer–Soviet Research*, 22 (1), pp. 91–102.

Razumovskiy, V.G., Ornatskiy, A.P., Maevskiy, E.M., and Igol'nikova, N.V., 1985. Determination of coefficients of hydraulic resistance at turbulent flow of water of supercritical pressure in smooth channels, (In Russian), *Applied Thermal Sciences*, (Промышленная Теплотехника, стр. 26–28), 7 (5), pp. 26–28.

RDIPE, 2004. N.A. Dollezhal Research and Development Institute of Power Engineering, Brochure, RDIPE, Moscow, Russia, 74 pages.

Renz, U. and Bellinghausen, R., 1986. Heat transfer in a vertical pipe at supercritical pressure, Proceedings of the 8th International Heat Transfer Conference, San Francisco, CA, USA, August 17–22, Vol. 3, pp. 957–962.

Retzlaff, K.M. and Ruegger, W.A., 1996. Steam turbines for ultrasupercritical power plants, Report GER-3945A, General Electric Company, Schenectady, NY, USA, 13 pages.

Rimpault, G., Maráczy, C., Kyrki-Rajamäki, R., et al., 2003. Core design feature studies and research needs for high performance light water reactors, Proceedings of the International Congress on Advances in Nuclear Power Plants (ICAPP'03), Cordoba, Spain, May 4–7, Paper No. 3194, 12 pages.

Rivkin, S.L., 1970. *Thermophysical Properties of Water in Critical Region. Handbook*, (In Russian), Izdatel'stvo Standartov Publishing House, Moscow, Russia, p. 635.

Rivkin, S.L. and Gukov, V.M., 1971. An experimental investigation of the isobaric specific heat-capacity of carbon dioxide containing a small quantity of nitrogen, at supercritical pressures, *Thermal Engineering* (Теплоэнергетика, стр. 82–83), 18 (10), pp. 120–122.

Robakidze, L.V., Miropol'skii, Z.L., and Khasanov-Agaev, L.R., 1983. Heat transfer to a medium of supercritical parameters with mixed convection in horizontal tubes, *Thermal Engineering* (Теплоэнергетика, стр. 71–72), 30 (6), pp. 368–369.

Rohatgi, U.S. and Duffey, R.B., 1998. Stability, DNB, and CHF in natural circulation two-phase flow, *International Communications in Heat and Mass Transfer*, 25 (2), pp. 161–174.

Rudyka, A.V., Safonof, A.G., and Kazarinov, S.I., 1971. Ensuring reliability of the boilers in supercritical units when operating at variable pressures, *Thermal Engineering* (Теплоэнергетика, стр. 45–49), 18 (10), pp. 67–72.

Rzaev, M.A., Kelbaliev, R.F., Bayramov, N.M., et al., 2003. Methods for estimation of heat transfer deterioration at turbulent flow and supercritical pressures of fluids, (In Russian), Transactions of the XIV School-Seminar of Young Scientists and Specialists under the Leadership of Academician RAS A.I. Leont'ev "Problems of Gas Dynamics and Heat-Mass-Transfer in Power

Installations," Rybinsk, Russia, May 26–30, Vol. 1, Moscow, MEI Publishing House, pp. 109–112.

Sabersky, R.H. and Hauptmann, E.G., 1967. Forced convection heat transfer to carbon dioxide near the critical point, *International Journal of Heat & Mass Transfer,* 10 (11), pp. 1499–1508.

Sagayama, Yu., 2005. Progress of the Generation IV nuclear energy system development project, Proceedings of the International Conference GLOBAL-2005 "Nuclear Energy Systems for Future Generation and Global Sustainability," Tsukuba, Japan, October 9–13, Paper No. 588, 6 pages.

Sakurai, K., Ko, H.S., Okamoto, K., and Madarame, H., 2000. Visualization study for pseudo-boiling in supercritical carbon dioxide under forced convection in rectangular channel, Proceedings of the 1st International Symposium on Supercritical Water-Cooled Reactor Design and Technology (SCR-2000), Tokyo, Japan, November 6–8, Paper No. 303.

Sakurai, S., Yoshida, N., Shiga, S., and Oka, Y., 2003. Development of supercritical-water cooled power reactor core design study with 3-D core simulator, Proceedings of the Joint International Conference Global Environment and Nuclear Energy Systems/Advanced Nuclear Power Plants (GENES4/ANP2003), Kyoto, Japan, September 15–19, Paper No. 1115, 7 pages.

Samoilov, A.G., 1985. *Heating Fuel Elements of Nuclear Reactors,* (In Russian), Energoatomizdat Publishing House, Moscow, Russia, 222 pages.

Samoilov, A.G., Kashtanov, A.I., and Volkov, V.S., 1982. *Dispersion Fuel Elements, Vol. 2, Design and Serviceability,* (In Russian), Energoizdat Publishing House, Moscow, Russia, 255 pages.

Samoilov, A.G., Pozdnyakova, A.V., and Volkov, V.S., 1976. Steam-superheating fuel elements of the reactors in the I.V. Kurchatov Beloyarsk nuclear power station, *Atomic Energy* (Атомная Энергия, стр. 371–377), 40 (5), pp. 451–457.

Sastry, V.S. and Schnurr, N.M., 1975. An analytical investigation of forced convection heat transfer to fluids near the thermodynamic critical point, *Journal of Heat Transfer,* Transactions of the ASME, 97 (2), pp. 226–230.

Scalabrin, G. and Piazza, L., 2002. Forced convection heat transfer to supercritical carbon dioxide inside tubes. Analysis through neural networks, Proceedings of the 12th International Heat Transfer Conference, Elsevier SAS, pp. 507–512.

Schmidt, E., 1960. Wärmetransport durch natürliche konvection in Stoffen bei Kritischen zustand, *International Journal of Heat & Mass Transfer,* 1 (1) pp. 92–101.

Schmidt, E., Eckert, E., and Grigull, V., 1946. Heat transfer by liquids near the critical state, AFF Translation, No. 527, Air Materials Command, Wright Field, Dayton, OH, USA, April.

Schnurr, N.M., 1969. Heat transfer to carbon dioxide in the immediate vicinity of the critical point, *Journal of Heat Transfer,* Transactions of the ASME, 91 (1), February, pp. 16–20.

Schnurr, N.M., Sastry, V.S., and Shapiro, A.B., 1976. A numerical analysis of heat transfer to fluids near the thermodynamic critical point including the thermal entrance region, *Journal of Heat Transfer,* Transactions of the ASME, 98 (4), pp. 609–615.

Seetharam, T.R. and Sharma, G.K., 1979. Free convective heat transfer to fluids in the near-critical region from vertical surfaces with uniform heat flux, *International Journal of Heat and Mass Transfer,* 22, pp. 13–20.

Sekimura, N., Kobayashi, K., and Morioka, T., 2001. Surface irradiation effects by STM observation; Surface damage in Si under energetic ion irradiation, Proceedings of the 1st International Symposium on Supercritical Water-Cooled Reactor Design and Technology (SCR-2000), Tokyo, Japan, November 6–8, Paper No. 207.

Selivanov, V.M. and Smirnov, A.M., 1984. Experimental investigation of heat transfer of water in a tube under supercritical pressure, (In Russian), ФЭИ-1602, Institute of Physics and Power Engineering (ФЭИ), Obninsk, Russia, 16 pages.

Semenovker, I.E. and Gol'dberg, Yu.A., 1970. Theoretical hydraulic and temperature stratification in waterfalls of supercritical boilers, *Thermal Engineering* (Теплоэнергетика, стр. 58–61), 17 (8), pp. 82–85.

Seo, K.W., Anderson, M.H., Oh, B.D., and Kim, M.H., 2005. Studies of supercritical heat transfer and flow phenomena, Proceedings of the International Topical Meeting on Nuclear Reactor Thermal Hydraulics (NURETH-11), Avignon, France, October 2–6, Paper No. 162, 17 pages.

Sevast'yanov, V.V., Sinitsin, A.T., and Yakaitis, F.L., 1980. Study of heat-exchange process in supercritical region of parameters under conditions of high-frequency pressure oscillations, *High Temperatures* (Теплофизика Высоких Температур, стр. 546–553), 18 (3), pp. 433–439.

Shadrin, A., Murzin, A., Kamachev, V., et al., 2005. Decontamination of metals, fabrics and soils in liquid CO_2, Proceedings of the International Conference GLOBAL-2005 "Nuclear Energy Systems for Future Generation and Global Sustainability," Tsukuba, Japan, October 9–13, Paper No. 125, 5 pages.

Sharma, G.K. and Protopopov, V.S., 1975. Experimental investigation of natural convection heat transfer from a vertical surface to carbon dioxide at super critical pressure, Proceedings of the 3rd National Heat and Mass Transfer Conference, Bombay, India, Vol. 1, Paper No. HMT–35–75.

Shenoy, S.U., Jagadish, B.S., and Sharma, G.K., 1985. Heat transfer by laminar forced convection to water in the near-critical region for flow through a horizontal tube under constant wall temperature conditions, Proceedings of the National Heat and Mass Transfer Conference, Bombay, India, Paper No. HMT–C19–85, pp. 305–309.

Shioiri, A., Moriya, K., Oka, Yo., et al., 2003, Development of supercritical water-cooled power reactor conducted by a Japanese joint team, Proceedings of the Joint International Conference Global Environment and Nuclear Energy Systems/Advanced Nuclear Power Plants (GENES4/ANP2003), Kyoto, Japan, September 15–19, Paper No. 1121, 9 pages.

Shiralkar, B. and Griffith, P., 1968. The deterioration in heat transfer to fluids at supercritical pressures and high heat fluxes, Report, Engineering Projects Laboratory, Department of Mechanical Engineering, Massachusetts Institute of Technology, Cambridge, MA, USA, June, 185 pages.

Shiralkar, B.S. and Griffith, P., 1969. Deterioration in heat transfer to fluids at supercritical pressures and high heat fluxes, *Journal of Heat Transfer*, Transactions of the ASME, 91 (1), February, pp. 27–36.

Shiralkar, B.S. and Griffith, P., 1970. The effect of swirl, inlet conditions, flow direction, and tube diameter on the heat transfer to fluids at supercritical pressure, *Journal of Heat Transfer*, Transactions of the ASME, 92 (3), August, pp. 465–474.

Shitsman, M.E., 1959. Heat transfer to water, oxygen and carbon dioxide in near critical region, (In Russian), *Thermal Engineering* (Теплоэнергетика, стр. 68–72), No. 1, pp. 68–72.

Shitsman, M.E., 1962. Investigation of heat transfer at water cooling in near critical region, (In Russian), *Thermal Engineering* (Теплоэнергетика, стр. 83–86), 9 (1), pp. 83–86.

Shitsman, M.E., 1963. Impairment of the heat transmission at supercritical pressures, *High Temperatures* (Теплофизика Высоких Температур, стр. 267–275), 1 (2), pp. 237–244.

Shitsman, M.E., 1966. The effect of natural convection on temperature conditions in horizontal tubes at supercritical pressures, *Thermal Engineering* (Теплоэнергетика, стр. 51–56), 13 (7), pp. 69–75.

Shitsman, M.E., 1967. Temperature conditions of evaporative surfaces at supercritical pressures, *Electrical Stations* (Электрические Станции, стр. 27–30), 38 (2), pp. 27–30.

Shitsman, M.E., 1968. Temperature conditions in tubes at supercritical pressures, *Thermal Engineering* (Теплоэнергетика, стр. 57–61), 15 (5), pp. 72–77.

Shitsman, M.W., 1974. Heat transfer to supercritical helium, carbon dioxide, and water: Analysis of thermodynamic and transport properties and experimental data, *Cryogenics*, 14 (2), February, pp. 77–83.

Shlykov, Yu.P., Koblyakov, A.N., and Leongardt, A.D., 1971a. Theoretical calculation of heat transfer during the turbulent flow of water at supercritical conditions, *High Temperatures* (Теплофизика Высоких Температур, стр. 765–770), 9 (4), pp. 694–698.

Shlykov, Yu.P., Koblyakov, A.N., and Leongardt, A.D., 1971b. Turbulent heat transfer in the supercritical region of state parameters, (In Russian), Transactions of the IVth All-Union Conference on Heat Transfer and Hydraulics at Movement of Two-Phase Flow inside Elements of Power Engineering Machines and Apparatuses, Leningrad, Russia, Part II, pp. 3–19.

Shvarts, A.L., Dik, E.P., Dudnikova, I.P., and Nadyrov, I., 1963. Investigation of the transient processes in once-through boiler of supercritical parameters, (In Russian), *Thermal Engineering* (Теплоэнергетика, стр. 35–41), No. 4, pp. 35–41.

Shvarts, A.L. and Glusker, B.N., 1976. A method of determining permissible flows with respect to conditions of stability in ∩- and U-shaped elements at supercritical pressures, *Thermal Engineering* (Теплоэнергетика, стр. 33–37), 23 (7), pp. 16–20.

Silin, V.A., 1973. Experimental investigation of deteriorated heat transfer regimes in turbulent flow of carbon dioxide of supercritical pressure, (In Russian), Ph.D. Thesis, Moscow Power Institute (МЭИ), Russia, 30 pages.

Silin, V.A., Voznesensky, V.A., and Afrov, A.M., 1993. The light water integral reactor with natural circulation of the coolant at supercritical pressure B-500 SKDI, *Nuclear Engineering and Design*, 144, pp. 327–336.

Simon, H.A. and Ecker, E.R.G., 1963. Laminar free convection in carbon dioxide near its critical point, *International Journal of Heat & Mass Transfer*, 6 (8), pp. 681–690.

Sinitsyn, A.T., 1980. Theoretical determination of the existence region of high-frequency oscillations in a liquid under supercritical pressure, *High Temperatures* (Теплофизика Высоких Температур, стр. 1211–1214), 18 (6), pp. 904–907.

Sirota, A.M., Latunin, V.I., and Belyaeva, G.M., 1974. Experimental investigation of the thermal conductivity maxima of water in the critical region, *Thermal Engineering* (Теплоэнергетика, стр. 52–58), 21 (10), pp. 70–78.

Skvortsov, S.A. and Feinberg, S.M., 1961. Use of supercritical parameters of steam in the pressurized water power reactors, (In Russian), Institute of Atomic Energy by the name of I.V. Kurchatov (ИАЭ АН СССР), Moscow, Russia, 18 pages.

Smirnov, A.M., 2000. Heat transfer and hydrodynamics at supercritical pressures (Bibliography), (In Russian and English), Obninsk, Russian Federation State Scientific Center Institute of Power Engineering by the name of A.I. Leypunskiy (ГНЦ РФ ФЭИ), 44 pages.

Smirnov, O.K. and Krasnov, S.N., 1978. An investigation of unsteady heat transfer to water at supercritical pressures, *Thermal Engineering* (Теплоэнергетика, стр. 86–87), 15 (5), pp. 70–72.

Smirnov, O.K. and Krasnov, S.N., 1979. Tube hot spots developing under transient loads and flow of supercritical water, *Heat Transfer–Soviet Research*, 11 (3), pp. 122–124.

Smirnov, O.K. and Krasnov, S.N., 1980. Investigation of unsteady heat transfer to supercritical water, *Heat Transfer–Soviet Research*, 12 (2), pp. 135–140.

Smith, D., 1999. Ultra-supercritical CHP: Getting more competitive, *Modern Power Systems*, January, pp. 21–32.

Smith, R.V., 1969. Review of heat transfer to helium, *Cryogenics*, Vol. 9, February, pp. 11–19.

Smolin, V.N. and Polyakov, V.K., 1965. Experimental investigation of heat transfer to water in tubes at supercritical pressure, (In Russian), *Transactions of TsKTI Boiler-Turbine Engineering* (Труды ЦКТИ Котло-Турбостроение, стр. 130–137), Issue 57, Leningrad, Russia, pp. 130–137.

Solomonov, V.M. and Lokshin, V.A., 1975. Temperature conditions and heat transfer in horizontal and inclined tubes of steam generators at supercritical pressure under conditions of joint free and forced convection, *Thermal Engineering* (Теплоэнергетика, стр. 52–55), 22 (7), pp. 74–77.

Sohn, M.S., Yoo, Yo.H., and Suh, K.Y., 2005. Recuperator for supercritical carbon dioxide Brayton cycle, Proceedings of the International Congress on Advances in Nuclear Power Plants (ICAPP'05), Seoul, Korea, May 15–19, Paper No. 5584, 5 pages.

Sorokin, V.G., Volosnikova, A.V., Vyatkin, S.A., et al., 1989. *Handbook of Steels and Alloys*, (In Russian), Mashinostroeniye Publishing House, Moscow, Russia.

Spalaris, C.N., Boyle, R.F., Evans, T.F., and Esch, E.L., 1961. Design, fabrication, and irradiation of superheat fuel element SH-4B in VBWR, Report GEAP-3796, Atomic Power Equipment Department, General Electric Company, San Jose, CA, USA, September, 70 pages.

Spinks, N.J., Pontikakis, N., and Duffey, R.B., 2002. Thermo-economic assessment of advanced, high-temperature CANDU reactors, Proceedings of the International Conference on Nuclear Engineering (ICONE-10), Arlington, VA, USA, April 14–18, Paper No. 22433, 11 pages.

Squarer, D., Bittermann, D., Schulenberg, T., et al., 2003a. Overview of the HPLWR project and future directions, Proceedings of the International Congress on Advances in Nuclear Power Plants (ICAPP'03), Córdoba, Spain, May 4–7, Paper No. 3137, 7 pages.

Squarer, D., Schulenberg, T., Struwe, D., et al., 2003b. High-performance light water reactor, *Nuclear Engineering and Design,* Vol. 221, pp. 167–180.

Sridharan, K., Zillmer, A., Licht, J.R., et al., 2004. Corrosion behaviour of candidate alloys for supercritical water reactors, Proceedings of the International Congress on Advances in Nuclear Power Plants (ICAPP'04), Pittsburgh, PA, USA, June 13–17, Paper No. 4136, pp. 527–536.

Starflinger, J., Aksan, N., Bittermann, D., et al., 2004. Status on R&D planning for supercritical water cooled reactor systems in the 6th European Frame Programme, Proceedings of the International Congress on Advances in Nuclear Power Plants (ICAPP'04), Pittsburgh, PA, USA, June 13–17, Paper No. 4285, pp. 459–469.

Stephan, K., Durst, M., and Windisch, R., 1985. Wärmeübertragung bei frier Strömung im überkritischen Gebiet reiner Fluide, *Wärme- und Stoffübertragung,* 19 (3), pp. 187–194.

Stewart, E., Stewart, P., and Watson, A., 1973. Thermo-acoustic oscillations in forced convection heat transfer to supercritical pressure water, *International Journal of Heat & Mass Transfer,* 16, pp. 257–270.

Styrikovich, M.A., Margulova, T.Kh., and Miropol'skii, Z.L., 1967. Problems in the development of designs of supercritical boilers, *Thermal Engineering* (Теплоэнергетика, стр. 4–7), 14 (6), pp. 5–9.

Styrikovich, M.A., Martynova, O.I., and Kurtova, I.S., 1966. The behaviour of feed water impurities in the circuits of supercritical units, *Thermal Engineering* (Теплоэнергетика, стр. 45–50), 13 (7), pp. 61–68.

Supercritical Fluids. Molecular Interactions, Physical Properties, and New Applications, 2002. Editors: Y. Arai, T. Sako, and Y. Takebayashi, Springer-Verlag, Berlin, Germany.

Supercritical pressure power reactor. A conceptual design, 1959. Report HW-59684, Prepared by the staff of Reactor Eng. Development Operation, Hanford Atomic Products Operation, Richland, WA, USA, March 18, 38 pages.

Sutotskii, G.P., Safonov, L.P., Kokoshkin, I.A., and Rabkina, M.B., 1989. Damages to power equipment of heat power plants with supercritical steam generators and its water-chemical regime, (In Russian), In book: *Transactions of TsKTI (Трулы ЦКТИ) Water-Chemical Regime and Corrosion of Power Equipment*, Leningrad, Russia, Issue 255, pp. 11–24.

Suzuki, Sh., 2001. Irradiation assisted stress corrosion cracking in light water reactors, Proceedings of the 1st International Symposium on Supercritical Water-Cooled Reactor Design and Technology (SCR-2000), Tokyo, Japan, November 6–8, Paper No. 206.

Swenson, H.S., Carver, J.R., and Kakarala, C.R., 1965. Heat transfer to supercritical water in smooth-bore tubes, *Journal of Heat Transfer,* Transactions of the ASME, Series C, 87 (4), pp. 477–484.

Szetela, E.J., 1962. Heat transfer to hydrogen including effects of varying fluid properties, *American Rocket Society Journal* (ARS), 32 (8), pp. 1289–1292.

Tanabe, T., Koshizuka, S., and Oka, Y., 2004. A subchannel analysis code for supercritical-pressure LWR with downward-flowing water rods, Proceedings of the International Congress on Advances in Nuclear Power Plants (ICAPP'04), Pittsburgh, PA, USA, June 13–17, pp. 502–508.

Tanaka, H., Nishiwaki, N., and Hirata, M., 1967. Turbulent heat transfer to supercritical carbon dioxide, Proceedings of the JSME Semi-International Symposium, Tokyo, Japan, September 4–8, Paper 267, pp. 127–134.

Tanaka, H., Nishiwaki, N., Hirata, M., and Tsuge, A., 1971. Forced convection heat transfer to fluid near critical point flowing in circular tube, *International Journal of Heat & Mass Transfer,* 14 (6), pp. 739–750.

Tanaka, H., Tsuge, A., Hirata, M., and Nishiwaki, N., 1973. Effects of buoyancy and acceleration owing to thermal expansion on forced turbulent convection in vertical circular tubes — criteria of the effects, velocity, and temperature profiles, and reverse transition from turbulent to laminar flow, *International Journal of Heat & Mass Transfer,* 16, pp. 1267–1288.

Tanger, G.E., Lytle, J.H., and Vachon, R.I., 1968. Heat transfer to sulphur hexafluoride near the thermodynamic critical region in a natural-circulation loop, *Journal of Heat Transfer,* Transactions of the ASME, 90 (1), pp. 37–42.

Tarasova, N.V. and Leont'ev, A.I., 1968. Hydraulic resistance during flow of water in heated pipes at supercritical pressures, *High Temperatures* (Теплофизика Высоких Температур, стр. 755–756), 6 (4), pp. 721–722.

The Flow and Level Handbook, 2000. OMEGA Engineering, Inc., 21st Century Edition, Vol. MM, (Transactions, Vol. 4, p. 22).

The Temperature Handbook, 2000. OMEGA Engineering, Inc., 21st Century Edition, Vol. MM.

Theory and Technique of Thermophysical Experiment. Textbook for Technical Universities, (In Russian), Yu.F. Gortyshov, F.N. Dresvyannikov, N.S. Idiatullin etc., Editor V.K. Shchukin, Energoatomizdat, Moscow, Russia, 360 pages.

Thermal and Nuclear Power Plants. Handbook, 2nd edition, 1988. (In Russian), Editors: V.A. Grigor'ev and V.M. Zorin, Energoatomizdat Publishing House, Moscow, Russia.

Thermal Calculations of Boiler Aggregates. (Standard Method), (In Russian), 1973. TsKTI (ЦКТИ), Leningrad, Russia, Issue 34, p. 354.

Thermal Power Engineering and Heat Engineering. General Aspects. Handbook, 1987. (In Russian), Editors: V.A. Grigor'ev and V.M. Zorin, 2nd edition, Energoatomizdat Publishing House, Moscow, Russia.

Thompson, W.R. and Geery, E.L., 1962. Heat transfer to cryogenic hydrogen at supercritical pressures, *Advances in Cryogenic Engineering,* Vol. 7, Editor K.D. Timmerhaus, Plenum Press, pp. 391–400.

Tkachev, V.A., 1981. Experimental study of free–convection heat transfer at supercritical pressures, *Heat Transfer–Soviet Research,* 13 (1), pp. 1–6.

Tomioka, O., Imai, T., Fujimoto, Sh., et al., 2005. Decontamination of uranium contained wastes of intricate structure using supercritical carbon dioxide, Proceedings of the International Conference GLOBAL-2005 "Nuclear Energy Systems for Future Generation and Global Sustainability," Tsukuba, Japan, October 9–13, Paper No. 318, 4 pages.

Topping, J., 1971. *Errors of Observation and Their Treatment,* 3rd edition, Chapman and Hall Ltd., London, UK, 119 pages.

Torgerson, D.F., Shalaby, B.A., and Pang, S., 2005. CANDU technology for Generation III+ and IV reactors, Proceedings of the International Conference on Nuclear Engineering (ICONE-13), Beijing, China, May 16–20, Paper No. 50855, 8 pages.

Treshchev, G.G. and Sukhov, V.A., 1977. Stability of flow in heated channels in the supercritical region of parameters of state, *Thermal Engineering* (Теплоэнергетика, стр. 79–81), 24 (5), pp. 68–71.

Treshchev, G.G., Sukhov, V.A., and Shevchenko, G.A., 1971. Flow autooscillations in a heated channel at the supercritical state parameters, Transactions of the IVth All-Union Conference on Heat Transfer and Hydraulics at Movement of Two-Phase Flow inside Elements of Power Engineering Machines and Apparatuses, Leningrad, Russia, pp. 167–175.

Tret'yakov, Yu.M., 1971. Investigation of the deposition rate of impurities in steam generating channels at supercritical pressure of the heat carrier, *Thermal Engineering* (Теплоэнергетика, стр. 44–47), 18 (2), pp. 65–69.

Trukhnii, A.D., 1998. A new power-generating unit for ultra-supercritical steam conditions, *Thermal Engineering* (Теплоэнергетика), 45 (5), p. 4128.

Tsao, D. and Gorzegno, W.P., 1981. Variable-pressure once-through steam generators—experience and development, Proceedings of the American Power Conference, Vol. 43, pp. 287–293.

UK Steam Tables in SI Units 1970, Edward Arnold (Publishers) Ltd., London, UK, 1970.

U.S. DOE Generation IV Nuclear Energy Systems Roadmap, 2001. Description and evaluation of candidate water-cooled reactor systems, Technical Working Group 1—Advanced Water Reactors, Report, September 12, 34 pages.

Uskenbaev, S. and Budnevich, S.S., 1972. Experimental study of heat transfer by nitrogen near the critical state under supercritical pressure, *Journal of Engineering-Physics* (Инженерно-Физический Журнал (ИФЖ), стр. 926–929), 22 (5), pp. 649–651.

Valueva, E.P. and Popov, V.N., 1985. Numerical modelling of mixed turbulent convection of water at subcritical and supercritical pressures, *Thermal Engineering* (Теплоэнергетика, стр. 62–65), 32 (9), pp. 525–529.

Valueva, E.P., Popov, V.N., and Filippovich, E.V., 1995. Heat transfer with transient and turbulent flow of hydrocarbon liquids under supercritical conditions in tubes, *Thermal Engineering* (Теплоэнергетика, стр. 30–36), 42 (3), pp. 211–220.

Valyuzhin, M.A. and Kuznetsov, Ye.V., 1986. Transfer of heat to supercritical helium in the region of a circular tube under highly nonisothermal conditions, *Heat Transfer–Soviet Research*, 18 (6), pp. 97–103.

Valyuzhinich, M.A., Eroshenko, V.M. and Kuznetsov, E.V., 1985a. Entrance section heat transfer in flow of helium at supercritical pressure, *Journal of Engineering-Physics* (Инженерно-Физический Журнал (ИФЖ), стр. 546–551), 48 (4), pp. 397–401.

Valyuzhinich, M.A., Eroshenko, V.M., and Kuznetsov, E.V., 1985b. Heat transfer in the flow of supercritical-pressure helium with high thermal loads, *Journal of Engineering-Physics* (Инженерно-Физический Журнал (ИФЖ), стр. 574–579), 49 (4), pp. 1158–1162.

Valyuzhinich, M.A., Eroshenko, V.M., and Kuznetsov, E.V., 1985c. Heat transfer to fluids at supercritical pressure in the entrance region under conditions of strong nonisothermicity, *Thermal Engineering* (Теплоэнергетика, стр. 64–66), 32 (1), pp. 48–51.

Valyuzhinich, M.A., Eroshenko, V.M., and Kuznetsov, E.V., 1986. Experimental study of heat transfer in the turbulent convection of helium at supercritical pressure under strongly nonisothermal conditions, *High Temperatures* (Теплофизика Высоких Температур, стр. 89–94), 24 (1), pp. 80–85.

Vargaftik, N.B., Vinogradov, Y.K., and Yargin, V.S., *Handbook of Physical Properties of Liquids and Gases. Pure Substances and Mixtures,* 3rd Augmented and revised edition, Begell House, New York, NY, USA, 1996.

Vasic, A. and Khartabil, H.F., 2005. Passive cooling of the CANDU SCWR fuel at LOCA/LOECC conditions, Proceedings of the International Conference GLOBAL-2005 "Nuclear Energy Systems for Future Generation and Global Sustainability," Tsukuba, Japan, October 9–13, Paper No. 184, 6 pages.

Vasilenko, G.V. and Sutotskii, G.P., 1980. The formation of magnetite in the feedwater of a supercritical boiler when ammonia-hydrazine treatment is practiced, *Thermal Engineering* (Теплоэнергетика, стр. 59–60), 27 (11), pp. 642–643.

Vasilenko, G.V., Zenkevich, Yu.V., and Mazurova, O.K., 1975. Precipitation of aluminium compounds in a supercritical steam generator, Thermal Engineering (Теплоэнергетика, стр. 62–63), 22 (6), pp. 83–84.

Verdiev, Ch.M., 2002. Specifics of heat transfer at supercritical pressures in regime of existence of high frequency thermoacoustic instabilities, (In Russian), Proceedings of the 3rd Russian National Heat Transfer Conference, Moscow, Russia, October, 21–25, MEI Publishing House, Moscow, Vol. 2, pp. 91–94.

Verdiev, Ch.M., Kelbaliev, R.F., Tavakkuli, D., and Valizade, Kh.M., 2001. Investigation of the process of improving heat transfer at laminar, transition and turbulent regimes of flow of hydrocarbon fluid of supercritical pressure, (In Russian), Transactions of the XIII School-Seminar of Young Scientists and Specialists under the Leadership of Academician RAS A.I. Leont'ev "Physical Basis of Experimental and Mathematical Modelling of Processes of Gas Dynamic and Heat-Mass-Transfer in Power Installations," St.-Petersburg, Russia, May 20–25, Vol. 1, MEI Publishing House, Moscow, pp. 43–45.

Vetrov, V.I., 1990. Calculation of frequencies of thermoacoustic oscillations, generated in heated channels for supercritical pressures of heat-transfer agents, *High Temperatures* (Теплофизика Высоких Температур, стр. 309–314), 28 (2), pp. 233–239.

Vetrov, V.I., Gerliga, V.A., and Razumovskiy, V.G., 1977. Experimental investigation of thermoacoustic oscillations, generated in heated channels at supercritical pressures of water, (In Russian), *Problems of Atomic Science and Techniques, Series Dynamics of Nuclear Power Installations*, Gor'kiy, Russia, Issue 2, No. 12, pp. 51–57.

Vikhrev, Yu.V., Barulin, Yu.D., and Kon'kov, A.S., 1967. A study of heat transfer in vertical tubes at supercritical pressures, *Thermal Engineering* (Теплоэнергетика, стр. 80–82), 14 (9), pp. 116–119.

Vikhrev, Yu.V., Kon'kov, A.S., Lokshin, V.A., et al., 1971. Temperature regime of steam generating tubes at supercritical pressure, (In Russian), Transactions of the IVth All-Union Conference on Heat Transfer and Hydraulics at Movement of Two-Phase Flow inside Elements of Power Engineering Machines and Apparatuses, Leningrad, Russia, pp. 21–40.

Vikhrev, Yu.V., Kon'kov, A.S., and Sinitsin, I.T., 1970. Temperature regime of horizontal tubes at supercritical pressure, (In Russian), *Electrical Stations* (Электрические Станции, стр. 35–38), (7), pp. 35–38.

Vikhrev, Yu.V., Kon'kov, A.S., Solomonov, V.M., and Sinitsyn, I.T., 1973. Heat transfer in horizontal and inclined steam-generating tubes at supercritical pressures, *High Temperatures* (Теплофизика Высоких Температур, стр. 1316–1318), 11 (6), pp. 1183–1185.

Vikhrev, Yu.V. and Lokshin, V.A., 1964. An experimental study of temperature conditions in horizontal tubes at supercritical pressures, *Thermal Engineering* (Теплоэнергетика, стр. 79–82), 11 (12), pp. 105–109.

Vlakhov, E.S., Miropol'skii, Z.L., and Khasanov-Agaev, L.R., 1981. Heat transfer to a supercritical medium with mixed convection and rising flow in heated tubes, *Thermal Engineering* (Теплоэнергетика, стр. 69–71), 28 (11), pp. 680–682.

Vukalovich, M.P. and Altunin, V.V., 1968. *Thermophysical Properties of Carbon Dioxide,* Collet's Publishers, Ltd., London & Wellingborough, UK, 463 pages.

Waata, Ch., Schulenberg, T., Cheng, Xu., and Starflinger, J., 2005. Results of a coupled neutronics and thermal-hydraulics analysis of a HPLWR fuel assembly, Proceedings of the International Congress on Advances in Nuclear Power Plants (ICAPP'05), Seoul, Korea, May 15–19, Paper No. 5064, 7 pages.

Walisch, T., Dörfler, W., and Trepp, Ch., 1997. Heat transfer to supercritical carbon dioxide in tubes with mixed convection, ASME Proceedings of the 32nd National Heat Transfer Conference, Vol. 12, Baltimore, MD, USA, HTD-Vol. 350, pp. 79–98.

Walker, B.J. and Harden, D.G., 1967. The "density effect" model: prediction and verification of the flow oscillation threshold in a natural-circulation loop operating near the critical point, ASME, Paper No. 67-WA/HT-23, 9 pages.

Wallin, W.R., Rice, R.E., and Novick, M., 1961. Borax V integral nuclear superheat reactor experiments, IAEA Symposium on Power Reactor Experiments, Vienna, Austria, SM – 21/10, pp. 9–26.

Was, G.S. and Alen, T.R., 2005. Time, temperature, and dissolved oxygen dependence of oxidation of austenitic and ferritic-martensitic alloys in supercritical water, Proceedings of the International Congress on Advances in Nuclear Power Plants (ICAPP'05), Seoul, Korea, May 15–19, Paper No. 5690, 24 pages.

Watanabe, Yu., Adschiri, T., and Kondo, T., 2001. Importance of pressure in corrosion and oxidation of alloys in supercritical water and superheated steam, Proceedings of the 1st International Symposium on Supercritical Water-Cooled Reactor Design and Technology (SCR-2000), Tokyo, Japan, November 6–8, Paper No. 402.

Watson, A., 1977. The influence of axial wall conduction in variable property convection — with particular reference to supercritical pressure fluids, *International Journal of Heat and Mass Transfer,* 20, pp. 65–71.

Watts, M.J. and Chou, C.T., 1982. Mixed convection heat transfer to supercritical pressure water, Proceedings of the 7th International Heat Transfer Conference, Munchen, Germany, Vol. 3, Paper No. 6–10, pp. 495–500.

Weinberg, R.S., 1972. Experimental and theoretical study of buoyancy effects in forced convection to supercritical pressure carbon dioxide, Ph.D. Thesis, University of Manchester, UK.

White, F.M., 1994. *Fluid Mechanics.* Third edition, McGraw-Hill, Inc., New York, NY, USA, pp. 364–365.

Winterton, R.H.S., 1998. Where did the Dittus and Boelter equation come from? *International Journal of Heat and Mass Transfer,* 41 (4–5), pp. 809–810.

Wood, R.D. and Smith, J.M., 1964. Heat transfer in the critical region—temperature and velocity profiles in turbulent flow, *AIChE Journal,* 10 (2), pp. 180–186.

Wright, J.H. and Patterson, J.F., 1966. Status and application of supercritical-water reactor coolant, Proceedings of the American Power Conference, Vol. 28, pp. 139–149.

Yamada, K. and Ka, Yo., 2005. Research and development of Supercritical Water-Cooled Reactor (SCWR) in Japan, Proceedings of the International Conference GLOBAL-2005 "Nuclear Energy Systems for Future Generation and Global Sustainability," Tsukuba, Japan, October 9–13, Paper No. 239, 6 pages.

Yamagata, K., Nishikawa, K., Hasegawa, S., et al., 1972. Forced convective heat transfer to supercritical water flowing in tubes, *International Journal of Heat & Mass Transfer*, 15 (12), pp. 2575–2593.

Yamaji, A., Oka, Yo., Ishiwatari, Yu., et al., 2005a. Rationalization of the fuel integrity and transient criteria for the super LWR, Proceedings of the International Congress on Advances in Nuclear Power Plants (ICAPP'05), Seoul, Korea, May 15–19, Paper No. 5538.

Yamaji, A., Oka, Yo., and Koshizuka, S., 2005b. Three-dimensional core design of high temperature supercritical-pressure light water reactor with neutronic and thermal-hydraulic coupling, *Journal of Nuclear Science and Technology*, 42 (1), pp. 8–19.

Yamaji, A., Tanabe, T., Oka, Yo., et al., 2005c. Evaluation of the peak cladding surface temperature of the Super LWR with subchannel analyses, Proceedings of the International Conference GLOBAL-2005 "Nuclear Energy Systems for Future Generation and Global Sustainability," Tsukuba, Japan, October 9–13, Paper No. 557, 6 pages.

Yamaji, A., Oka, Yo., Yang, J., et al., 2005d. Design and integrity analyses of the super LWR fuel rod, Proceedings of the International Conference GLOBAL-2005 "Nuclear Energy Systems for Future Generation and Global Sustainability," Tsukuba, Japan, October 9–13, Paper No. 556, 6 pages.

Yamaji, A., Kamei, K., Oka, Yo., and Koshizuka, S., 2004. Improved core design of high temperature supercritical-pressure light water reactor, Proceedings of the International Congress on Advances in Nuclear Power Plants (ICAPP'04), Pittsburgh, PA, USA, June 13–17, Paper No. 4331, pp. 509–517.

Yamaji, A., Oka, Yo., and Koshizuka, S., 2003a. Three-dimensional core design of SCLWR-H with neutronic and thermal-hydraulic coupling, Proceedings of the ANS/ENS International Winter Meeting and Nuclear Technology Expo, Embedded Topical Meeting GLOBAL 2003 "Advanced Nuclear Energy and Fuel Cycle Systems", New Orleans, LA, USA, November 16–20, pp. 1763–1771.

Yamaji, A., Oka, Yo., and Koshizuka, S., 2003b. Core design of a high-temperature reactor cooled and moderated by supercritical water, Proceedings of the Joint International Conference Global Environment and Nuclear Energy Systems/Advanced Nuclear Power Plants (GENES4/ANP2003), Kyoto, Japan, September 15–19, Paper No. 1041, 8 pages.

Yamaji, A., Oka, Yo., and Koshizuka, S., 2003c. Fuel design of high temperature reactors cooled and moderated by supercritical light water, Proceedings of the Joint International Conference Global Environment and Nuclear Energy Systems/Advanced Nuclear Power Plants (GENES4/ANP2003), Kyoto, Japan, September 15–19, Paper No. 1040, 8 pages.

Yamaji, A., Oka, Yo., and Koshizuka, S., 2003d. Core design of a high temperature reactor cooled and moderated by supercritical light water, Proceedings of the Joint International Conference Global Environment and Nuclear Energy Systems/Advanced Nuclear Power Plants (GENES4/ANP2003), Kyoto, Japan, September 15–19, Paper No. 1041, 10 pages.

Yamashita, T., Yoshida, S., Mori, H., et al., 2003. Heat transfer study under supercritical pressure conditions, Proceedings of the Joint International Conference Global Environment and Nuclear Energy Systems/Advanced Nuclear Power Plants (GENES4/ANP2003), Kyoto, Japan, September 15–19, Paper No. 1119, 8 pages.

Yang, S.K. and Khartabil, H.F., 2005. Normal and deteriorated heat transfer correlations for supercritical fluids, *Transaction of the American Nuclear Society*, November 13–17, Washington, D.C., USA, Vol. 93, pp. 635–637.

Yang, J., Oka, Yo., Liu, J., et al., 2005. Development on statistical thermal design procedure for Super LWR, Proceedings of the International Conference GLOBAL-2005 "Nuclear Energy Systems for Future Generation and Global Sustainability," Tsukuba, Japan, October 9–13, Paper No. 555, 6 pages.

Yang, W.S. and Zavaljevski, N., 2005. Effects of water rods on supercritical water reactor stability, Proceedings of the International Congress on Advances in Nuclear Power Plants (ICAPP'05), Seoul, Korea, May 15–19, Paper No. 5101, 11 pages.

Yanovskii, L.S., 1995. Heat transfer during forced flow of mixtures of organic liquids at supercritical pressures, *Heat Transfer Research*, 26 (3–8), pp. 405–410.

Yanovskiy, L.S., Kuznetsov, E.V., Myakichev, A.S., and Tikhonov, A.I., 1987. In book: *Modern Problems of Hydrodynamic and Heat Transfer in Power Units and Cryogenic Technique* (In Russian), Vsesouzniy Zaochniy Mashinostroitel'niy Institut, Moscow, Russia, p. 129.

Yaskin, L.A., 1980. Calculation of combined-convection heat transfer from supercritical steam in horizontal pipes, *Heat Transfer–Soviet Research*, 12 (3), pp. 43–44.

Yaskin, L.A., Jones, M.C., Yeroshenko, V.M., et al., 1977. A correlation for heat transfer to superphysical helium in turbulent flow in small channels, *Cryogenics*, 17 (10), pp. 549–552.

Yeroshenko, V.M., Starostin, A.D., and Yaskin, L.A., 1980. Calculation of visco-inertial flows of supercritical helium in heated pipes, *Heat Transfer–Soviet Research*, 12 (2), pp. 129–134.

Yeroshenko, V.M. and Yaskin, L.A., 1981. Applicability of various correlations for the prediction of turbulent heat transfer of supercritical helium, *Cryogenics*, 21 (2), pp. 94–96.

Yi, T.T., Ishiwatari, Yu., Liu, J., et al., 2005a. Thermal and stability considerations of super LWR during sliding pressure startup, *Journal of Nuclear Science and Technology*, 42 (6), pp. 537–548.

Yi, T.T., Ishiwatari, Yu., Liu, J., et al., 2005b. Thermal and stability considerations of super LWR during sliding pressure startup, Proceedings of the International Congress on Advances in Nuclear Power Plants (ICAPP'05), Seoul, Korea, May 15–19, Paper No. 5383.

Yi, T.T., Ishiwatari, Yu., Koshizuka, S., and Oka, Yo., 2004a. Startup thermal analysis of a high-temperature supercritical-pressure light water reactor, *Journal of Nuclear Science and Technology*, 41 (8), pp. 790–801.

Yi, T.T., Koshizuka, S., Oka, Yo., et al., 2004b. A linear stability analysis of supercritical water reactors, (I) Thermal-hydraulic stability, *Journal of Nuclear Science and Technology*, 41 (12), pp. 1166–1175.

Yi, T.T., Koshizuka, S., Oka, Yo., et al., 2004c. A linear stability analysis of supercritical water reactors, (II) Coupled neutronic thermal-hydraulic stability, *Journal of Nuclear Science and Technology*, 41 (12), pp. 1176–1186.

Yi, T.T., Ishiwatari, Yu., Koshizuka, S., and Oka, Yo., 2003. Startup of a high-temperature reactor cooled and moderated by supercritical-pressure light water, Proceedings of the Joint International Conference Global Environment and Nuclear Energy Systems/Advanced Nuclear Power Plants (GENES4/ANP2003), Kyoto, Japan, September 15–19, Paper No. 1036, 10 pages.

Yoo, J., Ishiwatari, Yu., Oka, Yo., and Liu, J., 2005. Composite core design of high power density Supercritical Water Cooled Fast reactor, Proceedings of the International Conference GLOBAL-2005 "Nuclear Energy Systems for Future Generation and Global Sustainability," Tsukuba, Japan, October 9–13, Paper No. 246, 6 pages.

Yoon, H.Y. and Bae, Y.Y., 2003. Feasibility study of a passive safety system for a supercritical water-cooled reactor, Proceedings of the Joint International Conference Global Environment and Nuclear Energy Systems/Advanced Nuclear Power Plants (GENES4/ANP2003), Kyoto, Japan, September 15–19, Paper No. 1022, 6 pages.

Yoon, S.H., Kim, J.H., Hwang, Y.W., et al., 2003. Heat transfer and pressure drop characteristics during the in-tube cooling process of carbon dioxide in the supercritical region, *International Journal of Refrigeration*, 26, pp. 857–864.

Yoshida, S. and Mori, H., 2000. Heat transfer to supercritical pressure fluids flowing in tubes, Proceedings of the 1st International Symposium on Supercritical Water-Cooled Reactor Design and Technology (SCR-2000), Tokyo, Japan, November 6–8, Paper No. 106.

Zhao, J., Saha, P., and Kazimi, M.S., 2005. Stability of supercritical water-cooled reactor during steady-state and sliding pressure start-up, Proceedings of the International Topical Meeting on Nuclear Reactor Thermal Hydraulics (NURETH-11), Avignon, France, October 2–6, Paper No. 106, 21 pages.

Zhou, N. and Krishnan, A., 1995. Laminar and turbulent heat transfer in flow of supercritical CO_2, HTD-Vol. 307, 1995 National Heat Transfer Conference—Vol. 5, ASME, pp. 53–63.

Zhukovskiy, A.V., Krasyakova, L.Yu., Belyakov, I.I., and Phephelova, N.D., 1971. Heat transfer in horizontal tube at supercritical pressures, (In Russian), *Soviet Energy Technology* (Энергомашиностроение), No. 2, pp. 23–26.

Zuber, N., 1966. An analysis of thermally induced flow oscillations in the near-critical and super-critical thermodynamic region, Report NASA-CR-80609, Research and Development Center, General Electric Company, Schenectady, NY, USA, May 25, 159 pages.

INDEX

H

I